T0254870

Applied Probability

Valérie Girardin • Nikolaos Limnios

Applied Probability

From Random Experiments to Random Sequences and Statistics

 Springer

Valérie Girardin
Laboratoire de Mathématiques Nicolas
Oresme
Université de Caen Normandie
Caen, France

Nikolaos Limnios
Laboratoire de Mathématiques Appliquées
de Compiègne
Université de Technologie de Compiègne
Compiègne, France

ISBN 978-3-030-97965-2   ISBN 978-3-030-97963-8   (eBook)
https://doi.org/10.1007/978-3-030-97963-8

Translation from the French language edition: Probabilités et introduction à la statistique by Valérie
Girardin, and Nikolaos Limnios, © Vuibert 2014. Published by Vuibert. All Rights Reserved.

This Springer imprint is published by the registered company Springer Nature Switzerland AG
The registered company address is: Gewerbestrasse 11, 6330 Cham, Switzerland

Preface

Some of the books on probability that have been published in the past years focus on theoretical developments, while others are oriented towards applications. Books written for beginners in probability, or dedicated to a particular application, usually contain only a limited number of theoretical notions, while others are very complete but very dense. We hope that a pragmatic book close to applications but without giving up mathematical rigor, serious yet friendly, can at the same time be useful and topical.

Random variables are defined by reference to a random experiment as functions whose values depend on the result of the experiment. They are called real random variables if they take values in the real line. A finite family of such variables is a random vector, while a denumerable family is a random sequence. All these random elements are studied together with notions necessary to their use in applied fields. To this end, we consider reliability linked to the lifetime of living and industrial systems, entropy linked to information theory, simulation methods, and we also present basics in inferential statistics.

Even if discrete finite probability spaces and random variables can be studied through elementary methods, the associated theory remains insufficient to model rigorously classical stochastic experiments; for instance, infinite sequences of tossing need a more sophisticated treatment. We have therefore chosen to include the discrete spaces in the more general framework of measured spaces in which all possible real situations are included. Nevertheless, we present measure and integration theory only as much as necessary to understand the spaces in which we actually work.

This book is addressed to advanced undergraduate students in mathematics and to postgraduate students in applied mathematics. It will also be of use to researchers and engineers working in other fields but interested by the rigorous mathematical basis of probability theory. To benefit from the reading, no notions of probability are required. The prerequisites are limited to some classical undergraduate courses in mathematical analysis.

Each chapter is illustrated by a great number of examples and exercises with their solutions. They are intended not only as solutions of classical problems in probability but also as complements to the main text together with an opening to applied fields. A table of notation and a detailed index are given for easy reference. A classified bibliography proposes further reading in theoretical and applied fields.

The volume is organized as follows.

In Chap. 1, the basic notions of Kolmogorov's system of axioms is presented, and completed by all intrinsic notions of probability theory, such as independence and conditioning of events. We state different formulae necessary for the effective calculus of probability of events. Primary principles of entropy end the chapter.

In Chap. 2, the discrete or continuous real random variables are presented, together with the tools necessary to their investigation, from probabilistic tools such as distributions and distribution functions to analytical tools such as moment generating functions of all kinds. First notions of reliability are given.

In Chap. 3, the simultaneous study of several real random variables is presented. The notions given in Chaps. 1 and 2 are extended to these real random vectors. Specific notions linked to relations between variables such as independence, order statistics, and entropy are given. Attention is focused on the effective calculus of distributions of random variables and vectors, in particular, Gaussian vectors.

In Chap. 4, elements of stochastic topology with the different types of convergence of random sequences are presented: almost sure, in mean, in square mean, in probability, and in distribution. We detail different laws of large numbers and central limit theorems, the most remarkable results of probability theory. Some basic stochastic simulation methods are developed.

In Chap. 5, basic notions of parametric and non-parametric inferential statistics are presented. Point estimation, confidence intervals, and statistical testing are a first step in the huge domain of mathematical statistics.

The volume is our own loose translation of the first of our two books published in French by Vuibert, Paris, which are in their third edition. The second volume, published in English by Springer in 2018, begins where this volume ends. The interested reader will find there complements on random sequences indexed by integers—such as martingales and Markov chains, together with an introduction to the random processes theory—especially jump Markov and semi-Markov processes. Application is proposed in many fields of applied probability, such as reliability, information theory, production, risk, seismic analysis, and queueing theory.

Caen, France   Valérie Girardin
Compiègne, France   Nikolaos Limnios
April 2022

Contents

Notation

$\mathbb{N}$	Set of integers
$\mathbb{Z}$	Set of rational integers
$[\![m, n]\!]$	Set of integers $\{m, \ldots, n\}$
$\mathbb{Q}$	Set of rational numbers
$\mathbb{R}$	Set of real numbers
$]a, b[$	Set of real numbers $a < x < b$
$[a, b]$	Set of real numbers $a \leq x \leq b$
$\overline{\mathbb{R}}$	Extended real line $[-\infty, +\infty]$
$\mathbb{R}_+$	Set of non negative numbers $[0, +\infty[$
$\mathbb{C}$	Set of complex numbers
$\mathcal{S}_n$	Set of permutations of $\{1, \ldots, n\}$
$\mathcal{R}e\, z$	Real part of the complex number z
$\mathcal{I}m\, z$	Imaginary part of the complex number z
i	Square root of -1, i.e., $i^2 = -1$
E^*	Set E minus 0
∂E	Boundary of the set E
$\overset{\circ}{E}$	Interior of E
$|E|$	Cardinal of E
$\prod_{i=1}^{n} E_i$	Cartesian product of the sets $E_1, \ldots, E_n$
E^n	Order n Cartesian product of E
$E^{\mathbb{N}}$	Set of sequences of elements of E
$a \vee b$	Maximum of the real numbers a and b
$a \wedge b$	Minimum of the real numbers a and b
$x \to x_0^+$	x tends to x_0, with $x > x_0$
$x \to x_0^-$	x tend to x_0, with $x < x_0$
$[x]$	Integer part of x
Δx_n	Increment of the sequence (x_n), i.e., $\Delta x_n = x_n - x_{n-1}$
$\log$	Neperian logarithm
Id	Identity function
$f \circ g$	Composite of the functions f and g
$f^{\circ n}$	Order n composite of f
$f(\cdot, y)$	Partial function $x \longrightarrow f(x, y)$ for a fixed y
f^+	Positive part of f, that is $\max(f, 0)$

f^-	Negative part of f, that is $\max(-f, 0)$
$\mathcal{C}^0(I)$	Set of continuous functions on $I \subset \mathbb{R}$
$\mathcal{C}^n(I)$	Set of n times differentiable functions on $I \subset \mathbb{R}$
$o(x)$	o of x at $x_0 \in \overline{\mathbb{R}}$, i.e., $\lim_{x \to x_0} o(x)/x = 0$
$O(x)$	O of x at $x_0 \in \overline{\mathbb{R}}$, i.e., $\lim_{x \to x_0} O(x)/x = c$, with $c \in \mathbb{R}$
$f(x) \approx g(x)$	Equivalent functions at $x_0 \in \overline{\mathbb{R}}$, i.e., $\lim_{x \to x_0} f(x)/g(x) = 1$
$f(x) \overset{\sim}{=} c$	$c \in \mathbb{R}$ is an approximated value of $f(x)$
$f(+\infty)$	Limit of the function f at $+\infty$
$\overline{\lim} f_n$	Superior limit of the sequence of functions (f_n), i.e., $\overline{\lim} f_n(x) = \inf_{n \geq 0} \sup_{k \geq n} f_k(x)$
$\underline{\lim} f_n$	Inferior limit of (f_n), i.e., $\underline{\lim} f_n(x) = \sup_{n \geq 0} \inf_{k \geq n} f_k(x)$
$\mathrm{Sp}(v_1, \ldots, v_n)$	Linear subspace spanned by $v_1, \ldots, v_n$
$\mathbf{1}_d$	d-dimensional column vector with components 1
$(v_1, \ldots, v_d)$	Row vector
$(v_1, \ldots, v_d)'$	Column vector
$< v, w >$	Scalar product of $v = (v_1, \ldots, v_d)$ and $w = (w_1, \ldots, w_d)$, I.e., $< v, w > = \sum_{i=1}^d v_i w_i$
$\|v\|$	Euclidean norm of $v = (v_1, \ldots, v_d)$, i.e., $\|v\| = (\sum_{i=1}^d |v_i|^2)^{1/2}$
I_d	d-dimensional identity matrix
$\det M$	Determinant of the matrix M
M'	Transpose of the matrix M
$\mathrm{diag}(a_i)$	Diagonal matrix whose i-th coefficient is a_i
$J_\varphi(u)$	Jacobian of the function φ at $u \in \mathbb{R}^d$
a.e.	Almost everywhere
a.s.	Almost surely
i.i.d.	Independent and identically distributed
Ω	Sample space of a random experiment
ω	Outcome of a random experiment
$\mathcal{F}$	σ-algebra
$\sigma(\mathcal{C})$	σ-algebra generated by $\mathcal{C}$
$\mathcal{B}(E)$	Borel σ-algebra of the topological space E
$\mathcal{P}(\Omega)$	Set of all subsets of Ω
μ	Positive measure
δ_ω	Dirac measure at ω
λ	Lebesgue measure on $\mathbb{R}$
λ_d	Lebesgue measure on $\mathbb{R}^d$
λ_F	Lebesgue-Stieltjes measure on $\mathbb{R}$ associated to the mass function F
$\mathbb{P}$	Probability measure
$(\Omega, \mathcal{F}, \mathbb{P})$	Probability space
$\Omega \setminus A = \overline{A}$	Complement of a subset A of Ω
$\mathbb{P}(B \mid A)$	Conditional probability of B given A
$n!$	Factorial n

$\binom{N}{n}$	Binomial coefficient
$\mathcal{I}(P)$	Entropy of P
$\mathcal{I}(P \mid Q)$	Entropy of P relative to Q
$\otimes_{i=1}^{n} \mathcal{F}_i$	Product of the σ-algebra s $\mathcal{F}_1, \ldots, \mathcal{F}_n$
$\mathcal{F}^{\otimes n}$	Order n product of $\mathcal{F}$
$\otimes_{i=1}^{n} \mathbb{P}_i$	Product of the probabilities $\mathbb{P}_1, \ldots, \mathbb{P}_n$
$\mathbb{P}^{\otimes n}$	Order n product of $\mathbb{P}$
$\mathbb{1}_A$	Indicator function of a subset A of Ω
$(f \in B)$	Inverse image of B by f
$(a \leq f < b)$	Inverse image of $[a, b[$ by f
$(f \in dx)$	Short for $(x < f \leq x + dx)$
$\sigma(f)$	σ-algebra generated by the measurable function f
$\sigma(f_i, i \in I)$	σ-algebra generated by the collection $\{f_i : i \in I\}$
X	Random variable
$\mathbb{P}_X$	Probability distribution of X
$X \sim P$	The variable X has the probability distribution P
$X \sim Y$	The variables X and Y have the same distribution
F_X	Distribution function of the variable X
I_α	Set of quantiles of order α
r_α	Quantile of order α
$\int f d\mu$	Lebesgue integral of the function f with respect to the measure μ
$\int_{\mathbb{R}} f dF$	Lebesgue-Stieltjes integral of f with respect to the mass function F
$\mathcal{L}^p(\Omega, \mathcal{F}, \mu)$	Set of $\mathcal{F}$-measurable functions whose p-th power is μ-integrable
$L^p(\Omega, \mathcal{F}, \mu)$	Set of the equivalence classes for the relation of a.e. equality in $\mathcal{L}^p(\Omega, \mathcal{F}, \mu)$
$\|f\|_p$	L^p norm
$\mathbb{E} X$	Expectation (or mean) of the variable X
$\operatorname{Var} X$	Variance of X
$\mathcal{B}(p)$	Bernoulli distribution
$\mathcal{B}(n, p)$	Binomial distribution
$\mathcal{B}_-(r, p)$	Negative binomial distribution
$\mathcal{G}(p)$	Geometric distribution
$\mathcal{P}(\lambda)$	Poisson distribution
$\mu \ll \nu$	The measure μ is absolutely continuous with respect to the measure ν
f_X	Density of the absolutely continuous variable X
$\mathcal{U}(a, b)$	Uniform distribution on the interval $[a, b]$
$\mathcal{N}(\mu, \sigma^2)$	Gaussian distribution with mean $\mu \in \mathbb{R}$ and variance $\sigma^2 \in \mathbb{R}_+$
$\gamma(a, b)$	Gamma distribution
$\chi^2(n)$	Chi square distribution with $n \in \mathbb{N}^*$ degrees of freedom
$\mathcal{E}(n, \lambda)$	Erlang distribution

$\mathcal{E}(\lambda)$	Exponential distribution
$\mathcal{W}(\alpha, \lambda)$	Weibull distribution
$\mathcal{LN}(a, b)$	Log-normal distribution
$\mathcal{N}^{-1}(\mu, \lambda)$	Inverse-Gaussian distribution
$\beta_1(p, q)$	First type beta distribution
$\beta_2(p, q)$	Second type beta distribution
$F(n, m)$	Fisher-Snedecor distribution
$\tau(n)$	Student distribution
$\mathcal{C}(\alpha)$	Cauchy distribution
g_X	Generating function of the variable X
$\hat{f}$	Fourier transform of the function f
$\hat{\mu}$	Fourier transform of the measure μ
φ_X	Characteristic function of the variable X
ψ_X	Laplace transform of the non negative variable X
M_X	Moments generating function of X
h_X	Cramér transform of X
R	Survival function or reliability of a system
h	Hazard rate function
$\mathbb{C}\text{ov}\,(X, Y)$	Covariance of the variables X and Y
$\rho_{X,Y}$	Correlation coefficient of X and Y
$\overset{st}{\geq}$	Stochastic order relation
$H(X)$	Entropy of the variable X
$H(X, Y)$	Joint entropy of the variables X and Y
$H(X \mid Y)$	Entropy of X conditional on Y
$K(X \mid Y)$	Entropy of X relative to Y
$(X_1, \ldots, X_d)'$	d-dimensional random vector
$\mathcal{M}(n; p_1, \ldots, p_r)$	Multinomial distribution
Γ_X	Variance-covariance matrix of the vector X
$Cov(X, Y)$	Covariance matrix of the vectors X and Y
$(X_{(1)}, \ldots, X_{(n)})$	Ordered sample of the variables $X_1, \ldots, X_n$
$f * g$	Convolution product of the functions f and g
$\mu * \nu$	Convolution product of the measures μ and ν
$f * F$	Lebesgue-Stieltjes convolution of the function f and of the mass function F
$F^{*(n)}$	Order n Lebesgue-Stieltjes convolution of F
$\mathcal{N}_d(M, \Gamma)$	d-dimensional Gaussian distribution with mean the vector M and covariance matrix Γ
(X_n)	Random sequence indexed by $\mathbb{N}$
$(\Omega, \mathcal{F}, \mathbb{P})^{\mathbb{N}}$	Infinite product space of $(\Omega, \mathcal{F}, \mathbb{P})$
$\overline{\lim} A_n$	Superior limit of the sequence of events (A_n)
$\underline{\lim} A_n$	Inferior limit of (A_n)
$\mathcal{CP}(\lambda, P)$	Compound Poisson distribution with parameter $\lambda \in \mathbb{R}_+$
$\xrightarrow{p.s.}$	Almost sure convergence
$\xrightarrow{P}$	Convergence in probability

$\xrightarrow{\mathcal{L}}$	Convergence in distribution
$\xrightarrow{L^p}$	Convergence in L^p-norm
$\widehat{P}_n$	Empirical distribution of an n-sample
$\widehat{F}_n$	Empirical distribution function
$\overline{X}_n$	Empirical mean
S_n^*	Empirical variance
I_n	Confidence interval
$\chi^2(P, Q)$	χ^2 distance between distributions P and Q
$L(\theta; x_1, \ldots, x_n)$	Likelihood of an n-sample with parameter θ
$\mathcal{I}_{(X_1,\ldots,X_n)}(\theta)$	Fisher information

Events and Probability Spaces

1

A random experiment is modeled in mathematics through a probability space, which is a particular type of measure space. General notions on measure spaces are here presented only as much as required for understanding the spaces described thereafter. The presentation is based on the fundamental principles of Kolmogorov's axioms, which—completed by the notions of independence and conditioning—constitute the basis of probability theory. This chapter also includes different formulas necessary for computing the probabilities of events.

1.1   Sample Space

Even if a random experiment is repeated under identical conditions, variations in the results are observed. In other words, identical conditions will not lead to a unique result but rather to a set of possible results.

▷ *Example 1.1 (Some Random Experiments)*

- Tossing a coin, rolling a die, playing a card game or the lottery, etc., called "games of chance."
- Observing the speed of a moving vehicle during a time interval $[s, t]$.
- Observing the lifetime without failure of a device, for instance a battery.
- Observing the number of incoming calls at a telephone exchange.
- Observing the growth of a population of bacteria.
- Observing the disintegration of some radioactive substance.   ◁

A random experiment is modeled in mathematics using the set of all possible results of the experiment, referred to as the sample space, and typically denoted by Ω. Elements ω of Ω are called issues, results, or outcomes.

V. Girardin, N. Limnios, *Applied Probability*,
https://doi.org/10.1007/978-3-030-97963-8_1

▷ *Example 1.2 (Coin Tossing)* The following sample spaces can be considered:

- For one toss, $\Omega = \{H, T\}$.
- For two tosses, $\Omega = \{(H, H), (H, T), (T, H), (T, T)\}$.
- For n tosses, $\Omega = \{(u_1, \ldots, u_n) : u_i \in \{H, T\}\}$.
- For an infinite number of tosses, $\Omega = \{H, T\}^{\mathbb{N}}$.   ◁

▷ *Example 1.3 (Urn Problem)* Suppose an urn contains four colored balls—blue, red, green, and white:

- If one ball is drawn at random, the sample space is $\Omega = \{B, R, G, W\}$.
- If two balls are drawn at random and without replacement, then
  $\Omega = \{(B, R), (B, G), (B, W), (R, B), (R, G), (R, W), (G, B), (G, R), (G, W)\}$.
- If balls are drawn with replacement until the red one shows, then
  $\Omega = \{R, (B, R), (G, R), (W, R), (B, B, R), (B, G, R), (B, W, R), \ldots \}$.   ◁

▷ *Example 1.4 (Speed of a Moving Vehicle)* A natural sample space for observing the speed of a moving vehicle in the time interval $[s, t]$ is $\Omega = C^0(]s, t[)$, the set of all continuous numerical functions defined on $]s, t[$.   ◁

▷ *Example 1.5 (Lifetime)* The sample space for investigating the lifetime of some device is either $\Omega = \mathbb{R}_+$ if the count of time is continuous, or $\Omega = \mathbb{N}$ if discrete.   ◁

A random event is an event linked to a random experiment, whose occurrence depends only on the result of that experiment. Mathematically, an event is a subset of Ω, that contains all the outcomes ω realizing this random event.

▷ *Example 1.6 (Continuation of Example 1.2)* For three tosses:

- "getting heads two times"$=\{(H, H, T), (H, T, H), (T, H, H), (H, H, H)\}$.
- "getting tails three times"$= \{(T, T, T)\}$.   ◁

▷ *Example 1.7 (Continuation of Example 1.4)* The event "the speed is less than v_o" is represented in Ω by the set $\{\omega \in \Omega : \sup_{u \in [s,t]} \omega(u) < v_o\}$.   ◁

Historically, the rigorous definition of a probability as a mathematical object emerged through statistical frequencies of events. Suppose we observe the occurrence of an event A, in a series of N independent repetitions (called trials) of the same random experiment. If $n(A)$ denotes the number of occurrences of A, the frequency of A is $n(A)/N$. As N becomes large, the frequency of the event fluctuates less and less, so we can write

$$\frac{n(A)}{N} \longrightarrow p_A, \quad N \to +\infty,$$

and call p_A the probability of the observed event. This convergence—whose exact meaning will be specified in Chap. 4, observed on experimental data, constitutes the empirical law of large numbers. This law has led to extract the notion of probability from the experiment. The set function $A \rightarrow n(A)/N$ satisfies both following properties:

- $0 \leq n(A)/N \leq 1$;
- If $A_1, \ldots, A_k$ are incompatible events—that cannot occur simultaneously—linked to the experiment with sample space Ω, then

$$\frac{n(A_1 \cup \cdots \cup A_k)}{N} = \frac{n(A_1) + \cdots + n(A_k)}{N}.$$

Every event of a random experiment with sample space Ω is a subset of Ω. On the contrary, a subset of Ω is not always an event; it has to belong to a collection of interesting events, referred to as a σ-algebra and denoted by $\mathcal{F}$. Furthermore, the means of measuring the events may vary, which leads to the mathematical notion of measure.

1.2 Measure Spaces

A random experiment is modeled in mathematics through a probability space, that is a measure space whose total measure is equal to 1. Precisely, a measure space is a triple composed of a sample space, a σ-algebra, and a measure. We present most results in relation to these spaces without proofs.

1.2.1 σ-Algebras

Different kinds of collections of subsets of the sample space can be specified.

Definition 1.8 Let Ω be any set.

1. A collection $\mathcal{A}$ of subsets of Ω is called a π-system if $\mathcal{A}$ is closed under complement and finite intersection, that is:

    $a.$ If $A \in \mathcal{A}$, then $\Omega \backslash A = \overline{A} \in \mathcal{A}$.
    $b.$ If $A \in \mathcal{A}$ and $B \in \mathcal{A}$, then $A \cap B \in \mathcal{A}$.

    If, moreover, $\Omega \in \mathcal{A}$, this collection is called an algebra (or a Boolean field).
2. A collection $\mathcal{C}$ of subsets of Ω is called a λ-system (or monotone class) if $\mathcal{C}$ contains Ω and is closed under monotone limits—in mathematical words, if for all non-decreasing (increasing) sequence (A_n) of elements of $\mathcal{C}$, we have $\cup_{n \geq 0} A_n \in \mathcal{C}$ ($\cap_{n \geq 0} A_n \in \mathcal{C}$).

3. A collection $\mathcal{F}$ of subsets of Ω is called a σ-algebra (or σ-field) on Ω if $\mathcal{F}$ is closed under complement and countable unions, that is

> *i.* $\Omega \in \mathcal{F}$.
> *ii. If* $A \in \mathcal{F}$, *then* $\overline{A} \in \mathcal{F}$.
> *iii. If* $A_n \in \mathcal{F}$ *for* $n \in \mathbb{N}$, *then* $\cup_{n \geq 0} A_n \in \mathcal{F}$.

The pair $(\Omega, \mathcal{F})$ is referred to as a measurable space and the elements of $\mathcal{F}$ are the measurable sets, called events in probability theory. The space is said to be discrete or countable if it is finite or denumerable, that is $\Omega = \{\omega_i : i \in I\}$ with $I \subset \mathbb{N}$. It is denumerable if $I = \mathbb{N}$.

Properties of σ-Algebras

1. The empty set $\emptyset$ belongs to $\mathcal{F}$, by *i.* and *ii.*.
2. $\mathcal{F}$ is closed under countable intersections—namely, if $A_n \in \mathcal{F}$ for all integers n, then $\cap_{n \geq 0} A_n \in \mathcal{F}$. This is a consequence of *ii.* and *iii.*, since $\overline{\cap_{n \geq 0} A_n} = \cup_{n \geq 0} \overline{A_n}$.
3. $A \subset B$ with $B \in \mathcal{F}$ implies neither $A \in \mathcal{F}$ nor $B \setminus A \in \mathcal{F}$.

▷ *Example 1.9 (Smallest and Biggest σ-Algebras)* The set of all subsets of Ω, typically denoted by $\mathcal{P}(\Omega)$, is the largest possible σ-algebra on Ω, in the sense of inclusion. In the same way, the trivial σ-algebra $\{\emptyset, \Omega\}$ is the smallest one.   ◁

Exercise 1.1 shows that no enumerable σ-algebra exists. A σ-algebra included in another one is studied in Exercise 1.5.

Any intersection of σ-algebras is a σ-algebra. Since $\mathcal{P}(\Omega)$ is a σ-algebra, generated σ-algebras can be defined using intersections.

Definition 1.10 Let $\mathcal{C} \subset \mathcal{P}(\Omega)$. The σ-algebra generated by $\mathcal{C}$ on Ω is the intersection of all σ-algebras on Ω containing $\mathcal{C}$; in mathematic words

$$\mathcal{F} = \sigma(\mathcal{C}) \iff \begin{cases} \mathcal{F} \; \sigma\text{-algebra on } \Omega, \\ \mathcal{C} \subset \mathcal{F}, \\ \forall \mathcal{F}' \; \sigma\text{-algebra on } \Omega, \quad \mathcal{C} \subset \mathcal{F}' \Rightarrow \mathcal{F} \subset \mathcal{F}'. \end{cases}$$

The σ-algebra generated by $\mathcal{C}$ is also referred to as the minimal σ-algebra over $\mathcal{C}$. Note that if $\mathcal{C} \subset \mathcal{D} \subset \mathcal{P}(\Omega)$, then $\sigma(\mathcal{C}) \subset \sigma(\mathcal{D})$. Besides, for all subsets $\mathcal{C}$ and $\mathcal{D}$ of $\mathcal{P}(\Omega)$,

$$\sigma(\mathcal{C} \cup \mathcal{D}) = \sigma(\{C \cup D : C \in \mathcal{C}, \, D \in \mathcal{D}\}) = \sigma(\{C \cap D : C \in \mathcal{C}, \, D \in \mathcal{D}\}),$$

also denoted by $\sigma(\mathcal{C}, \mathcal{D})$.

▷ *Example 1.11 (A Union of σ-Algebras that Is Not a σ-Algebra)* For all $A \in \mathcal{P}(\Omega)$ and $B \in \mathcal{P}(\Omega)$, first, $\sigma(A) = \{\emptyset, A, \overline{A}, \Omega\}$, and second,

$$\sigma(A, B) = \sigma(\{A, B\})$$

$$= \{\emptyset, A, B, \Omega, \overline{A}, \overline{B}, A \cap B, \overline{A} \cap \overline{B}, \overline{A} \cap B, A \cap \overline{B}, A \cup B, \overline{A} \cup \overline{B},$$

$$\overline{A} \cup B, A \cup \overline{B}, (A \cap B) \cup (\overline{A} \cap \overline{B}), (A \cap \overline{B}) \cup (\overline{A} \cap B)\}.$$

Clearly, $\sigma(A, B) \not\subset \sigma(A) \cup \sigma(B)$, which means that the union of σ-algebras is not a σ-algebra in general.   ◁

The next classical result, stated without proof, induces that the smallest σ-algebra containing a π-system is the smallest λ-system containing it.

Theorem 1.12 (Monotone Class) *If a π-system $\mathcal{A}$ is included in a λ-system $\mathcal{C}$, then its generated σ-algebra $\sigma(\mathcal{A})$ is also included in $\mathcal{C}$.*

If $(\Omega, \mathcal{O})$ is a topological space, then $\sigma(\mathcal{O})$ is referred to as the Borel σ-algebra of Ω, classically denoted by $\mathcal{B}(\Omega)$. Its elements are the Borel sets.

The σ-algebra $\mathcal{B}(\mathbb{R})$ is the smallest σ-algebra of $\mathbb{R}$ that contains all intervals. It is also the set of all parts of $\mathbb{R}$ with a length. Exercise 1.2 shows that all subsets of $\mathbb{R}$ are not Borel sets, that is $\mathcal{B}(\mathbb{R}) \neq \mathcal{P}(\mathbb{R})$.

Theorem 1.13 *The Borel σ algebra of $\mathbb{R}$ is generated by any of the following collections of intervals of $\mathbb{R}$:*

1. $\mathcal{B}(\mathbb{R}) = \sigma(\{]x, y[: (x, y) \in \mathbb{R}^2\})$.   *2.* $\mathcal{B}(\mathbb{R}) = \sigma(\{] - \infty, x[: x \in \mathbb{R}\})$.
3. $\mathcal{B}(\mathbb{R}) = \sigma(\{]x, y] : (x, y) \in \mathbb{R}^2\})$.   *4.* $\mathcal{B}(\mathbb{R}) = \sigma(\{] - \infty, x] : x \in \mathbb{R}\})$.

Proof We prove only the two first points.

1. A subset $\mathcal{O}$ of $\mathbb{R}$ is an open set of the topological space $\mathbb{R}$ if and only if for all $r \in \mathcal{O}$, some $(a, b) \in \mathbb{R}^2$ exists such that $r \in]a, b[\subset \mathcal{O}$, from which the first equality follows.
2. First, $] - \infty, x[= \cup_{n \geq 0}] - n, x[$ is a denumerable union of open sets, and hence $\sigma(\{] - \infty, x[: x \in \mathbb{R}\}) \subset \mathcal{B}(\mathbb{R})$.
   Conversely, for all real numbers $a < b$, we have $]a, b[=] - \infty, b[\cap]a, +\infty[$ and $]a, +\infty[= \overline{] - \infty, a]} = \cup_{n \geq 0}] - \infty, a + 1/n[$, so that $\sigma(\{]a, b[: (a, b) \in \mathbb{R}^2\}) \subset \sigma(\{] - \infty, x[: x \in \mathbb{R}\})$, and the second equality follows.
   □

In the same way, the Borel σ-algebra of $\mathbb{R}^d$, say $\mathcal{B}(\mathbb{R}^d)$, can be shown to be generated by either all the rectangles, the balls, or the rectangles with rational ends, etc. In particular, $\mathcal{B}(\mathbb{R}^2)$ is the set of all subsets of $\mathbb{R}^2$ with an area, and $\mathcal{B}(\mathbb{R}^3)$ of all subsets of $\mathbb{R}^3$ with a volume.

1.2.2 Measures

Let us begin with some definitions and main properties of positive measures, which will be called measures for the sake of simplicity.

Definition 1.14 Let $(\Omega, \mathcal{F})$ be a measurable space. A set function $\mu : \mathcal{F} \to \overline{\mathbb{R}}_+$ is a (positive) measure on $(\Omega, \mathcal{F})$ if $\mu(\emptyset) = 0$ and

$$\mu\left(\bigcup_{n \geq 0} A_n\right) = \sum_{n \geq 0} \mu(A_n), \quad A_n \in \mathcal{F}, \ A_i \cap A_j = \emptyset, \ i \neq j,$$

a property called the axiom of countable additivity (or σ-additivity).

We will consider here only positive measures, which we will simply call measures. The triple $(\Omega, \mathcal{F}, \mu)$ is referred to as a measure space. The quantity $\mu(\Omega)$ is called the total mass of the space. Where no ambiguity may arise, we will also say that μ is a measure on Ω, without specifying the underlying σ-algebra.

Proposition 1.15 *The axiom of countable additivity can alternatively be stated as follows:*

1. For any disjoint measurable sets A and B,

$$\mu(A \cup B) = \mu(A) + \mu(B).$$

2. **(Axiom of countable subadditivity)** *For any sequence (A_n) of measurable sets,*

$$\mu\left(\bigcup_{n \geq 0} A_n\right) \leq \sum_{n \geq 0} \mu(A_n).$$

The axiom of countable subadditivity is also referred to as Boole's inequality.

Proof If μ satisfies the axiom of countable additivity, then *1.* clearly holds.
Let (A_n) be any sequence of measurable sets in $\mathcal{F}$. Setting $B_n = A_n \backslash \cap_{i=0}^{n-1} A_i$ defines a sequence of pairwise disjoint measurable sets of $\mathcal{F}$. Since $\cup_{n \geq 0} A_n = \cup_{n \geq 0} B_n$, *2.* holds.
The proof of the converse is omitted. □

Properties of Measures
1. If $A \subset B$, then $\mu(A) \leq \mu(B)$ and $\mu(B) = \mu(B \backslash A) + \mu(A)$, since $B = (B \backslash A) \cup A$ is a union of disjoint sets.
2. $\mu(A \cup B) = \mu(A) + \mu(B) - \mu(A \cap B)$, since $A \cup B = (A \backslash A \cap B) \cup B$.

If $N \in \mathcal{F}$ is such that $\mu(N) = 0$, then N is said to be μ-null. If $N \in \mathcal{P}(\Omega)$ and if some $A \in \mathcal{F}$ exists such that $N \subset A$ and $\mu(A) = 0$, then N is also said to be μ-null. A measure μ is said to be complete if every μ-null set is measurable.

Let $(\Omega, \mathcal{F}, \mu)$ be any measure space and let (P) denote a property concerning the elements of Ω. If (P) holds except on a μ-null set, then (P) is said to hold almost everywhere with respect to μ (μ-a.e.). If $S \subset \Omega$ and $\mu(\omega) \neq 0$ only for all $\omega \in S$, then S is called the support of μ.

Definition 1.16 Two measure spaces $(\Omega, \mathcal{F}, \mu)$ and $(\Omega', \mathcal{F}', \mu')$ are said to be isomorphic if a μ-null set $N \in \mathcal{F}$ and a bijective function $\varphi \colon \Omega \setminus N \to \Omega'$ exist satisfying $\mu(A) = \mu'(\varphi(A))$ for all $A \in \mathcal{F}$ such that $A \cap N = \emptyset$.

A collection $\{A_i \in \mathcal{P}(\Omega) \setminus \emptyset : i \in I\}$, with $I \subset \mathbb{N}$, is a partition (also called dissection if all A_i are measurable sets) of Ω if $\cup_{i \in I} A_i = \Omega$ and $A_i \cap A_j = \emptyset$ for $i \neq j$.

Definition 1.17 A measure μ on a measurable space $(\Omega, \mathcal{F})$ is said to be

1. Continuous (or diffuse) if $\{\omega\} \in \mathcal{F}$ and $\mu(\{\omega\}) = 0$ for all $\omega \in \Omega$.
2. Discrete if $\mu(\Omega \setminus S) = 0$ for some finite or countable measurable set $S \in \mathcal{F}$.
3. Arithmetic if $(\Omega, \mathcal{F}) = (\mathbb{R}_+, \mathcal{B}(\mathbb{R}_+))$ and $\mu(\{x\}) = 0$ for all $x \notin \{x_0 + n\delta : n \in \mathbb{N}\}$, where $x_0 \in \mathbb{R}_+$ is fixed. The constant $\delta \in \mathbb{R}_+^*$ is called the period of μ.
4. Σ-finite, if a countable partition (A_n) of Ω exists such that $\mu(A_n) < +\infty$ for all n
5. Finite (or bounded) if its total mass is finite, that is $\mu(\Omega) < +\infty$.

If μ is discrete, then

$$\mu(A) = \sum_{\omega \in A} \mu(\{\omega\}), \quad A \in \mathcal{F}.$$

Exercise 2.5 shows that measures that are neither continuous nor discrete do exist.

▷ *Example 1.18* All measures defined on countable spaces are clearly discrete. All discrete measures defined on $\mathbb{N}$ are arithmetic, with $x_0 = 0$ and $\delta = 1$. ◁

The next proposition is an immediate consequence of the theorem of the monotone class.

Proposition 1.19 *Two σ-finite measures that are identical on a π-system $\mathcal{A}$ are also identical on the σ-algebra generated by $\mathcal{A}$.*

Below we present some general positive measures of use in probability theory.

Dirac Measure

For any $\omega \in \Omega$, the Dirac measure (or unit mass) concentrated at ω is the discrete finite measure on $(\Omega, \mathcal{F})$ typically denoted by δ_ω and defined by

$$\delta_\omega(A) = \begin{cases} 1 & \text{if } \omega \in A, \\ 0 & \text{otherwise.} \end{cases}$$

Note that when both A and ω are fixed, $\delta_\omega(A) = \mathbb{1}_A(\omega)$.

Counting Measure

The counting measure is the measure μ defined on any countable measurable space $(\Omega, \mathcal{F})$ by

$$\mu(A) = \sum_{\omega \in \Omega} \delta_\omega(A), \quad \text{for all } A \in \mathcal{F}.$$

This measure takes values in $\overline{\mathbb{N}}$ and is finite whenever Ω is a finite set.

Lebesgue Measure

Using the theorem of the monotone class, a unique measure λ on $(\mathbb{R}, \mathcal{B}(\mathbb{R}))$ such that

$$\lambda([x, y]) = \lambda([x, y[) = \lambda(]x, y]) = \lambda(]x, y[) = y - x, \quad x \le y,$$

can be proven to exist. This measure is continuous and σ-finite; moreover, it is invariant with respect to translation.

In the same way, a unique measure λ_d on $(\mathbb{R}^d, \mathcal{B}(\mathbb{R}^d))$, for $d > 1$, can be proven to exist such that the measure of any parallelepiped is its volume, namely,

$$\lambda_d\left(\prod_{i=1}^d [x_i, y_i]\right) = \prod_{i=1}^d \lambda([x_i, y_i]) = \prod_{i=1}^d (y_i - x_i), \quad x_i \le y_i, \ i \in [\![1, d]\!].$$

The boundary of such a parallelepiped adds nothing to its measure. Therefore, two parallelepipeds differing only by a part of their boundary have the same measure. Where no ambiguity may arise, we will denote λ_d by λ too.

Definition 1.20 A function $F \colon \mathbb{R} \to \mathbb{R}$ that is non-decreasing and continuous from the right is called a mass function.

If $\lim_{x \to +\infty} F(x) < 1$, then F is said to be improper or defective.

If the support of F is $\{x_0 + nh : n \in \mathbb{Z}\}$, where the shift $x_0 \in \mathbb{R}$ and the span $h > 0$ are fixed, then F is said to be arithmetic.

As shown below in Chap. 2, a mass function F such that $\lim_{x\to-\infty} F(x) = 0$ and $\lim_{x\to+\infty} F(x) = 1$ is a distribution function. Arithmetic distribution functions are also called lattice distribution functions.

▷ *Example 1.21 (Lebesgue–Stieltjes Measures)* For any mass function F, a unique measure λ_F such that $\lambda_F(]x, y]) = F(y) - F(x)$ for all $x \leq y$, can be proven to exist, by using the theorem of the monotone class. This defines a one-to-one correspondence—up to an additive constant—between all mass functions and all measures on $(\mathbb{R}, \mathcal{B}(\mathbb{R}))$ such that the measure of any bounded interval of $\mathbb{R}$ is bounded.

Note that $\lambda_F(\{x\}) = \lim_{n\to+\infty} \lambda_F(]x - 1/n, x]) = 0$ if and only if F is a continuous function. Furthermore, the Lebesgue measure on $\mathbb{R}$ is obtained for the identity function $F(x) = x$ for all $x \in \mathbb{R}$.

Lebesgue–Stieltjes measures on $(\mathbb{R}^d, \mathcal{B}(\mathbb{R}^d))$ are similarly defined. ◁

1.3 Probability Spaces

We present the notion of probability through Kolmogorov's system of axioms, with detailed formulas for computing probability of events.

1.3.1 General Case

Let us first present general probability spaces. The discrete spaces will be especially studied in Sec. 1.3.3.

Definition 1.22 Let $(\Omega, \mathcal{F})$ be a measurable space. A function $\mathbb{P}: \mathcal{F} \to \mathbb{R}$ is a probability (or probability measure) on $(\Omega, \mathcal{F})$ if it is a positive measure with total mass equal to 1—in other words, satisfying Kolmogorov axioms:

Axiom 1 $\mathbb{P}(A) \geq 0$, for all $A \in \mathcal{F}$.
Axiom 2 $\mathbb{P}(\Omega) = 1$.
Axiom 3 $\mathbb{P}(\cup_{n\geq 0} A_n) = \sum_{n\geq 0} \mathbb{P}(A_n)$ for any sequence (A_n) of pairwise disjoint elements of $\mathcal{F}$.

The triple $(\Omega, \mathcal{F}, \mathbb{P})$ is referred to as a probability space.

▷ *Example 1.23 (A Probability Derived from a Measure)* Let μ be any finite measure defined on a measurable space $(\Omega, \mathcal{F})$. The measure $\mathbb{P}$ defined on $(\Omega, \mathcal{F})$ by $\mathbb{P}(A) = \mu(A)/\mu(\Omega)$ is a probability and $(\Omega, \mathcal{F}, \mathbb{P})$ is a probability space. ◁

The elements of Ω are the results of the experiment and the elements of $\mathcal{F}$ are the events. Disjoint events are incompatible. An event $A \in \mathcal{F}$ such that $\mathbb{P}(A) = 0$ is null. If $\mathbb{P}(A) = 1$, then A is almost sure. If $\mathbb{P}(A)$ is close to 0, then A is rare. The space Ω is sure, and the empty set $\emptyset$ is impossible. A property that is true $\mathbb{P}$-almost everywhere is said to be almost sure, or $\mathbb{P}$-a.s.

Thanks to additivity, $\mathbb{P}((\Omega \setminus A) \cup A) = \mathbb{P}(\Omega) = 1$, and hence the probability of the complement (or complementary event) $\overline{A} = \Omega \setminus A$ of A is $\mathbb{P}(\overline{A}) = 1 - \mathbb{P}(A)$.

The uniform probability defined in the following example is a basic one.

▷ *Example 1.24 (Uniform Probability)*   If Ω is a finite space, the measure $\mathbb{P}$ defined by $\mathbb{P}(A) = |A|/|\Omega|$ for all $A \subset \Omega$ is a probability on $(\Omega, \mathcal{P}(\Omega))$, referred to as uniform probability. All outcomes $\omega \in \Omega$ have the same probability $\mathbb{P}(\{\omega\}) = 1/|\Omega|$.   ◁

The following formulas prove to be useful tools for computing probability of unions and intersections of events.

Proposition 1.25 (Law of Total Probability)   *For any partition* $\{B_i : i \in I\} \subset \mathcal{F}$ *of* Ω, *with* $I \subset \mathbb{N}$,

$$\mathbb{P}(A) = \sum_{i \in I} \mathbb{P}(A \cap B_i), \quad A \in \mathcal{F}.$$

Proof   Since the events $A \cap B_i$ are pairwise disjoint, we have

$$\mathbb{P}(A) = \mathbb{P}(A \cap \Omega) = \mathbb{P}\Big[A \cap \Big(\bigcup_{i \in I} B_i\Big)\Big] = \mathbb{P}\Big[\bigcup_{i \in I}(A \cap B_i)\Big],$$

and the result follows by countable additivity.   □

Proposition 1.26 (Disjunction Rules)
Disjunction rule *For any events A and B,*

$$\mathbb{P}(A \cup B) = \mathbb{P}(A) + \mathbb{P}(B) - \mathbb{P}(A \cap B).$$

Inclusion–exclusion formula *For any finite family of events $A_1, \ldots, A_n$,*

$$\mathbb{P}\Big(\bigcup_{i=1}^{n} A_k\Big) = \sum_{k=1}^{n}(-1)^{k+1} \sum_{1 \le i_1 < \cdots < i_k \le n} \mathbb{P}\Big(\bigcap_{j=1}^{k} A_{i_j}\Big).$$

Proof   Let us prove the proposition by induction.
For any events A and B, we have $A \cup B = A \cup (B \cap \overline{A})$, a disjoint union. By the axiom of additivity, $\mathbb{P}(A \cup B) = \mathbb{P}(A \cup (B \cap \overline{A})) = \mathbb{P}(A) + \mathbb{P}(B \cap \overline{A})$. Again, the union $B = (B \cap A) \cup (B \cap \overline{A})$ is a disjoint union, so $\mathbb{P}(B) = \mathbb{P}(B \cap A) + \mathbb{P}(B \cap \overline{A})$, yielding the disjunction rule.

Suppose the inclusion–exclusion formula holds for some $n \geq 1$. We have

$$\bigcup_{i=1}^{n+1} A_i = \left(\bigcup_{i=1}^{n} A_i \right) \cup \left(A_{n+1} \cap \overline{\bigcup_{i=1}^{n} A_i} \right),$$

with

$$A_{n+1} \cap \overline{\bigcup_{i=1}^{n} A_i} = A_{n+1} \setminus \left(A_{n+1} \cap \bigcup_{i=1}^{n} A_i \right),$$

and hence,

$$\mathbb{P}\left(\bigcup_{i=1}^{n+1} A_i \right) = \mathbb{P}\left(\bigcup_{i=1}^{n} A_i \right) + \mathbb{P}(A_{n+1}) - \mathbb{P}\left[A_{n+1} \cap \left(\bigcup_{i=1}^{n} A_i \right) \right].$$

Since

$$A_{n+1} \cap \bigcup_{i=1}^{n} A_i = \bigcup_{i=1}^{n} (A_{n+1} \cap A_i),$$

the inclusion–exclusion formula follows from the induction hypothesis. □

Exercise 1.3 shows an application of the inclusion–exclusion formula to the matching problem.

Proposition 1.27 *For any finite family of events* $A_1, \ldots, A_n$,

$$\mathbb{P}\left(\bigcup_{i=1}^{n} A_i \right) = \mathbb{P}(A_1) + \sum_{k=2}^{n-1} \mathbb{P}\left(\left(\bigcap_{i=1}^{k-1} \overline{A_i} \right) \cap A_k \right).$$

Proof The additivity of $\mathbb{P}$ applied to the transformation of the union into a union of disjoint events, say

$$A_1 \cup \cdots \cup A_n = A_1 \cup (\overline{A_1} \cap A_2) \cup (\overline{A_1} \cap \overline{A_2} \cap A_3) \cup \cdots \cup (\overline{A_1} \cap \cdots \cap \overline{A_{n-1}} \cap A_n),$$

yields the result. □

The notion of continuity for probability reads as follows for sequences that are monotone in the sense of inclusion.

Proposition 1.28 *If* (A_n) *is a monotone sequence of events, then*

$$\mathbb{P}(\lim_{n\to+\infty} A_n) = \lim_{n\to+\infty} \mathbb{P}(A_n),$$

where $\lim_{n\to+\infty} A_n = \cup_{n\geq 0} A_n$ $(\lim_{n\to+\infty} A_n = \cap_{n\geq 0} A_n)$ *for a non-decreasing (increasing) sequence.*

Proof Let us suppose that (A_n) is non-decreasing, and set $B_n = A_n \cap (\Omega \setminus A_{n-1})$. We have $\cup_{n\geq 0} B_n = \cup_{n\geq 0} A_n$ and the events B_n are pairwise disjoint, so the conclusion follows by countable additivity.   □

Note that the above notion of limit loses all meaning for non monotonic sequences.

1.3.2   Conditional Probabilities

When a fixed event is known to have occurred, the probability is to be modified accordingly.

Theorem-Definition 1.29 *Let* $(\Omega, \mathcal{F}, \mathbb{P})$ *be a probability space and let* A *be an event with positive probability. The function* $\mathbb{P}(\cdot \mid A)$ *defined on* $\mathcal{F}$ *by*

$$\mathbb{P}(B \mid A) = \frac{\mathbb{P}(B \cap A)}{\mathbb{P}(A)}, \quad B \in \mathcal{F},$$

is a probability. It is called probability conditional on (or assuming) A.

Proof Since $\mathbb{P}$ is a probability, the set function $\mathbb{P}(\cdot \mid A)$ clearly takes values in $[0, 1]$, with $\mathbb{P}(\Omega \mid A) = 1$. If (A_n) is a sequence of pairwise disjoint events, then

$$\mathbb{P}\left(\bigcup_{n\geq 0} A_n \mid A\right) = \frac{\mathbb{P}[(\cup_{n\geq 0} A_n) \cap A]}{\mathbb{P}(A)} = \frac{\mathbb{P}[\cup_{n\geq 0}(A_n \cap A)]}{\mathbb{P}(A)}.$$

Since $\mathbb{P}$ is countably additive, $\mathbb{P}[\cup_{n\geq 0}(A_n \cap A)] = \sum_{n\geq 0} \mathbb{P}(A_n \cap A)$, and hence we get $\mathbb{P}(\cup_{n\geq 0} A_n \mid A) = \sum_{n\geq 0} \mathbb{P}(A_n \mid A)$.   □

Note that if A and B are incompatible (disjoint) events, then $\mathbb{P}(B \cap A) = 0$, and hence $\mathbb{P}(B \mid A) = 0$; if B implies A—that is $B \subset A$, then $\mathbb{P}(B \mid A) = 1$.

▷ *Example 1.30 (Rolling Fair Dice)* Rolling one fair die, the natural sample space is $\Omega = [\![1, 6]\!]$ equipped with the σ-algebra $\mathcal{P}(\Omega)$ and the uniform probability $\mathbb{P}$.

Set $A =$"the score is even" and $B =$"the score is less than 3." We have $A \cap B =$ "the score is 2," $\mathbb{P}(A \cap B) = 1/6$ and $\mathbb{P}(A) = 1/2$, and hence

$$\mathbb{P}(B \mid A) = \frac{1/6}{1/2} = \frac{1}{3}.$$

In the same way, $\mathbb{P}(A \mid B) = 1/2$.

Rolling two good dice—say a black one and a white one, the sample space of is $\Omega = [\![1, 6]\!]^2$ equipped with the σ-algebra $\mathcal{P}(\Omega)$ and the uniform probability.

Let us compute the probability of the sum of scores being more than 6 assuming the score of the black die is 3. The event "the score of the black die is 3" can be written $C = \{(3, x) : x \in [\![1, 6]\!]\}$, and "the sum is more than 6" is $D = \{(x, y) \in [\![1, 6]\!]^2 : x + y > 6\}$. Therefore, $C \cap D = \{(3, 4), (3, 5), (3, 6)\}$, from which we deduce

$$\mathbb{P}(D \mid C) = \frac{\mathbb{P}(C \cap D)}{\mathbb{P}(C)} = \frac{\mid C \cap D \mid}{\mid C \mid} = \frac{3}{6},$$

that is $\mathbb{P}(D \mid C) = 1/2$, as expected.   ◁

▷ *Example 1.31 (Conditional Probability and Children)* A family has two children. Let us compute the conditional probability of both children being boys assuming:

• One of them at least is a boy.

   The natural sample space is $\Omega = \{(b, g), (b, b), (g, g), (g, b)\}$, equipped with the σ-algebra $\mathcal{P}(\Omega)$ and the uniform probability $\mathbb{P}$. The associated events are $C = \{(b, b)\}$ and $D = \{(b, g), (b, b), (g, b)\}$. Since $C \subset D$, we have $\mathbb{P}(C \mid D) = \mathbb{P}(C)/\mathbb{P}(D) = 1/3$.

   An alternative sample space would be $\widetilde{\Omega} = \{\{b, b\}, \{b, g\}, \{g, g\}\}$ equipped with $\mathcal{P}(\widetilde{\Omega})$ and the probability $\widetilde{\mathbb{P}}$ defined by $\widetilde{\mathbb{P}}(\{b, g\}) = 1/2$ and $\widetilde{\mathbb{P}}(\{b, b\}) = \widetilde{\mathbb{P}}(\{g, g\}) = 1/4$. Then $\widetilde{\mathbb{P}}(C) = 1/4$ and $\widetilde{\mathbb{P}}(D) = 3/4$. Again, $\widetilde{\mathbb{P}}(C \mid D) = 1/3$.

• The eldest is a boy.

   Considering $(\Omega, \mathcal{P}(\Omega), \mathbb{P})$, we get $E = \{(b, g), (b, b)\}$, and hence we compute

$$\mathbb{P}(E \mid D) = \frac{\mathbb{P}(E)}{\mathbb{P}(D)} = \frac{1/2}{3/4} = \frac{2}{3}.$$

The alternative space is no more pertinent.   ◁

The definition of a conditional probabilities implies straightforwardly that for all events A and B with positive probabilities,

$$\mathbb{P}(A \mid B)\mathbb{P}(B) = \mathbb{P}(B \mid A)\mathbb{P}(A) = \mathbb{P}(A \cap B).$$

This generalizes to any finite collection of events as follows.

Theorem 1.32 (Compound Probabilities) *If $A_1, \dots, A_n$ are a collection of events such that $\mathbb{P}(A_1 \cap \dots \cap A_{n-1}) > 0$, then*

$$\mathbb{P}\left(\bigcap_{i=1}^{n} A_i\right) = \mathbb{P}(A_1)\mathbb{P}(A_2 \mid A_1)\mathbb{P}(A_3 \mid A_1 \cap A_2) \dots \mathbb{P}(A_n \mid A_1 \cap A_2 \cap \dots \cap A_{n-1}).$$

Proof By induction, since $\mathbb{P}(A_1 \cap A_2) = \mathbb{P}(A_1)\mathbb{P}(A_2 \mid A_1)$ and

$$\mathbb{P}(A_1 \cap \dots \cap A_n) = \mathbb{P}(A_n \mid A_1 \cap \dots \cap A_{n-1})\mathbb{P}(A_1 \cap \dots \cap A_{n-1}),$$

by definition of conditional probabilities.   □

▷ *Example 1.33 (Urn Problem)* An urn contains 15 balls: 4 black ones, 6 red ones, and 5 white ones. Three balls are drawn at random successively and without replacement.

The natural sample space is $\Omega = \{(\omega_1, \omega_2, \omega_3) : \omega_i \in \{W, R, B\}\}$, equipped with the σ-algebra $\mathcal{P}(\Omega)$. Let us show how a probability $\mathbb{P}$ can be defined on this space by computing the probability of the event $E=$ "the first ball is red, the second is white, and the third is black".

Set $A =$ "the first ball is red," $B=$ "the second ball is white," and $C =$ "the third ball is black". Since $E = A \cap B \cap C$, thanks to the compound probabilities theorem, we get $\mathbb{P}(A \cap B \cap C) = \mathbb{P}(A)\mathbb{P}(B \mid A)\mathbb{P}(C \mid A \cap B)$.

First, $\mathbb{P}(A) = 6/15$. Once A has occurred, the urn contains 4 black balls, 5 red balls, and 5 white balls, from which we deduce that $\mathbb{P}(B \mid A) = 5/14$. Once both A and B have occurred —that is $A \cap B$, the urn contains 4 black balls, 5 red balls, and 4 white balls, and hence $\mathbb{P}(C \mid A \cap B) = 4/13$. Therefore, $\mathbb{P}(E) = (6/15) \times (5/14) \times (4/13)$.

Giving a general formula for the probability of any event would be much more intricate, and is not compulsory.   ◁

The next formula can be used to compute the probability of the cause when the probability of the effect is known.

Proposition 1.34 (Bayes' Theorem) *Let the family $\{B_i \in \mathcal{F} : i \in I\}$ be a partition of Ω into events with positive probabilities. For any event $A \in \mathcal{F}$ with positive probability,*

$$\mathbb{P}(B_{i_0} \mid A) = \frac{\mathbb{P}(A \mid B_{i_0})\mathbb{P}(B_{i_0})}{\sum_{i=1}^{n} \mathbb{P}(A \mid B_i)\mathbb{P}(B_i)}, \quad i_0 \in I.$$

Proof By definition of conditional probabilities,

$$\mathbb{P}(B_{i_0} \mid A) = \frac{\mathbb{P}(A \cap B_{i_0})}{\mathbb{P}(A)} = \frac{\mathbb{P}(A \mid B_{i_0})\mathbb{P}(B_{i_0})}{\mathbb{P}[\cup_{i=1}^{n}(A \cap B_i)]}.$$

Since the events are pairwise disjoint, thanks to the law of total probability,

$$\mathbb{P}\left[\bigcup_{i=1}^{n}(A \cap B_i)\right] = \sum_{i=1}^{n} \mathbb{P}(A \cap B_i) = \sum_{i=1}^{n} \mathbb{P}(A \mid B_i)\mathbb{P}(B_i),$$

and the result follows. □

▷ *Example 1.35 (Screening Test for a Disease)* Suppose we know the proportion of positive tests within those infected, the proportion of negative tests among the healthy ones, and the proportion of ill individuals within the population.

Let us compute the probability that a positively tested individual (T) is ill (M),

$$\mathbb{P}(M \mid T) = \frac{\mathbb{P}(T \mid M)\mathbb{P}(M)}{\mathbb{P}(T \mid M)\mathbb{P}(M) + \mathbb{P}(T \mid \overline{M})\mathbb{P}(\overline{M})}$$

$$= \frac{\mathbb{P}(T \mid M)\mathbb{P}(M)}{\mathbb{P}(T \mid M)\mathbb{P}(M) + [1 - \mathbb{P}(\overline{T} \mid \overline{M})][1 - \mathbb{P}(M)]}.$$

For example:

- If $\mathbb{P}(T \mid M) = 0.95$, $\mathbb{P}(\overline{T} \mid \overline{M}) = 0.95$ and $\mathbb{P}(M) = 0.001$, then $\mathbb{P}(M \mid T) \cong 0.02$.
- If $\mathbb{P}(T \mid M) = 0.99$, $\mathbb{P}(\overline{T} \mid \overline{M}) = 0.99$ and $\mathbb{P}(M) = 0.001$, then $\mathbb{P}(M \mid T) \cong 0.1$.
- If $\mathbb{P}(T \mid M) = 0.999$, $\mathbb{P}(\overline{T} \mid \overline{M}) = 0.999$ and $\mathbb{P}(M) = 0.001$, then $\mathbb{P}(M \mid T) \cong 0.5$.

This is an easy illustration of the important issue of accuracy of tests. ◁

1.3.3   Discrete Case: Combinatorial Analysis and Entropy

As long as the sample space is countable, the set $\mathcal{P}(\Omega)$ of all subsets of the sample space is a convenient σ-algebra. Then, any subset of the sample space is an event. A natural characterization of probabilities follows.

Theorem 1.36 *If Ω is countable, then any probability $\mathbb{P}$ on $(\Omega, \mathcal{P}(\Omega))$ is characterized by a sequence of real numbers $(p_i)_{i \in I}$ such that $p_i \geq 0$ for $i \in I$, and $\sum_{i \in I} p_i = 1$.*

We can write $\Omega = \{\omega_i : i \in I\}$ with $I \subset \mathbb{N}$, and $\mathbb{P} = \sum_{i \in I} p_i \delta_{\omega_i}$, where δ_{ω_i} is the unit mass at ω_i.

Proof First, setting $\mathbb{P}(\{\omega_i\}) = p_i$ for $\omega_i \in \Omega$, and $\mathbb{P}(A) = \sum_{\omega_i \in A} \mathbb{P}(\{\omega_i\})$ for $A \in \mathcal{P}(\Omega)$, clearly defines a probability on $(\Omega, \mathcal{P}(\Omega))$.
Conversely, if $\mathbb{P}$ is a probability, set $p_i = \mathbb{P}(\{\omega_i\})$ for $\omega_i \in \Omega$. Clearly, $p_i \geq 0$ for $i \in I$ and $\sum_{i \in I} p_i = \mathbb{P}(\Omega) = 1$ by definition of probabilities.   □

When Ω is finite, the uniform probability $\mathbb{P}$ is defined by $\mathbb{P}(\{\omega\}) = 1/|\Omega|$ for $\omega \in \Omega$; see Example 1.24. Then $\mathbb{P}(A) = |A|/|\Omega|$, for $A \in \mathcal{P}(\Omega)$, so that computing the probability of any event amounts to computing its cardinal, that is to say to combinatorial analysis.

In this aim, let us present some typical occupancy problems through randomly placing n elements into N cells. Let the set of these elements be $\Omega = \{\omega_1, \ldots, \omega_n\}$ and the set of cells $C = \{c_1, \ldots, c_N\}$.

- For $N = n$ and indistinguishable elements, placing exactly 1 element in each cell:

    The first element can be placed into n cells, the second element into $(n - 1)$ cells, and so on until 2 possible choices remain for the $n-1$-th element and only 1 choice for the last element. Therefore, the number of possible orderings—called permutations —of n elements is

$$n! = n(n - 1) \ldots 1.$$

  This counts the possible outcomes in a contest without equal ranking. This is also the number of bijections from Ω to C.
- For $N \geq n$ and distinguishable elements, placing at most 1 element in each cell:

    There are N possible choices for the first element, and so on, until $(N - n + 1)$ choices remain for the last element. Therefore, the number of permutations without repetitions of n elements in N cells is

$$\frac{N!}{(N - n)!}.$$

This counts the possible rankings (without equal ranking) of the n chosen individuals in a contest involving N candidates. This is also the number of injections from Ω to C.

- For $N \geq n$ and indistinguishable elements, placing at most 1 element in each cell:

  There are $N!/(N-n)!$ possible ranked drawings. Within each of these drawings, exactly $n!$ permutations of elements are possible. Therefore, the number of combinations without repetition of n elements chosen among N elements is the binomial coefficient

$$\binom{N}{n} = \frac{N!}{n!(N-n)!}.$$

This counts the possible aggregates of n candidates passing an exam with N candidates. Clearly, $\binom{n}{0} = \binom{n}{n} = 1$, and $\binom{n}{p} = \binom{n}{n-p}$. The following Pascal's triangle formula is well-known,

$$\binom{n}{p} = \binom{n-1}{p-1} + \binom{n-1}{p},$$

as well as the binomial theorem,

$$(a+b)^n = \sum_{k=0}^{n} \binom{n}{k} a^k b^{n-k}.$$

- For any N and n, and distinguishable elements, placing any number of elements in each cell:

  This is the number of functions from Ω to C, namely $|C|^{|\Omega|} = N^n$. Especially, $|\mathcal{P}(\Omega)| = 2^{|\Omega|}$ because $\sum_{k=0}^{n} \binom{n}{k} = 2^n$.

- For any N and n, for indistinguishable elements, placing any number of elements in each cell:

  We state without proof that the number of such placements is $\binom{N+n-1}{n}$. This counts all possible ways for distributing n sheets of paper to N students.

- For any N and n, and distinguishable elements, placing k_i elements in the cell c_i:

  Clearly, a necessary condition is $\sum_{i=1}^{N} k_i = n$. There are $\binom{n}{k_1}$ possible choices for the k_1 elements in c_1, then $\binom{n-k_1}{k_2}$ choices for the k_2 elements in c_2, and so on, until only k_N elements remain to be put in c_N. Therefore the number of possibilities is the so-called multinomial coefficient

$$\binom{n}{k_1 \ldots k_n} = \frac{n!}{k_1! \ldots k_N!} = \binom{n}{k_1} \cdot \binom{n-k_1}{k_2} \cdots \binom{n-k_1-\cdots-k_{N-1}}{k_N}.$$

This is for example the number of possible distributions of n sheets of paper to N students, with k_i sheets given to the i-th student. See Chap. 3 for links to the multinomial distribution.

▷ *Example 1.37 (Continuation of Example 1.3)* Drawing at random and without replacement all the balls, the probability of obtaining (W, G, R, B) is $1/24$, since the number of permutations of 4 balls is $4! = 1 \times 2 \times 3 \times 4 = 24$.

Drawing 2 balls, the probability of obtaining (W, R) is $1/12$, since the number of permutations without repetition of 2 balls among 4 is $4!/(4-2)! = 12$.   ◁

▷ *Example 1.38 (Urn Problem)* An urn contains m white balls and n black balls. Let us draw without replacement r balls $(r \leq m)$. The probability that only white balls show is $\binom{m}{r}/\binom{m+n}{r}$, since $\binom{m}{r}$ possible drawings of r balls among m exist and $\binom{m+n}{r}$ of r balls among $m + n$.

This leads to the hyper-geometric distribution to be defined in Chap. 2.   ◁

Let us now present the notion of entropy of a random system. Entropy measures the quantity of information contained in the system, that is to say the uncertainty of a random phenomenon.

▷ *Example 1.39 (A Source in Terms of Information Theory)* A source of information produces random sequences of symbols drawn from a given set. The source is said to be without memory if none of occurring symbols depends on the preceding ones. A source is modeled by the set of all possible outcomes and is identified to a probability space. For a discrete source, the sample space is called the alphabet, an outcome is a letter, and an event is a word.   ◁

Definition 1.40 Let $\Omega = \{\omega_i : i \in I\}$ be a discrete sample space and let P be any probability on $(\Omega, \mathcal{P}(\Omega))$. Set $p_i = P(\{\omega_i\})$. The Shannon entropy (or quantity of information) of P is defined by

$$\mathbb{S}(P) = -\sum_{i \in I} p_i \log p_i,$$

with the convention $0 \log 0 = 0$.

The definition holds for any base of logarithm, although the binary logarithm is usual in information theory. The quantity of information linked to $A \in \mathcal{P}(\Omega)$ is $-\log P(A)$. In other words, P associates to each outcome ω_i the quantity of information $-\log p_i$ and $\mathbb{S}(P)$ is the weighted mean of these quantities according to their probability of occurrence. The same procedure operates on a partition $\mathcal{A} = (A_i)_{i=1}^{n}$ of any sample space Ω; then, the system $(\mathcal{A}, P)$ associated with the entropy $\mathbb{S}(\mathcal{A}, P) = \mathbb{S}(p_1, \ldots, p_n) = -\sum_{i=1}^{n} p_i \log p_i$ has to be considered, with here $p_i = P(A_i)$.

▷ *Example 1.41 (A Binary Information System)* Suppose a system contains two elements.

If $P(A_1) = 1 - P(A_2) = 1/200$, then $\mathbb{S}(\mathcal{A}, P) \cong 0.0215$ (for the natural logarithm). It is easy to guess which of the two events will occur.

If $P(A_1) = 1 - P(A_2) = 1/20$, then $\mathbb{S}(\mathcal{A}, P) = 0.101$, prediction is less easy.

If $P(A_1) = 1 - P(A_2) = 1/2$, then $\mathbb{S}(\mathcal{A}, P) = 0.693$, uncertainty is maximum and prediction impossible.   ◁

Properties of Shannon Entropy
1. $\mathbb{S}(P) \geq 0$, since all $p_i \in [0, 1]$, with equality if $p_{i_0} = 1$ for some i_0.
2. **(Continuous)** $\mathbb{S}(p_1, \ldots, p_n)$ is continuous with respect to each p_i.
3. **(Symmetric)** $\mathbb{S}(p_1, \ldots, p_n) = \mathbb{S}(p_{\sigma(1)}, \ldots, p_{\sigma(n)})$ for any permutation σ of $[\![1, n]\!]$.
4. **(Maximum)** For a finite space, the entropy is maximum for the uniform probability. Indeed, $\mathbb{S}(1/n, \ldots, 1/n) = \log n$ and

$$\mathbb{S}(p_1, \ldots, p_n) - \log n = -\sum_{i=1}^{n} p_i \log p_i + \log \frac{1}{n} = \sum_{i=1}^{n} p_i \log \frac{1}{np_i}.$$

Since $\log x \leq x - 1$ for all positive x, we get

$$\mathbb{S}(p_1, \ldots, p_n) - \log n \leq \sum_{i=1}^{n} p_i \left(\frac{1}{np_i} - 1 \right) = 0.$$

5. **(Non-decreasing)** $\mathbb{S}(1/n, \ldots, 1/n)$ increases when n increases (the uncertainty of a system increases with the number of its elements).

A functional satisfying 1, 2, and 4 can be proven to be the Shannon entropy up to a multiplicative constant (depending on the chosen base of logarithm). These conditions are linked to some physical interpretation of the systems.

The uniform probability has maximum entropy among all probabilities defined on a finite set. Further, the method of maximum entropy consists of choosing among all probabilities satisfying a finite number of constraints the solution that maximizes the entropy—meaning the one linked to the most uncertain system.

Proposition 1.42 (Additivity) *If $p_n = \sum_{k=1}^{m} q_k$, then*

$$\mathbb{S}(p_1, \ldots, p_{n-1}, q_1, \ldots, q_m) = \mathbb{S}(p_1, \ldots, p_n) + p_n \mathbb{S}\left(\frac{q_1}{p_n}, \ldots, \frac{q_m}{p_n} \right).$$

Proof We compute

$$\mathbb{S}(p_1, \ldots, p_{n-1}, q_1, \ldots, q_m) - \mathbb{S}(p_1, \ldots, p_n) =$$

$$= p_n \log p_n - \sum_{k=1}^{m} q_k \log q_k = p_n \sum_{k=1}^{m} \frac{q_k}{p_n} \log p_n - \sum_{k=1}^{m} q_k \log q_k$$

$$= -p_n \sum_{k=1}^{m} \frac{q_k}{p_n} \log \frac{q_k}{p_n} = p_n \mathbb{S}\Big(\frac{q_1}{p_n}, \ldots, \frac{q_m}{p_n}\Big),$$

which gives the searched result.   □

For studying two distributions of events simultaneously, relative entropy is a necessary tool.

Definition 1.43 Let P and Q be two probabilities defined on the same discrete sample space $\Omega = \{\omega_i : i \in I\}$, such that $p_i = 0$ when $q_i = 0$, where $p_i = P(\{\omega_i\})$ and $q_i = Q(\{\omega_i\})$. The relative entropy of P with respect to Q—also called Kullback–Leibler information or divergence—is

$$\mathbb{K}(P \mid Q) = \sum_{i \in I} p_i \log \frac{p_i}{q_i}.$$

The solution minimizing Kullback–Leibler information with respect to a reference probability among probabilities satisfying smooth constraints will be the closest possible to the reference in terms of information—behaving as a projection. Indeed, the Kullback–Leibler information is a measure of distance between distributions or probabilities; it is nonnegative and $\mathbb{K}(P \mid Q) = 0$ if and only if $P = Q$. Still, it is not a distance in the sense of metric spaces because it is not symmetric and the triangular inequality does not hold true. Nevertheless, a Pythagorean theorem for projection can be proven to hold under reasonable hypothesis on the involved constraints.

1.4 Independence of Finite Collections

All notions of independence are based on the independence of σ-algebras.

Definition 1.44 Any σ-algebras $\mathcal{F}_1, \ldots, \mathcal{F}_n$ included in $\mathcal{F}$ are said to be (mutually) independent if

$$\mathbb{P}(B_1 \cap \cdots \cap B_n) = \mathbb{P}(B_1) \ldots \mathbb{P}(B_n), \quad B_1 \in \mathcal{F}_1, \ldots, B_n \in \mathcal{F}_n.$$

Contrary to incompatibility that is intrinsic to events, independence depends on $\mathbb{P}$: Two σ-algebras can be independent with respect to a probability $\mathbb{P}$ and dependent with respect to another probability $\mathbb{P}'$.

Intuitively, events are independent if the occurrence of one does not depend on the occurrence of the others.

Definition 1.45 The events $A_1, \ldots, A_d$ of $\mathcal{F}$ are said to be (mutually) independent if their generated σ-algebras are independent, that is if

$$\mathbb{P}(A_{k_1} \cap \cdots \cap A_{k_n}) = \prod_{i=1}^{n} \mathbb{P}(A_{k_i}), \quad 2 \leq n \leq d, \ 1 \leq k_1 < \cdots < k_n \leq d.$$

They are pairwise independent if $\mathbb{P}(A_i \cap A_j) = \mathbb{P}(A_i)\mathbb{P}(A_j)$ for any $1 \leq i \neq j \leq d$.

For two events, because $\sigma(A_i) = \{\Omega, \emptyset, A_i, \overline{A_i}\}$, independence reduces to a simple criterion: Two events A_1 and A_2 of $\mathcal{F}$ are independent if and only if

$$\mathbb{P}(A_1 \cap A_2) = \mathbb{P}(A_1)\mathbb{P}(A_2).$$

Two disjoint events may be independent only if one is null, and a null event is independent of any other event; see Exercise 1.9.

Pairwise independence does not imply mutual independence, as shown by the next example; see also Exercise 1.10.

▷ *Example 1.46 (Mutual and Pairwise Independence)* Suppose a random experiment is modeled by the sample space $\Omega = \{abc, acb, bac, bca, cab, cba, aaa, bbb, ccc\}$, equipped with the σ-algebra $\mathcal{F} = \mathcal{P}(\Omega)$ and the uniform probability $\mathbb{P}$, namely $\mathbb{P}(\{\omega\}) = 1/9$ for any $\omega \in \Omega$. Set $A_k=$"the k-th letter is a", for $k = 1, 2, 3$.

We compute $\mathbb{P}(A_1) = \mathbb{P}(\{abc, acb, aaa\}) = 3/9$, $\mathbb{P}(A_2) = \mathbb{P}(\{bac, cab, aaa\}) = 3/9$ and $\mathbb{P}(A_1 \cap A_2) = \mathbb{P}(\{aaa\}) = 1/9$, and hence $\mathbb{P}(A_1 \cap A_2) = \mathbb{P}(A_1)\mathbb{P}(A_2)$. In the same way, $\mathbb{P}(A_3) = 1/3$, $\mathbb{P}(A_2 \cap A_3) = \mathbb{P}(A_2)\mathbb{P}(A_3)$, and $\mathbb{P}(A_3 \cap A_1) = \mathbb{P}(A_3)\mathbb{P}(A_1)$, but $\mathbb{P}(A_1 \cap A_2 \cap A_3) = \mathbb{P}(\{aaa\}) = 1/9$. Therefore, $\mathbb{P}(A_1 \cap A_2 \cap A_3) \neq \mathbb{P}(A_1)\mathbb{P}(A_2)\mathbb{P}(A_3)$. The events A_1, A_2, and A_3 are pairwise independent, but are not mutually independent.   ◁

The next proposition, a straightforward consequence of the definitions, is a useful tool for proving that two σ-algebras are independent.

Proposition 1.47 *Let $\mathcal{C}$ and $\mathcal{D}$ be stable under finite intersections subsets of $\mathcal{F}$. If all $C \in \mathcal{C}$ and $D \in \mathcal{D}$ are independent, then the generated σ-algebras $\sigma(\mathcal{C})$ and $\sigma(\mathcal{D})$ are independent.*

Independence can be written in terms of conditional probabilities as follows, again a simple consequence of the definitions.

Proposition 1.48 *Let $(\Omega, \mathcal{P}(\Omega), \mathbb{P})$ be a countable probability space. Two events A and B with positive probabilities are independent if and only if*

$$\mathbb{P}(B \mid A) = \mathbb{P}(B) \quad or \quad \mathbb{P}(A \mid B) = \mathbb{P}(A).$$

Further, independence conditional on an event—or on the σ-algebra generated by this event—reads as follows.

Definition 1.49 Let B be an event with positive probability. The events $A_1, \ldots, A_d$ of $\mathcal{F}$ are said to be independent conditional on B if the σ-algebras they generate are independent for the probability conditional on B, in other words $\mathbb{P}(A_1 \cap \cdots \cap A_d \mid B) = \mathbb{P}(A_1 \mid B) \ldots \mathbb{P}(A_d \mid B)$.

1.5   Exercises

∇ **Exercise 1.1 (No Enumerable σ-Algebra Exists)**   Let $(\Omega, \mathcal{F})$ be a measurable space such that the cardinal of $\mathcal{F}$ is finite. Let $A(\omega)$ denote the intersection of all events containing ω.

1. A binary relation is defined on Ω by setting: $\omega \sim \omega'$ if $A(\omega) = A(\omega')$. Prove that this relation is an equivalence relation.
2. Let $\mathcal{C}$ denote the partition of $\mathcal{P}(\Omega)$ into equivalence classes linked to this relation. Prove that the cardinal of $\mathcal{C}$ is finite and that its generated σ-algebra is exactly $\mathcal{F}$.
3. Application: Prove that no discrete non-finite σ-algebra exists.

Solution
1. The union of the $A(\omega)$ is clearly Ω. If $A(\omega) \cap A(\omega')$ were non-empty, then:
   Either $\omega \in A(\omega) \cap A(\omega') \subset A(\omega)$, and $A(\omega) \cap A(\omega')$ would be contained in all the events containing ω, which contradicts the definition of $A(\omega)$, unless $A(\omega) = A(\omega')$;
   Or $\omega \in A(\omega) \setminus [A(\omega) \cap A(\omega')] \subset A(\omega)$, and $A(\omega) \setminus [A(\omega) \cap A(\omega')]$ would be contained in all the events containing ω, which is not possible.
   Clearly, the relation is reflexive, symmetric, and transitive, and its classes are the sets $A(\omega)$.
2. We have $\mathcal{C} = \{A(\omega) \in \mathcal{F} : \omega \in \Omega\}$. Each $A(\omega)$ is an event—as a finite intersection of events, and hence $\mathcal{C} \subset \mathcal{F}$. Moreover, $\mathcal{C}$ is finite as a subset of $\mathcal{F}$. Using Definition 1.10, since $\mathcal{C} \subset \mathcal{F}$ and $\mathcal{F}$ is a σ-algebra, $\sigma(\mathcal{C}) \subset \mathcal{F}$. Finally, for any event $A \in \mathcal{F}$, we can write $A = \cup_{\omega \in A} A(\omega)$, where the union is finite since $\mathcal{C}$ is finite. Therefore $A \in \sigma(\mathcal{C})$, and hence $\sigma(\mathcal{C}) = \mathcal{F}$.
3. Questions 1 and 2 apply to an enumerable σ-algebra, but then the partition $\mathcal{C}$ would be enumerable too, and the set of all countable unions of elements of $\mathcal{C}$, that is $\sigma(\mathcal{C})$, would no more be enumerable, hence a contradiction. $\triangle$

∇ **Exercise 1.2 (A Subset of $\mathbb{R}$ that Is Not a Borel Set)** An equivalence relation is defined on $[0, 1]$ by setting: $x \sim y$ if $x - y \in \mathbb{Q}$. We define a set $E \subset [0, 1]$ by taking exactly one point in each equivalence class—thanks to the axiom of choice. Set $E + r = \{x + r : x \in E\}$, for $r \in \mathbb{R}$.

1. Prove that for all $x \in [0, 1]$, there exist $r \in \mathbb{Q} \cap [-1, 1]$ such that $x \in E + r$.
2. Prove that if $r \neq s$ are not rational numbers, then $E + r$ and $E + s$ are disjoint.
3. Application: Prove that E is not a Borel set.

Solution
1. For all $x \in [0, 1]$, $y \in E$ exist such that $x \sim y$. Clearly, setting $r = x - y$ yields $x \in E + r$.
2. If $x \in E + r \cap E + s$, then $x = y + r = z + s$ with $y \in E$ and $z \in E$, or $y - z = s - r$. Since $r \neq s$, this would imply $y \sim z$, hence a contradiction.
3. If $E \in \mathcal{B}(\mathbb{R})$, then E would be measurable for the Lebesgue measure. Let us set $\alpha = \lambda(E)$ and

$$S = \bigcup_{r \in \mathbb{Q} \cap [-1,1]} (E + r).$$

From 2, we get $E + r \cap E + s = \emptyset$ for all r and s in $\mathbb{Q} \cap [-1, 1]$ such that $r \neq s$. From the invariance of λ with respect to translation, we get $\lambda(E + r) = \alpha$. Since S is a countable union of disjoint sets, we have

$$\lambda(S) = \sum_{r \in \mathbb{Q} \cap [-1,1]} \lambda(E + r) = \sum_{r \in \mathbb{Q} \cap [-1,1]} \alpha.$$

We have $\lambda(S) \leq 3$ because $S \subset [-1, 2]$, therefore $\alpha = 0$ and $\lambda(S) = 0$. But $[0, 1] \subset S$, and hence $\lambda(S) \geq 1$, a contradiction.

Note that this shows also that no measure that is invariant with respect to translation may exist on $\mathcal{P}(\mathbb{R})$.   △

∇ **Exercise 1.3 (Matching Problem)** Suppose n sportsmen leave their bags in the changing-room of their club and all take one at random when returning home. Compute the probability that no matches occur.

Solution
The sample space is the set $\mathcal{S}_n$ of permutations of $[\![1, n]\!]$ equipped with $\mathcal{P}(\mathcal{S}_n)$ and the uniform probability.

Set $A_i = $ "the i-th sportsman takes his own bag," for $i \in [\![1, n]\!]$. Then $A_1 \cup \cdots \cup A_n = $ "at least one sportsman takes his own bag," and $\overline{A_1 \cup \cdots \cup A_n} = $ "none of them takes his own bag." Since

$$\mathbb{P}(A_{i_1} \cap \cdots \cap A_{i_k}) = \mathbb{P}(A_1 \cap \cdots \cap A_k) = \frac{(n-k)!}{n!},$$

we get using the inclusion–exclusion formula—see Proposition 1.26,

$$P\left(\bigcup_{k=1}^{n} A_k\right) = \sum_{k=1}^{n}(-1)^{k-1} \binom{n}{k} \frac{(n-k)!}{n!} = \sum_{k=1}^{n} \frac{(-1)^{k-1}}{k!}.$$

Therefore, the probability that no match occurs is

$$1 - \sum_{k=1}^{n} \frac{(-1)^{k-1}}{k!} = \sum_{k=0}^{n} \frac{(-1)^k}{k!}.$$

This quantity tends to $1/e$ when n tends to infinity. $\triangle$

▽ Exercise 1.4 (Convex Combination of Probabilities)

1. Let $(\mathbb{P}_n)$ be a sequence of probabilities on $(\Omega, \mathcal{F})$. Let (a_n) be a sequence of positive real numbers whose sum is 1. Prove that the set function $\mathbb{P}$ is a probability, where

$$\mathbb{P}(A) = \sum_{i \geq 1} a_i \mathbb{P}_i(A), \quad A \in \mathcal{F}.$$

2. Application: Let $\{B_i : i \in I\}$ be a partition of Ω with positive probabilities, and let P be the set function defined on $\mathcal{P}(\Omega)$ by

$$P(A) = \sum_{i \in I} a_i \frac{|B_i \cap A|}{|B_i|}.$$

Give a necessary and sufficient condition on the real numbers a_i for P to be a probability.

Solution
1. Let (A_n) be a sequence of pairwise independent events. We have

$$\mathbb{P}\left(\bigcup_{n \geq 0} A_n\right) = \sum_{i \geq 1} a_i \mathbb{P}_i\left(\bigcup_{n \geq 0} A_n\right) \overset{(1)}{=} \sum_{i \geq 1} a_i \sum_{n \geq 0} \mathbb{P}_i(A_n).$$

(1) by countable additivity, because the set functions $\mathbb{P}_i$ are probabilities; see Definition 1.22. Hence, since the terms of the series are nonnegative,

$$\mathbb{P}\left(\bigcup_{n \geq 0} A_n\right) = \sum_{n \geq 0} \sum_{i \geq 1} a_i \mathbb{P}_i(A_n) = \sum_{n \geq 0} \mathbb{P}(A_n).$$

Finally, $\mathbb{P}(\Omega) = \sum_{i \in I} a_i = 1$ by hypothesis.

2. Necessarily $P(A) \geq 0$ for all $A \in \mathcal{P}(\Omega)$, and hence $a_i \geq 0$ for all $i \in I$. From $P(\Omega) = 1$, we get $\sum_{i \in I} a_i = 1$.

   In order to apply 1, it is sufficient to show that the set function P_i defined on $\mathcal{P}(\Omega)$ by $\mathbb{P}_i(A) = |B_i \cap A|/|B_i|$ is a probability for all $i \in I$. Clearly, we have $P_i(A) \geq 0$ and $P_i(\Omega) = 1$. Let now (A_n) be a sequence of pairwise independent events. We have

$$P_i\left(\bigcup_{n \geq 0} A_n\right) = \frac{|B_i \cap (\cup_{n \geq 0} A_n)|}{|B_i|} = \frac{|\cup_{n \geq 0} (B_i \cap A_n)|}{|B_i|}$$

$$\stackrel{(1)}{=} \frac{\sum_{n \geq 0} |B_i \cap A_n|}{|B_i|} = \sum_{n \geq 0} P_i(A_n).$$

   (1) by pairwise independence. $\triangle$

∇ **Exercise 1.5 (Restriction of σ-Algebras and Probabilities)** Let $(\Omega, \mathcal{F}, \mathbb{P})$ be a probability space and let A be an event.

1. Prove that $\mathcal{F}_A = \{A \cap B : B \in \mathcal{F}\}$ is a σ-algebra on A.
2. Give a condition for the restriction of $\mathbb{P}$ on $(A, \mathcal{F}_A)$ to be a probability.

Solution
1. Since $A = A \cap A$, we have $A \in \mathcal{F}_A$. If $C \in \mathcal{F}_A$, then $C = A \cap B$ with $B \in \mathcal{F}$, hence $A \backslash C = A \backslash (A \cap B) = A \cap (\Omega \backslash B)$ and $\Omega \backslash B \in \mathcal{F}$, so $A \backslash C \in \mathcal{F}_A$. Finally, let (C_n) be a sequence of elements of $\mathcal{F}_A$. We have $\cup_{n \geq 0} C_n = \cup_{n \geq 0} (A \cap B_n) = A \cap (\cup_{n \geq 0} B_n)$ and, since $\mathcal{F}$ is a σ-algebra, $\cup_{n \geq 0} B_n \in \mathcal{F}$. Therefore, thanks to Definition 1.8, $\mathcal{F}_A$ is indeed a σ-algebra on A.
2. The restriction of $\mathbb{P}$ to $\mathcal{F}_A$ is defined for $C = B \cap A$ by $\mathbb{P}_{|\mathcal{F}_A}(C) = \mathbb{P}(B \cap A)$. Let $(C_n) = (A \cap B_n)$ be a sequence of pairwise disjoint elements of $\mathcal{F}_A$. Since $\mathbb{P}$ is countably additive and the events of the sequence $(A \cap B_n)$ are pairwise disjoint, we have

$$\mathbb{P}_{|\mathcal{F}_A}\left(\bigcup_{n \geq 0} C_n\right) = \mathbb{P}_{|\mathcal{F}_A}\left[\bigcup_{n \geq 0} (A \cap B_n)\right] = \sum_{n \geq 0} \mathbb{P}(A \cap B_n) = \sum_{n \geq 0} \mathbb{P}_{|\mathcal{F}_A}(C_n).$$

   Moreover, $\mathbb{P}_{|\mathcal{F}_A}(A) = \mathbb{P}(A)$. Therefore, $\mathbb{P}_{|\mathcal{F}_A}$ is a probability if and only if $\mathbb{P}(A) > 0$. $\triangle$

∇ **Exercise 1.6 (Reliability of a Hi-Fi System)** A hi-fi system includes a tuner (1), a CD player (2), an amplifier (3), and two speakers (4) and (5); see Fig. 1.1. The system is said to be functioning whenever the tuner or the CD player, the amplifier, and at least one of the two speakers are functioning. All the components are supposed to function independently of each other.

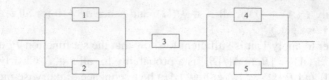

Fig. 1.1 Diagram for the hi-fi system—Exercise 1.6

1. Determine the probability space associated to this experiment.
2. Compute the probability of event A = "the system is functioning," in terms of probabilities of events C_i = "(i) is functioning."
3. Numerical application: $\mathbb{P}(C_1) = 0.9$, $\mathbb{P}(C_2) = 0.95$, $\mathbb{P}(C_3) = 0.8$, and $\mathbb{P}(C_4) = \mathbb{P}(C_5) = 0.7$.

Solution

1. This kind of systems is called a series–parallel system. The space is $\Omega = \prod_{j=1}^{5}\{u_j, d_j\}$ equipped with $\mathcal{P}(\Omega)$ and the product probability, where u_i corresponds to the functioning of component i and d_i to breakdown.
2. We can write $C_i = \{u_i\} \times \prod_{j\neq i}\{u_j, d_j\}$. Moreover, $A = (C_1 \cup C_2) \cap C_3 \cap (C_4 \cup C_5)$, and hence

$$\mathbb{P}(A) =$$

$$= \mathbb{P}(C_1 \cup C_2) \cap C_3 \cap [C_4 \cup C_5)] = \mathbb{P}(C_1 \cup C_2)\mathbb{P}(C_3)\mathbb{P}(C_4 \cup C_5),$$

$$= [\mathbb{P}(C_1) + \mathbb{P}(C_2) - \mathbb{P}(C_1)\mathbb{P}(C_2)]\,\mathbb{P}(C_3)\,[\mathbb{P}(C_4) + \mathbb{P}(C_5) - \mathbb{P}(C_4)\mathbb{P}(C_5)].$$

3. For the given data, we compute $\mathbb{P}(A) = [0.9 + 0.95 - 0.9 \times 0.95] \times 0.8 \times [0.7 + 0.7 - 0.7 \times 0.7] = 0.72436$. $\triangle$

∇ **Exercise 1.7 (A Telecom Problem)** A binary message is sent under the form of 0 and 1. During the transfer, 2/5-th of the 0 are changed into 1 and 1/3-rd of 1 are changed into 0. The message is 3 times 1 for 5 times 0.

1. Give the sample space of this experiment. Compute the probability that the received signal is the same as the transmitted message, assuming that the received message is first 0 and then 1.
2. The reliability of the transmission being unsatisfactory, the sender tries to improve it by sending three times the same message. The message is supposed to be the number received at least two times. Compute the probability of error assuming that the message is 0 and then 1.

Solution

1. The sample space is $\Omega = \{(0, 0), (0, 1), (1, 0), (1, 1)\}$ equipped with $\mathcal{P}(\Omega)$ and with the probability defined above.

The events $R_i =$ "the received signal is i" and $T_i =$ "the transmitted message is i" can be written $A_i = \{(0, i), (1, i)\}$ and $B_i = \{(i, 0), (i, 1)\}$, for $i = 0, 1$. Since $\mathbb{P}(T_1)/\mathbb{P}(T_0) = 3/5$ and $\mathbb{P}(T_1) + \mathbb{P}(T_0) = 1$, we have $\mathbb{P}(T_0) = 5/8$ and $\mathbb{P}(T_1) = 3/8$. We also know that $\mathbb{P}(R_1 \mid T_0) = 2/5 = 1 - \mathbb{P}(R_0 \mid T_0)$ and $\mathbb{P}(R_0 \mid T_1) = 1/3 = 1 - \mathbb{P}(R_1 \mid T_1)$. Bayes' theorem gives

$$\mathbb{P}(T_0 \mid R_0) = \frac{\mathbb{P}(T_0)\mathbb{P}(R_0 \mid T_0)}{\mathbb{P}(T_0)\mathbb{P}(R_0 \mid T_0) + \mathbb{P}(T_1)\mathbb{P}(R_0 \mid T_1)} = \frac{\frac{5}{8} \cdot \frac{3}{5}}{\frac{3}{8} + \frac{1}{8}} = \frac{3}{4}.$$

We compute in the same way $\mathbb{P}(T_1 \mid R_1) = 1/2$.

2. If the transmitted message is 0, the probability of error is

$$\mathbb{P}(E_0) = \binom{3}{2}\left(\frac{2}{5}\right)^2 \frac{3}{5} + \left(\frac{2}{5}\right)^3 = \frac{56}{125} < \frac{2}{5}.$$

If the transmitted message is 1,

$$\mathbb{P}(E_1) = \binom{3}{2}\left(\frac{1}{3}\right)^2 \frac{2}{3} + \left(\frac{1}{3}\right)^3 = \frac{7}{27} < \frac{1}{3}.$$

The reliability of transmission increases, but the process is three times more costly. △

▽ Exercise 1.8 (Laplace's Rule of Succession)

1. Suppose $N + 1$ urns are numbered from 0 to N. The k-th urn contains k/N white balls. An urn is chosen at random and then $n + 1$ balls are drawn from it, successively and with replacement.
   (a) Give the sample space associated with this experiment.
   (b) Compute the probability that the first n drawn balls are white.
   (c) Compute the probability that the $(n + 1)$-th ball is white assuming the first n ones were white.
2. Application: An event occurs with an unknown probability. Assuming it occurred n times at n trials, compute the probability that it occurs at the $(n + 1)$-th trial, when N tends to infinity.

Solution
1.
   (a) The sample space is $\Omega = [\![1, N]\!] \times \{0, 1\}^{n+1}$ equipped with $\mathcal{P}(\Omega)$ and with the probability $\mathbb{P}$ defined by

$$\mathbb{P}[\{(k, i_1, \ldots, i_{n+1})\}] = \frac{1}{N+1} \times \left(\frac{k}{N}\right)^{\sum_{j=1}^{n+1} i_j} \times \left(\frac{N-k}{N}\right)^{n+1 - \sum_{j=1}^{n+1} i_j},$$

for $k \in [\![1, N]\!]$ and $(i_1, \ldots, i_{n+1}) \in \{0, 1\}^{n+1}$.

(b) Set $A_n =$ "the first n balls are white." The collection of events $U_k =$ "the k-th urn is chosen" for $k \in [\![1, N]\!]$ is a partition of Ω, with the same probability $\mathbb{P}(U_k) = 1/(N+1)$. Therefore

$$\mathbb{P}(A_n) = \sum_{k=0}^{N} \mathbb{P}(U_k)\mathbb{P}(A_n \mid U_k) = \frac{1}{N+1} \sum_{k=0}^{N} \left(\frac{k}{N}\right)^n.$$

(c) Set $B_{n+1} =$ "the $(n+1)$-th ball is white." We have $B_{n+1} \cap A_n = A_{n+1}$, from which we get

$$\mathbb{P}(B_{n+1} \mid A_n) = \frac{\mathbb{P}(A_{n+1})}{\mathbb{P}(A_n)} = \frac{\sum_{k=0}^{N} k^{n+1}}{N \sum_{k=0}^{N} k^n}.$$

2. This experiment can be modeled by 1, with N tending to infinity. Since

$$\frac{1}{N} \sum_{k=0}^{N} \left(\frac{k}{N}\right)^n \longrightarrow \int_0^1 x^n dx,$$

we get

$$\mathbb{P}(B_{n+1} \mid A_n) \longrightarrow \frac{\int_0^1 x^{n+1} dx}{\int_0^1 x^n dx} = \frac{n+1}{n+2}.$$

This gives an approximation of the probability that the sun will rise tomorrow assuming it has risen daily until today. $\triangle$

∇ **Exercise 1.9 (Null Events and Independence)** Let $(\Omega, \mathcal{F}, \mathbb{P})$ be a probability space and let A and B be two events. Study the equivalence of each pair of propositions:

1. "A has probability 1," and "A is independent of itself."
2. "A is null," and "A is independent of any event."
3. "A and B are null," and "$A \cup B$ is null."
4. "A and B are independent and disjoint," and "A or B is null."

Solution
Note that if A and B are two events, then $A \cap B$ and $A \cup B$ are also events.

1. Let A be such that $\mathbb{P}(A) = 1$. Then $\mathbb{P}(A \cap A) = \mathbb{P}(A) = 1 = \mathbb{P}(A)\mathbb{P}(A)$, and hence A is independent of itself.

   Conversely, since $\mathbb{P}(\emptyset \cap \emptyset) = \mathbb{P}(\emptyset) = 0 = \mathbb{P}(\emptyset)\mathbb{P}(\emptyset)$, the empty set is independent of itself but is not an event with probability 1.

2. Since $A \cap B \subset A$, if $\mathbb{P}(A) = 0$, then $\mathbb{P}(A \cap B) = 0 = \mathbb{P}(A)\mathbb{P}(B)$, so A and B are independent.

   Conversely, since $\mathbb{P}(A \cap \Omega) = \mathbb{P}(A) = \mathbb{P}(A)\mathbb{P}(\Omega)$, the space Ω is independent of any other event, but is not null.
3. Since $\mathbb{P}(A \cup B) \leq \mathbb{P}(A) + \mathbb{P}(B)$, if A and B are null, then $A \cup B$ is null.

   Conversely, since $A \subset A \cup B$ and $B \subset A \cup B$, if $\mathbb{P}(A \cup B) = 0$, then A and B are null.
4. If A and B are independent and disjoint, then $\mathbb{P}(A)\mathbb{P}(B) = \mathbb{P}(A \cap B) = 0$, and hence A or B is null.

   Conversely, since $\mathbb{P}(A \cap B) \leq \mathbb{P}(A)$, if A is null, then $A \cap B$ is null but is not necessarily empty.

$\triangle$

∇ **Exercise 1.10 (Independence and Dice)** Two fair dice are rolled, a black one and a white one. Study the independence of the events A="the score of the black die is even," B= "the score of the white die is odd," and C= "both scores are either even or odd."

Deduce that two events independent for one probability are not necessarily independent for another one.

Solution
The sample space is $\Omega = [\![1, 6]\!]^2$, equipped with $\mathcal{P}(\Omega)$ and with the uniform probability, that is $\mathbb{P}(\{\omega\}) = 1/36$ for all $\omega \in \Omega$.

We have $\mathbb{P}(A) = |A|/|\Omega| = 18/36 = 1/2$. In the same way, we compute $\mathbb{P}(B) = \mathbb{P}(C) = 1/2$. We have $A \cap B$ = "the score of the black die is even and the score of the white die is odd," hence $\mathbb{P}(A \cap B) = 1/4 = \mathbb{P}(A)\mathbb{P}(B)$ and $\mathbb{P}(A \mid B) = \mathbb{P}(A)$. The events A and B are independent.

On the one hand, $A \cap C$="both scores are even" and $B \cap C$="both scores are odd." So $|A \cap C| = |B \cap C| = 9$, and $\mathbb{P}(A \cap C) = \mathbb{P}(B \cap C) = 1/4 = \mathbb{P}(A)\mathbb{P}(C) = \mathbb{P}(B)\mathbb{P}(C)$. Hence A, B, and C are pairwise independent.

On the other hand, since $A \cap B \cap C = \emptyset$, we have $\mathbb{P}(A \cap B \cap C) \neq \mathbb{P}(A)\mathbb{P}(B)\mathbb{P}(C)$. The three events are not mutually independent.

Finally, $\mathbb{P}(A \mid C) = \mathbb{P}(A) = 1/2 = \mathbb{P}(B \mid C)$ but $\mathbb{P}(A \cap B \mid C) = 0$, so A and B are not independent given C. $\triangle$

Random Variables

2

In this chapter, all the tools necessary to investigate one random variable at a time are presented. They include specific probabilistic tools such as distributions, distribution functions, densities, reliability, as well as analytical tools, such as Lebesgue integrals, generating functions, Fourier transform.

Relations between random variables, necessary to simultaneously investigate several random variables, will be presented in Chap. 3.

2.1 Random Variables

A random variable is defined with respect to a random experiment as a function whose value depends on the result of the experiment. In mathematical words, a random variable is a measurable function defined on a probability space. Therefore, we first present properties of general measurable functions, with a particular attention to Borel functions.

2.1.1 Measurable Functions

Let us begin with basics on direct and inverse images, indicator and elementary functions.

Let $f : \Omega \to \Omega'$ be any function. If $A \subset \Omega$, the image of A by f is the set $f(A) = \{\omega' \in \Omega' : \exists \omega \in A, \ f(\omega) = \omega'\}$. If $A' \subset \Omega'$, the inverse image of A' by f is defined as $f^{-1}(A') = \{\omega \in \Omega : f(\omega) \in A'\}$, a well-defined set even if f is not bijective. We have $A \subset f^{-1}(f(A))$ and $f(f^{-1}(A')) \subset A'$; further, for any family I of indices, $f^{-1}(\cup_{i \in I} A'_i) = \cup_{i \in I} f^{-1}(A'_i)$ and $f^{-1}(\cap_{i \in I} A'_i) = \cap_{i \in I} f^{-1}(A'_i)$.

Definition 2.1 The indicator function of $A \subset \Omega$ is defined by

$$\mathbb{1}_A(\omega) = \begin{cases} 1 & \text{if } \omega \in A, \\ 0 & \text{if } \omega \in \Omega \backslash A. \end{cases}$$

Note that:

- $\mathbb{1}_{\overline{A}} = 1 - \mathbb{1}_A$, • $\mathbb{1}_{A \cup B} = \mathbb{1}_A + \mathbb{1}_B - \mathbb{1}_{A \cap B} = \max(\mathbb{1}_A, \mathbb{1}_B)$,
- $\sigma(\mathbb{1}_A) = \sigma(A)$, • $\mathbb{1}_{A \cap B} = \mathbb{1}_A \times \mathbb{1}_B = \min(\mathbb{1}_A, \mathbb{1}_B)$.

Definition 2.2 A function f such that $f = \sum_{i=1}^{n} \alpha_i \mathbb{1}_{A_i}$, where $\alpha_i \in \mathbb{R}$ and $A_i \in \mathcal{P}(\Omega)$ for $i \in [\![1, n]\!]$, with $\cup_{i=1}^{n} A_i = \Omega$, is called an elementary—or step—function.

This additive decomposition is unique for a non-null function if we impose $A_i \cap A_j = \emptyset$ and $\alpha_i \neq \alpha_j$ for $i \neq j$.

After the σ-algebras generated by events in Chap. 1, let us now define σ-algebras generated by functions.

Theorem-Definition 2.3 *Let Ω be a set, $(\Omega', \mathcal{F}')$ a measurable set, and let $f : \Omega \to \Omega'$ be a function. The set $f^{-1}(\mathcal{F}')$ is a σ-algebra on Ω, called the σ-algebra generated by f and denoted by $\sigma(f)$. In other words,*

$$\sigma(f) = \{A \in \mathcal{P}(\Omega) : \exists A' \in \mathcal{F}' \text{ such that } A = f^{-1}(A')\}.$$

Proof Let us prove that $\sigma(f)$ is a σ-algebra.

First, since $\Omega = f^{-1}(\Omega')$, we have $\Omega \in \sigma(f)$.

Second, if $A \in \sigma(f)$, we can write $A = f^{-1}(A')$ with $A' \in \mathcal{F}'$, and we have

$$\Omega = f^{-1}(A' \cup \overline{A'}) = f^{-1}(A') \cup f^{-1}(\overline{A'}) = A \cup f^{-1}(\overline{A'}).$$

Since the two latter sets are disjoint, $\overline{A} = f^{-1}(\overline{A'})$, and hence belongs to $\sigma(f)$.

Finally, if (A'_n) is a sequence of elements of $\mathcal{F}'$, we have $\cup_{n \geq 0} f^{-1}(A'_n) = f^{-1}(\cup_{n \geq 0} A'_n)$ and $\cup_{n \geq 0} A'_n \in \mathcal{F}'$, so $\cup_{n \geq 0} f^{-1}(A'_n) \in f^{-1}(\mathcal{F}')$. □

In the same way, let $\{f_i : i \in I\}$, where $I \subset \mathbb{N}$, be a collection of functions from Ω into some measurable spaces $(\Omega'_i, \mathcal{F}'_i)$. The σ-algebra on Ω generated by $\cup_{i \in I} f_i^{-1}(\mathcal{F}'_i)$ is referred to as the σ-algebra generated by the collection and is denoted by $\sigma(f_i, i \in I)$. This σ-algebra is the smallest that makes measurable all the functions of the collection.

Definition 2.4 Let $(\Omega, \mathcal{F})$ and $(\Omega', \mathcal{F}')$ be two measurable spaces. A function $f : \Omega \to \Omega'$ is said to be $(\mathcal{F}, \mathcal{F}')$-measurable (or $\mathcal{F}$-measurable) if $\sigma(f) \subset \mathcal{F}$. When $(\Omega', \mathcal{F}') = (\mathbb{R}^d, \mathcal{B}(\mathbb{R}^d))$, measurable functions are called Borel functions.

Equivalently, f is measurable if and only if $f^{-1}(A') \in \mathcal{F}$ for all $A' \in \mathcal{F}'$. On the one hand, every function is measurable for the σ-algebra $\mathcal{P}(\Omega)$. On the other hand, the only measurable functions for the σ-algebra $\{\Omega, \varnothing\}$ are the constant ones.

The set $f^{-1}(A')$ is usually denoted by $(f \in A')$; in the same way, $f^{-1}(\{\omega'\}) = (f = \omega')$ and, for $(\Omega', \mathcal{F}') = (\mathbb{R}, \mathcal{B}(\mathbb{R}))$, we set $(f \in [a, b]) = (a \leq f \leq b)$ and $(f \in dx) = (x \leq f \leq x + dx)$.

If $\mathcal{C}'$ is a set of subsets of Ω', then $\sigma(f^{-1}(\mathcal{C}')) = f^{-1}(\sigma(\mathcal{C}'))$. Consequently, if $\mathcal{F}' = \sigma(\mathcal{C}')$, then $f : \Omega \to \Omega'$ is $(\mathcal{F}, \mathcal{F}')$-measurable if and only if $f^{-1}(\mathcal{C}') \subset \mathcal{F}$. Thus, according to Theorem 1.13, $f : \Omega \to \mathbb{R}$ is a Borel function if $(f \leq x) \in \mathcal{F}$ for all real numbers x.

▷ *Example 2.5 (Indicator Functions of Events)* If $A \subset \Omega$ belongs to the σ-algebra $\mathcal{F}$, then the indicator function of A is a Borel function.   ◁

Most of the following properties of Borel functions are proven in Exercise 2.2.

Properties of Measurable Functions

1. The composites of measurable functions are measurable functions.
2. The absolute value of any Borel function is a Borel function.
3. The set of all finite-valued Borel functions is an $\mathbb{R}$-linear space.
4. The product, the quotient—when well-defined, the maximum and the minimum of Borel functions are Borel functions.
5. The supremum and infimum limits of sequences of Borel functions are Borel functions—when finite.
6. All increasing or continuous real functions are Borel functions.
7. Every Borel function defined on a finite set is a step function. Conversely, a step function $f = \sum_{i=1}^{n} \alpha_i \mathbb{1}_{A_i}$ is a Borel function if and only if $A_i \in \mathcal{F}$ for $i \in [\![1, n]\!]$. Moreover, $\sigma(f) = \sigma(A_1, \dots, A_n)$.

Theorem 2.6

1. *Every Borel function is the difference of two nonnegative Borel functions.*
2. *Every nonnegative Borel function is the limit of an increasing sequence of step Borel functions.*

Proof

1. Let us set $f^+ = \sup(f, 0) = f\mathbb{1}_{(f \geq 0)}$ for the positive part of f, and $f^- = \sup(-f, 0) = -f\mathbb{1}_{(f \leq 0)}$ for its negative part. Then $f = f^+ - f^-$ and both these functions are nonnegative Borel functions. Note that $|f| = f^+ + f^-$.

2. The proof is made by constructing such a sequence. Set

$$f_n = n \, \mathbb{1}_{(f \geq n)} + \sum_{k=0}^{n2^n - 1} \frac{k}{2^n} \mathbb{1}_{(k/2^n \leq f < (k+1)/2^n)}.$$

The function f_n is indeed a step Borel function for every integer n.

Let $\omega \in \Omega$. Since f is finite, for n big enough, $f(\omega) \leq n$.

Moreover, since $\cup_{k=0}^{n2^n-1}[k/2^n, (k+1)/2^n[= [0, n[$, some $k_0 \in [\![0, n2^n - 1]\!]$ exists such that $k_0/2^n \leq f(\omega) < (k_0 + 1)/2^n$ and we have $|f_n(\omega) - f(\omega)| < 1/2^n$. Obviously, (f_n) is increasing and converges to f. □

The above theorem is the basis of a classical scheme of proof of which the following is an example.

Theorem 2.7 *Let f and g be two Borel functions defined on $(\Omega, \mathcal{F})$ taking values in $(\mathbb{R}^d, \mathcal{B}(\mathbb{R}^d))$. The function g is $(\sigma(f), \mathcal{B}(\mathbb{R}^d))$-measurable if and only if a Borel function $\varphi \colon \mathbb{R}^d \to \mathbb{R}^d$ exists such that $g = \varphi \circ f$.*

Proof Suppose that $g = \varphi \circ f$ for some Borel function φ. If B is a Borel set of $\mathcal{B}(\mathbb{R}^d)$, then $g^{-1}(B) = (\varphi \circ f)^{-1}(B) = f^{-1}(\varphi^{-1}(B))$. Since φ is a Borel function, $\varphi^{-1}(B) \in \mathcal{B}(\mathbb{R}^d)$, and since f is $\sigma(f)$-measurable, we obtain $g^{-1}(B) \in \sigma(f)$.
Conversely,

- if $g = \mathbb{1}_A$ is $\sigma(f)$-measurable, then necessarily $A \in \sigma(f)$, that is $A = f^{-1}(A')$ with $A' \in \mathcal{B}(\mathbb{R}^d)$. If $\omega \in \Omega$, we have $\mathbb{1}_A(\omega) = \mathbb{1}_{(f \in A')}(\omega) = \mathbb{1}_{A'}(f(\omega))$, so $g = \mathbb{1}_{A'} \circ f$, and hence $\varphi = \mathbb{1}_{A'}$ is a Borel function.
- if $g = \sum_{i=1}^n \alpha_i \mathbb{1}_{A_i}$ where $A_1, \ldots, A_n$ are events, then $g = \varphi \circ f$ for $\varphi = \sum_{i=1}^n \alpha_i \mathbb{1}_{A_i'}$, with $A_i' = f(A_i)$.
- if g is a nonnegative Borel function, then a sequence (g_n) of step Borel functions converging to g exists, with $g_n = \varphi_n \circ f$ for all n, for some Borel functions φ_n. Set $C = \{x \in \mathbb{R}^d : \varphi_n(x) \text{ converges}\}$. Since g_n converges to g, we have $f(\Omega) \subset C$. Moreover, $C = (\underline{\lim}\varphi_n = \overline{\lim}\varphi_n) \cap (\overline{\lim}\varphi_n < +\infty)$, so that $C \in \mathcal{B}(\mathbb{R}^d)$. Then $g = \varphi \circ f$ with $\varphi(x) = \mathbb{1}_C(x) \lim_n \varphi_n(x)$.
- finally, if g is a general Borel function, it can be written as the difference of two nonnegative Borel functions, which yields the conclusion.

□

Note that if $f \colon (\Omega, \mathcal{F}) \to (\Omega', \mathcal{F}')$ is a measurable function and if $\varphi \colon (\Omega', \mathcal{F}') \to (\mathbb{R}, \mathcal{B}(\mathbb{R}))$ is a Borel function, then $\sigma(\varphi \circ f) \subset \sigma(f)$. Precisely, $\sigma(f) = \cup_\varphi \sigma(\varphi \circ f)$.

The next defined image measures play an essential part in probability, as will be shown in the following section.

Theorem-Definition 2.8 *Let* $h\colon (\Omega, \mathcal{F}) \to (\mathbb{R}^d, \mathcal{B}(\mathbb{R}^d))$ *be a Borel function and let* μ *be a measure on* $(\Omega, \mathcal{F})$. *The function* $\mu_h = \mu \circ h^{-1}$ *defined by*

$$\mu_h(B) = \mu[h^{-1}(B)] = \mu[\{\omega \in \Omega : h(\omega) \in B\}], \quad B \in \mathcal{B}(\mathbb{R}^d),$$

is a measure on $(\mathbb{R}^d, \mathcal{B}(\mathbb{R}^d))$, *called the image measure of* μ *by* h.

Moreover, if μ is a probability, then μ_h is also a probability.

2.1.2 Distributions and Distribution Functions

In probability theory, a measurable function defined on a probability space is referred to as a random variable. We will study here only real random variables—taking values in $\mathbb{R}$, and extended real random variables—taking values in $\overline{\mathbb{R}}$; when no confusion may arise, we will simply call them random variables. Random variables taking values in $\mathbb{R}^d$ with $d > 1$ will be studied in Chap. 3. Note that a function $Z\colon \Omega \to \mathbb{C}$ whose both real and imaginary parts are real random variables is called a complex random variable.

Definition 2.9 Let X be a random variable defined on a probability space $(\Omega, \mathcal{F}, \mathbb{P})$. The probability $\mathbb{P} \circ X^{-1}$ on $(\mathbb{R}, \mathcal{B}(\mathbb{R}))$ is the image of $\mathbb{P}$ by X, referred to as the (probability) distribution or law of X, and typically denoted by $\mathbb{P}_X$; in mathematical words,

$$\mathbb{P}_X(B) = \mathbb{P}[X^{-1}(B)] = \mathbb{P}(X \in B), \quad B \in \mathcal{B}(\mathbb{R}).$$

If $\mathbb{P}_X = P$, then X is said to have the distribution P, and we will write $X \sim P$. If X and Y are two random variables with the same distribution, we will also write $X \sim Y$. Note that if $X \sim -X$, then X is said to be symmetric, and if $X = Y$ a.s., then Y is referred to as a version of X.

The notion of distribution is linked to the notion of isomorphism of models: if two given probability spaces are isomorphic—see Definition 1.16, then any random variable can equivalently be considered as defined either on one or on the other, as shown in the next example.

▷ *Example 2.10 (Coding)* Suppose a true (or symmetric) coin is tossed. A random variable X is defined by setting $X=0$ if the coin falls on tails and $X=1$ if it falls on heads. The variable X may equivalently be considered as defined on the space $(\Omega, \mathcal{P}(\Omega), \mathbb{P})$ with $\Omega = \{0, 1\}$ and $\mathbb{P}(\{0\}) = \mathbb{P}(\{1\}) = 1/2$, or on the isomorphic space $(\Omega', \mathcal{P}(\Omega'), \mathbb{P}')$ with $\Omega' = \{\text{heads, tails, edge}\}$, $\mathbb{P}'(\{\text{edge}\}) = 0$ and $\mathbb{P}'(\{\text{heads}\}) = \mathbb{P}'(\{\text{tails}\}) = 1/2$. The distribution is the same in both cases, that is $\mathbb{P}_X(\{0\}) = \mathbb{P}_X(\{1\}) = 1/2$.

Actually, the distribution is again the same if a die is rolled and a random variable X is defined by $X=0$ if the result is even and $X=1$ if odd, or if X measures the success of any experiment with two possible outcomes with the same probability of occurrence. $\lhd$

Consequently, we can speak of distributions—in Example 2.10, a Bernoulli distribution with parameter $1/2$—without specifying the experiment. In particular, two a.s. equal random variables have the same distribution. Conversely, if any probability P is given on $(\mathbb{R}, \mathcal{B}(\mathbb{R}))$, then a random variable defined on some probability space $(\Omega, \mathcal{F}, \mathbb{P})$ exists with distribution P; indeed, it is enough to consider $(\Omega, \mathcal{F}, \mathbb{P}) = (\mathbb{R}, \mathcal{B}(\mathbb{R}), P)$ and the identity function.

Since $\mathcal{B}(\mathbb{R}) = \sigma(\{] - \infty, x] : x \in \mathbb{R}\})$—see Theorem 1.13, the quantities $P(] - \infty, x])$ for all $x \in \mathbb{R}$ characterize any probability P defined on $(\mathbb{R}, \mathcal{B}(\mathbb{R}))$.

Definition 2.11 Let X be a random variable on $(\Omega, \mathcal{F}, \mathbb{P})$. The distribution function of $\mathbb{P}_X$, that is $F_X : \mathbb{R} \to [0, 1]$, defined by

$$F_X(x) = \mathbb{P}_X(] - \infty, x]) = \mathbb{P}(X \leq x),$$

is called the distribution function of X.

Properties of Distribution Functions
1. F_X is increasing on $\mathbb{R}$.
2. F_X is continuous from the right on $\mathbb{R}$; indeed, by Proposition 1.28, since $(X \leq x + 1/n)$ is a decreasing sequence converging to $(X \leq x)$, we have

$$\lim_{n \to +\infty} F_X\left(x + \frac{1}{n}\right) = \lim_{n \to +\infty} \mathbb{P}\left(X \leq x + \frac{1}{n}\right) = \mathbb{P}\left[\bigcap_{n \geq 0}\left(X \leq x + \frac{1}{n}\right)\right]$$

$$= \mathbb{P}(X \leq x) = F_X(x).$$

3. $\lim_{x \to -\infty} F_X(x) = \mathbb{P}(\emptyset) = 0$ and $\lim_{x \to +\infty} F_X(x) = \mathbb{P}(\Omega) = 1$.
4. If $x < y$, then $\mathbb{P}(x < X \leq y) = F_X(y) - F_X(x)$. Indeed,

$$(x < X \leq y) = (x < X) \cap (X \leq y) = (\overline{X \leq x}) \cap (X \leq y)$$

$$= (X \leq y) \backslash (X \leq x).$$

5. F_X characterizes $\mathbb{P}_X$: two random variables with the same distribution share the same distribution function and conversely.
6. F_X is continuous if and only if $\mathbb{P}(X = x) = 0$ for all $x \in \mathbb{R}$; indeed, $\lim_{h \to 0}[F_X(x + h) - F_X(x - h)] = \mathbb{P}(X = x)$ for all $x \in \mathbb{R}$.

Every function $F : \mathbb{R} \to [0, 1]$ satisfying properties 1, 2 and 3 is the distribution function of a unique distribution on $\mathbb{R}$—the associated Lebesgue-Stieltjes measure

Fig. 2.1 An example of quantiles

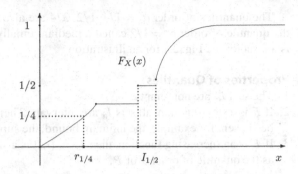

λ_F, and hence of a random variable X defined on some probability space. To be exact, F_X is, among all mass functions associated with the Lebesgue-Stieltjes measure $\mathbb{P}_X$, the one satisfying property 3. A one-to-one correspondence is thus established between probabilities on $(\mathbb{R}, \mathcal{B}(\mathbb{R}))$ and distribution functions. This justifies the use of the Borel σ-algebra $\mathcal{B}(\mathbb{R})$ instead of the set $\mathcal{P}(\mathbb{R})$ of all subsets of $\mathbb{R}$, on which all the probabilities are discrete.

If F_X is not continuous at x, then the discontinuity jump equals $\mathbb{P}(X = x)$ and x is referred to as a mass point or atom of F_X (or of $\mathbb{P}_X$). Further, if F_X is a step function, it is said to be purely atomic.

▷ *Example 2.12 (Integer Part)* Let X be any nonnegative random variable. The distribution function of the random variable $[X]$ (the integer part of X) is

$$F_{[X]}(x) = \mathbb{P}([X] \le x) = \mathbb{P}(X < [x]+1) = F_X([x]+1) - \mathbb{P}(X = [x]+1),\ x \in \mathbb{R}.$$

If $y \in [0, 1[$, then $(X - [X] \le y) = \cup_{n \ge 0}(n \le X \le n + y)$, and hence the distribution function of the random variable $Y = X - [X]$ is

$$F_Y(y) = \sum_{n \ge 0}[F_X(n + y) - F_X(n) + \mathbb{P}(X = n)] = 1 + \sum_{n \ge 0}[F_X(n + y) - F_X(n)];$$

finally, $F_Y(y) = 0$ for $y \in \mathbb{R}_-$ and $F_Y(y) = 1$ for $y \in]1, +\infty[$.   ◁

Note that if X is an extended random variable taking values in $\overline{\mathbb{R}}$, with $\mathbb{P}(X = -\infty) + \mathbb{P}(X = +\infty) > 0$, then its distribution function is a defective mass function, called defective distribution or subdistribution function. In that case, $\lim_{t \to +\infty} F(t) < 1$.

Definition 2.13 Let $\alpha \in [0, 1]$. Let X be a random variable with distribution P and distribution function F. The set of quantiles of order α of X (or of P) is the interval I_α of all real numbers x such that $\lim_{h \to 0+} F(x - h) \le \alpha \le F(x)$.

The quantiles of order $\alpha = 1/4, 1/2, 3/4$ are also referred to as quartiles, with the quantile of order $\alpha = 1/2$ called a median. Finally, a quantile of rational order is a fractile; see Fig. 2.1 for an illustration.

Properties of Quantiles
1. The sets I_α are non-empty.
2. If F is a step function, all sets I_α are intervals. Then, special points of I_α have to be chosen, for example, the infimum bound, the supremum bound, or the middle.
3. If F is an increasing function, then $I_\alpha = \{r_\alpha\}$ is a singleton, and r_α is referred to as the quantile of order α of P.

2.2   Expectation

The expectation of a random variable is its mean value. The most rigorous definition is based on the definition of Lebesgue integral.

2.2.1   Lebesgue Integral

Let us present the main properties of Lebesgue integral. Note that results will be presented without proofs; we refer the interested reader to a book on integration theory for details.

Definition 2.14 Let f be a Borel function defined on a measurable space $(\Omega, \mathcal{F}, \mu)$.

1. **(Integral of step functions)** The integral with respect to μ of a step Borel function $f = \sum_{i=1}^{n} \alpha_i \mathbb{1}_{A_i}$ is defined by

$$\int_\Omega f(\omega) d\mu(\omega) = \sum_{i=1}^{n} \alpha_i \mu(A_i),$$

with the convention $0 \times \infty = 0$.
2. **(Integral of nonnegative Borel functions)** The integral with respect to μ of a nonnegative Borel function f is defined by

$$\int_\Omega f(\omega) d\mu(\omega) = \lim_{n \to +\infty} \int_\Omega f_n(\omega) d\mu(\omega),$$

where (f_n) is a non-decreasing sequence of step functions converging to f. The function f is said to be μ-integrable if the limit is finite.

3. **(Integral of general Borel functions)** A Borel function f is said to be μ-integrable if $\int_\Omega |f(\omega)| d\mu(\omega)$ is finite, and then

$$\int_\Omega f(\omega) d\mu(\omega) = \int_\Omega f^+(\omega) d\mu(\omega) - \int_\Omega f^-(\omega) d\mu(\omega)$$

is well-defined.

The sequence (f_n) in Point 2. is known to exist thanks to Theorem 2.6 but is not unique. Nevertheless, the limit can be proven to be independent of the chosen sequence.

When no ambiguity may arise, we will write indifferently

$$\int_\Omega f(\omega) d\mu(\omega), \quad \int_\Omega f(\omega) \mu(d\omega), \quad \int_\Omega f d\mu, \text{ or } \int f d\mu.$$

For all $A \in \mathcal{F}$, the integral $\int_A f(\omega) \mu(d\omega) = \int_\Omega f(\omega) \mathbb{1}_A(\omega) \mu(d\omega)$ is referred to as the integral of f on A. Moreover, when $(\Omega, \mathcal{F}, \mu) = (\mathbb{R}, \mathcal{B}(\mathbb{R}), \lambda)$, we will write by analogy to Riemann integral, for $-\infty \le a < b \le +\infty$,

$$\int_{[a,b]} f d\lambda = \int_a^b f(x)\, dx = -\int_b^a f(x)\, dx.$$

Properties of Lebesgue Integrals

Let f_1, f_2, and f be real integrable functions defined on $(\Omega, \mathcal{F}, \mu)$.

1. If $\int_\Omega |f| d\mu = 0$, then $f = 0$ μ-a.e.; therefore, if $f_1 = f_2$ μ-a.e., then $\int_\Omega f_1 d\mu = \int_\Omega f_2 d\mu$.
2. **(Linear)** For any real number c, the function cf is integrable on Ω, with

$$\int_\Omega cf d\mu = c \int_\Omega f d\mu.$$

Also, $f_1 + f_2$ is integrable on Ω, with

$$\int_\Omega (f_1 + f_2) d\mu = \int_\Omega f_1 d\mu + \int_\Omega f_2 d\mu.$$

3. **(Non decreasing)** If $f_1 \le f_2$ μ-a.e., then $\int_\Omega f_1 d\mu \le \int_\Omega f_2 d\mu$. In particular, if f is a nonnegative function, its integral is also nonnegative.
4. If (A_n) is a sequence of pairwise disjoint elements of $\mathcal{F}$, then

$$\int_{\cup_{n \ge 0} A_n} f d\mu = \sum_{n \ge 0} \int_{A_n} f d\mu.$$

5. If δ_{ω_0} is the Dirac measure at the point ω_0, then $\int_\Omega f d\delta_{\omega_0} = f(\omega_0)$. More generally, if $\mu = \sum_{i \in I} \delta_{\omega_i}$ is the counting measure on some discrete space $\Omega = \{\omega_i : i \in I\}$, with $I \subset \mathbb{N}$, then the integral takes the form of a sum, that is

$$\int_\Omega f d\mu = \sum_{i \in I} f(\omega_i).$$

6. The integral of a measurable function f taking complex values is well-defined if both the real and imaginary parts of f are integrable, and then

$$\int_\Omega f d\mu = \int_\Omega \mathcal{R}e(f) d\mu + i \int_\Omega \mathcal{I}m(f) d\mu.$$

Integration with respect to mass functions is defined as integration with respect to Lebesgue-Stieltjes measures.

Definition 2.15 Let λ_F be the Lebesgue-Stieltjes measure on $\mathbb{R}$ associated with a mass function F. Let $f: \mathbb{R} \to \mathbb{R}$ be a Borel function. The integral $\int_\mathbb{R} f d\lambda_F$ is called the Lebesgue-Stieltjes integral of f with respect to F.

We will write in short $\int_\mathbb{R} f d\lambda_F = \int_\mathbb{R} f dF$ or $\int_\mathbb{R} f(x) dF(x)$ and F may in this respect be seen either as a measure or a function. If F is null on $\mathbb{R} \setminus [a, b]$ and piecewise differentiable on $[a, b]$ with jumps at $c_0 = a < c_1 < \cdots < c_{n+1} = b$, then

$$\int_\mathbb{R} f(x) d\lambda_F(x) = \int_\mathbb{R} f(x) F'(x) dx + f(a)[F(a^+) - F(a)]$$

$$+ \sum_{i=1}^n f(c_i)[F(c_i^+) - F(c_i^+)] + f(b)[F(b) - F(b^-)].$$

Let us now state some of the most celebrated theorems of integration theory.

Theorem 2.16 (Lebesgue Monotone Convergence) *Let (f_n) be a non-decreasing sequence of nonnegative Borel functions converging μ-a.e. to a function f. Then $\int_\Omega f_n d\mu$ converges to $\int_\Omega f d\mu$.*

Note that the limit f need not be finite, and also that this is sometimes called the Beppo-Levi convergence theorem.

Theorem 2.17 (Fatou's Lemma) *Let* (f_n) *be a sequence of nonnegative Borel functions. Then*

$$\int_\Omega \underline{\lim} f_n d\mu \leq \underline{\lim} \int_\Omega f_n d\mu.$$

Theorem 2.18 (Lebesgue Dominated Convergence) *Let* (f_n) *be a sequence of Borel functions such that* $f_n \longrightarrow f$ μ-*a.e.. If* $|f_n| \leq g$ μ-*a.e. for some* μ-*integrable function g and all integers n, then f is integrable and* $\int_\Omega f_n d\mu$ *converges to* $\int_\Omega f d\mu$. *Moreover* $\int_\Omega |f_n - f| d\mu$ *converges to 0.*

This result applies to integrals depending on real parameters as follows.

Proposition 2.19 *Let* $f : I \times \Omega \rightarrow \mathbb{R}$ *be a function, where I is any interval of* $\mathbb{R}$.

1. **(Continuity)** *If* $f(t, \cdot)$ *is measurable for all* $t \in I$, *if* $f(\cdot, \omega)$ *is continuous at* t_0 μ-*a.e., and if* $|f(t, \omega)| \leq g(\omega)$ μ-*a.e. for all* $t \in I$, *where g is some* μ-*integrable function, then* $t \rightarrow \int_\Omega f(t, \omega) d\mu(\omega)$ *is continuous at* t_0.
2. **(Differentiability)** *If* $f(t, \cdot)$ *is integrable for all* $t \in I$, *if* $f(\cdot, \omega)$ *is differentiable on I* μ-*a.e., and if* $\left|\frac{\partial f}{\partial t}(t, \omega)\right| \leq g(\omega)$ *for all* $t \in I$ μ-*a.e., where g is some* μ-*integrable function, then* $\int_\Omega f(\cdot, \omega) d\mu(\omega)$ *is differentiable,* $\frac{\partial f}{\partial t}(t, \cdot)$ *is integrable, and*

$$\frac{d}{dt} \int_\Omega f(t, \omega) d\mu(\omega) = \int_\Omega \frac{\partial f}{\partial t}(t, \omega) d\mu(\omega).$$

The following result, easily proven for step functions, then for nonnegative functions and finally for general functions by difference of nonnegative functions, is the basis of the transfer theorem to be stated in the next section.

Theorem 2.20 *Let* $(\Omega, \mathcal{F}, \mu)$ *be a measure space,* $h \colon \Omega \rightarrow \Omega'$ *a measurable function and* $\mu' = \mu \circ h^{-1}$ *the image measure of* μ *by h. Let* $f \colon \Omega' \rightarrow \mathbb{R}$ *be a Borel function. Then*

$$\int_\Omega (f \circ h) d\mu = \int_{\Omega'} f d\mu'.$$

The following result is one of the most famous concerning functions with several variables.

Theorem 2.21 (Fubini) *Let* $(\Omega, \mathcal{F}, \mu)$ *and* $(\Omega', \mathcal{F}', v)$ *be two measure spaces. If* $f \colon (\Omega \times \Omega', \mathcal{F} \times \mathcal{F}', \mu \otimes v) \rightarrow (\mathbb{R}, \mathcal{B}(\mathbb{R}))$ *is a Borel function, then both partial functions* $f(\cdot, \omega')$ *and* $f(\omega, \cdot)$ *are measurable on their respective spaces.*

If moreover f is nonnegative or $\mu \otimes \nu$-integrable, then

$$\iint_{\Omega \times \Omega'} f(\omega, x) d(\mu \otimes \nu)(\omega, \omega') = \int_{\Omega} \left(\int_{\Omega'} f(\omega, \omega') d\nu(\omega') \right) d\mu(\omega)$$

$$= \int_{\Omega'} \left(\int_{\Omega} f(\omega, \omega') d\mu(\omega) \right) d\nu(\omega').$$

Note that if f is not either nonnegative or $\mu \otimes \nu$-integrable, the right-hand sides of both above equalities may be defined while the left-hand side is not.

The next change of variables formula is especially of use for computing distributions of functions of random variables. Let us recall that a C^1-diffeomorphism is a continuously differentiable mapping ψ whose inverse is continuously differentiable too. Its Jacobian is defined as

$$J_\psi(u) = \det \begin{pmatrix} \frac{\partial \psi_1}{\partial x_1}(u) & \cdots & \frac{\partial \psi_1}{\partial x_d}(u) \\ \vdots & & \vdots \\ \frac{\partial \psi_d}{\partial x_1}(u) & \cdots & \frac{\partial \psi_d}{\partial x_d}(u) \end{pmatrix}.$$

Note that $J_{\psi^{-1}}(u) = 1/J_\psi(\psi^{-1}(u))$.

Theorem 2.22 (Change of Variable Formula) *Let D_1 and D_2 be two open sets of $\mathbb{R}^d$. Let $f: D_2 \to \mathbb{R}^d$ be a Borel function. Let $\psi: D_1 \to D_2$ be a C^1-diffeomorphism with Jacobian $J_\psi(u)$. The function $x \to f(x)$ is integrable on D_2 if and only if the function $u \to f \circ \psi(u) |J_\psi(u)|$ is integrable on D_1, and we have*

$$\int_{D_2} f(x)\, dx = \int_{D_1} f \circ \psi(u) |J_\psi(u)| du. \tag{2.1}$$

For $d = 1$ and $D_1 = [a, b]$, this amounts to the well-known formula

$$\int_{\psi(a)}^{\psi(b)} f(x)\, dx = \int_a^b f \circ \psi(u) \psi'(u) du. \tag{2.2}$$

Let us now present shortly the spaces L^p, that constitute a natural framework for many of the notions investigated in this book.

Let $(\Omega, \mathcal{F}, \mu)$ be any measure space. Let p be a positive integer. The linear space of $\mathcal{F}$-measurable real-valued functions whose p-th power is μ-integrable is typically denoted by $\mathcal{L}^p(\Omega, \mathcal{F}, \mu)$. The quantity

$$\|f\|_p = \left(\int_\Omega |f(\omega)|^p d\mu(\omega) \right)^{1/p}$$

is referred to as the L^p-norm. The space $\mathcal{L}^\infty(\Omega, \mathcal{F}, \mu)$ is defined as the set of all $\mathcal{F}$-measurable functions f whose sup-norm is finite, where

$$\|f\|_\infty = \sup\{x \geq 0 : \mu(|f| \geq x) > 0\} \tag{2.3}$$
$$= \inf\{x \geq 0 : \mu(|f| \geq x) = 0\} < +\infty.$$

If f and g are μ-integrable and if $f = g$ μ–a.e., then $\int_\Omega f d\mu = \int_\Omega g d\mu$, and hence $\| \cdot \|_p$ is a semi-norm on $\mathcal{L}^p(\Omega, \mathcal{F}, \mu)$, for $p \in \overline{\mathbb{N}}^*$. Note that the triangular inequality is Minkowski's inequality; see Theorem 2.26 below.

Consider now the equivalence relation defined on $\mathcal{L}^p(\Omega, \mathcal{F}, \mu)$ by $f \sim g$ if $f = g$ μ-a.e.. The value of $\|f\|_p$ depends only on the class of f. The quotient space is typically denoted by

$$L^p(\Omega, \mathcal{F}, \mu) = \mathcal{L}^p(\Omega, \mathcal{F}, \mu)/\sim,$$

or simply L^p when no ambiguity may arise. Each class is identified to one of its elements, and hence speaking of μ-a.e. equality remains possible in L^p. We will say that a function is L^p if its p-th power is integrable, that is if $\|f\|_p < +\infty$.

Proposition 2.23 If $\mu(\Omega) < +\infty$, then $L^{p'} \subset L^p$ for all $1 \leq p \leq p' \leq +\infty$.

In particular, the above inclusion holds true for any probability measure μ.

The following inequalities are often of use.

Theorem 2.24 (Markov's Inequality) Let $p \in \mathbb{N}^*$ and let $\varepsilon > 0$. If $f \in L^p(\mu)$, then

$$\mu(|f| \geq \varepsilon) \leq \frac{1}{\varepsilon^p}\|f\|_p^p.$$

For $p = q = 2$, the next inequality is also referred to as Cauchy's or Schwarz's inequality.

Theorem 2.25 (Hölder's Inequality) Let $p \in \mathbb{N}^*$ and $q \in \mathbb{N}^*$ be such that $1/p + 1/q = 1$. If $f \in L^p$ and $g \in L^q$, then

$$\|fg\|_1 \leq \|f\|_p\|g\|_q.$$

Theorem 2.26 (Minkowski's Inequality) If $f \in L^p$ and $g \in L^p$, then

$$\|f + g\|_p \leq \|f\|_p + \|g\|_p.$$

Theorem 2.27 The space $L^2(\Omega, \mathcal{F}, \mu)$, equipped with the scalar product $< f, g >_2 = \int_\Omega fg d\mu$ is a Hilbert space.

Proof Due to Hölder's inequality, $< f, g >_2$ is well-defined. It is easy to check that $< f, g >_2 = < g, f >_2$, that $< f, f >_2$ is nonnegative, and that the function $(f, g) \rightarrow < f, g >_2$ is bilinear. The associated norm is $\| f \|_2 = (\int |f|^2 d\mu)^{1/2}$.

We admit that L^2 is complete for this norm. $\qquad\qquad\qquad\qquad\qquad\qquad\square$

All L^p spaces for $p \geq 1$ are Banach spaces. The space L^2 is the only Hilbert space in the collection.

2.2.2 Expectation

The (mathematical) expectation of a random variable can now be defined.

Definition 2.28 Let $(\Omega, \mathcal{F}, \mathbb{P})$ be a probability space.

1. Let X be a nonnegative random variable on $(\Omega, \mathcal{F}, \mathbb{P})$. The expected value (or mean) $\mathbb{E} X$ of X is defined by

$$\mathbb{E} X = \int_\Omega X(\omega) d\mathbb{P}(\omega)$$

   if this quantity is finite, and as $\mathbb{E} X = +\infty$ otherwise.
2. Let X be a random variable $(\Omega, \mathcal{F}, \mathbb{P})$. Set $X^+ = \sup(X, 0)$ and $X^- = -\inf(X, 0)$. The expected value of X is defined by

$$\mathbb{E} X = \mathbb{E} X^+ - \mathbb{E} X^-$$

   if both expected values are finite. If one is finite and the other is infinite (negative infinite), then $\mathbb{E} X = +\infty \ (-\infty)$. If none of them is finite, the expected value of X is not defined.

Since $|X| = X^+ + X^-$, the expected value of X is finite if and only if the expected value of $|X|$ is finite, and then X is a.s. finite.

The transfer theorem—an immediate consequence of Theorem 2.20—allows an integral on Ω to be transformed into an integral on $\mathbb{R}$ via the image probability, that is the distribution of the variable.

Theorem 2.29 (Transfer) *Let X be a random variable and let $h : \mathbb{R} \rightarrow \mathbb{R}$ be a Borel function. If $h \circ X$ is integrable, then*

$$\int_\Omega h(X(\omega)) d\mathbb{P}(\omega) = \mathbb{E}[h(X)] = \int_\mathbb{R} h(x) d\mathbb{P}_X(x).$$

We can also write

$$\mathbb{E}\,[h(X)] = \int_{\mathbb{R}} h(x)\mathbb{P}_X(dx) = \int_{\mathbb{R}} h(x)\mathbb{P}(X \in dx) = \int_{\mathbb{R}} h(x)d\,F_X(x).$$

The expected value of a random variable thus appears as an average of its values weighted by their distribution values. Most properties of the expectation operator derive straightforwardly from the properties of the integral operator.

In particular, the expectation is linear and increasing. In other words, if X and Y are two real random variables defined on the same space, then

$$\mathbb{E}\,(aX + bY) = a\mathbb{E}\,X + b\mathbb{E}\,Y \quad \text{for all } (a, b) \in \mathbb{R}^2.$$

If $X \geq Y$, then $\mathbb{E}\,X \geq \mathbb{E}\,Y$, with $\mathbb{E}\,X \geq 0$ for all nonnegative X. Similarly, if X is a.s. constant, then its expectation is equal to this constant.

If $X \in L^k(\Omega, \mathcal{F}, \mathbb{P})$, where $k \in \mathbb{N}^*$, then $\mathbb{E}\,(X^k)$ is referred to as the k-th (order) moment of X. By Proposition 2.23, all the l-th order moments of X are finite for $l \leq k$, and the finite quantities $\mathbb{E}\,[X(X - 1)\ldots(X - k + 1)]$ are called the factorial moments of X.

The first order moment of X is its expectation (or expected value, or mean); if the expected value of $|X|$ is finite, we will say that X is integrable or L^1; if the second order moment is finite, X is square integrable, or L^2.

The variance of a random variable is more often of use than its second order moment.

Definition 2.30 The variance of a square integrable random variable X is

$$\mathrm{Var}\,X = \mathbb{E}\,[(X - \mathbb{E}\,X)^2].$$

Note that $\mathrm{Var}\,(aX + b) = a^2\mathrm{Var}\,(X)$, for all $(a, b) \in \mathbb{R}^2$, the variance is quadratic. It is a dispersion parameter: the closer to its mean are the values of X, the smaller is its variance. The variance is usually denoted by σ_X^2. The nonnegative quantity σ_X is referred to as the standard deviation of X, measured with the same unities as the variable, which makes it an attractive quantity. Finally, X is said to be centered if $\mathbb{E}\,X = 0$, and standard if, moreover, $\mathrm{Var}\,X = 1$.

The following formula is more often used for computing variances than the definition.

Proposition 2.31 (König-Huygens Formula) *For any random variable* X *with finite variance,*

$$\mathrm{Var}\,X = \mathbb{E}\,(X^2) - (\mathbb{E}\,X)^2.$$

Proof Since the expectation is linear, we compute

$$\mathbb{E}\,[(X - \mathbb{E}\,X)^2] = \mathbb{E}\,[X^2 + (\mathbb{E}\,X)^2 - 2X\mathbb{E}\,X]$$
$$= \mathbb{E}\,(X^2) + (\mathbb{E}\,X)^2 - 2(\mathbb{E}\,X)(\mathbb{E}\,X),$$

and the result follows.   □

Since the moments of a random variable depend only on its distribution (by the transfer theorem), the moments of X are also referred to as the moments of its distribution.

Hölder's, Minkowski's, and Markov's inequalities are especially useful tools in investigating random variables. For example, for square integrable random variables X and Y, Schwarz's inequality reads

$$\mathbb{E}\,(|XY|)^2 \le \mathbb{E}\,(X^2)\mathbb{E}\,(Y^2).$$

The following inequality, referred to as Bienaymé-Chebyshev or Chebyshev's inequality, the simplest of Chebyshev type inequalities, is induced by Markov's inequality for $p = 2$. It will be completed by Poly-Zygmund inequality in Exercise 2.12.

Corollary 2.32 (Chebyshev's Inequality) *Let X be a square integrable random variable. If $\varepsilon > 0$, then*

$$\mathbb{P}(|X - \mathbb{E}\,X| \ge \varepsilon) \le \frac{1}{\varepsilon^2}\mathrm{Var}\,X.$$

Thus, the smaller is the variance of X, the closer to the expected value of X are its values.

Proof For any square integrable random variable Y and any $\varepsilon > 0$,

$$\mathbb{1}_{[\varepsilon,+\infty[}(Y(\omega)) \le \frac{Y(\omega)^2}{\varepsilon^2}, \quad \text{a.s..}$$

Integrating both sides of the above inequality yields

$$\mathbb{P}(|Y| \ge \varepsilon) \le \frac{\mathbb{E}\,Y^2}{\varepsilon^2},$$

from which the desired inequality follows by setting $Y = X - \mathbb{E}\,X$.   □

▷ *Example 2.33 (Deviation from the Mean)* Let X be any random variable with finite variance σ^2. We obtain from Chebyshev's inequality for $\varepsilon = k\sigma$ that

$$\mathbb{P}(|X - \mu| \geq k\sigma) \leq \frac{1}{k^2}.$$

For instance, the probability that the deviation from X to its mean be larger than 2σ (3σ) is less than $1/4$ ($1/9$), and so on.   ◁

The next inequality involving convex functions is also a classical one.

Theorem 2.34 (Jensen's Inequality) *Let X be a random variable taking values in an interval I of $\mathbb{R}$ and let $\varphi \colon I \to \mathbb{R}$ be a convex function. If both X and $\varphi(X)$ are integrable, then*

$$\varphi(\mathbb{E} X) \leq \mathbb{E}[\varphi(X)].$$

Moreover, if φ is strictly convex and X a.s. not constant, the inequality is strict.

Proof Since the function φ is convex, for every $z \in I$, some $\lambda \in \mathbb{R}$ exists such that

$$\varphi(x) \geq \varphi(z) + \lambda(x - z), \quad x \in I.$$

Setting $z = \mathbb{E} X$ and $x = X$ and taking expectation of both sides of the above inequality yields the result.   □

In particular, $|\mathbb{E} X| \leq \mathbb{E}(|X|)$ and $(\mathbb{E} X)^2 \leq \mathbb{E}(X^2)$, and also $\exp(\mathbb{E} X) \leq \mathbb{E}(\exp X)$.

2.3 Discrete Random Variables

A random variable is said to be discontinuous if its values constitute a discontinuous set. More particularly, the random variables involved in numerous applications take values in a countable set—among which the case $\mathbb{N}$ is of paramount importance. Such variables are said to be discrete.

2.3.1 General Properties

Let us begin with a rigorous definition of discrete random variables.

Definition 2.35 A random variable $X \colon (\Omega, \mathcal{F}, \mathbb{P}) \to (E, \mathcal{E})$ is said to be discrete if $X(\Omega)$ is $\mathbb{P}$-a.s. a countable set, that is if a collection $\{x_i \in E : i \in I\}$, where $I \subseteq \mathbb{N}$, exists such that $\mathbb{P}(X(\Omega) = \{x_i \in E : i \in I\}) = 1$.

We can write $X = \sum_{i \in I} x_i \mathbb{1}_{(X=x_i)}$. The distribution of X is determined by the set $\{\mathbb{P}(X = x_i) : i \in I\}$, with

$$\sum_{i \in I} \mathbb{P}(X = x_i) = 1 \quad \text{and} \quad \mathbb{P}(X = x_i) \geq 0, \quad i \in I.$$

If moreover $\mathbb{P}(X = x_i) > 0$ for all $i \in I$, we will say that X takes the values $\{x_i : i \in I\}$.

Note that a real random variable is discrete if and only if its distribution function is a step function—that is $F_X(x) = \sum_{x_i \leq x} \mathbb{P}(X = x_i)$, with jumps of heights $\mathbb{P}(X = x_i)$, an alternative definition.

▷ *Example 2.36 (Rolling Fair Dice)* Rolling one fair die, the probability space is $\Omega = [\![1, 6]\!]$ equipped with the σ-algebra $\mathcal{P}(\Omega)$. Let $X : \Omega \to \mathbb{R}$ be the variable defined by $X(\omega) = \omega$. Its distribution is determined by $\mathbb{P}(X = \omega) = 1/6$ for all $\omega \in \Omega$ and is referred to as a uniform distribution on $[\![1, 6]\!]$.

Rolling two fair dice, the space is $\Omega^2 = [\![1, 6]\!]^2$ equipped with $\mathcal{P}(\Omega^2)$. Let $Y : \Omega^2 \to \mathbb{R}$ be the random variable defined by $Y((\omega_1, \omega_2)) = \omega_1 + \omega_2$. Its distribution is given in the following table.

x	2	3	4	5	6	7	8	9	10	11	12
$\mathbb{P}(Y = x)$	$\frac{1}{36}$	$\frac{2}{36}$	$\frac{3}{36}$	$\frac{4}{36}$	$\frac{5}{36}$	$\frac{6}{36}$	$\frac{5}{36}$	$\frac{4}{36}$	$\frac{3}{36}$	$\frac{2}{36}$	$\frac{1}{36}$

◁

▷ *Example 2.37 (Integer Part Function)* Let $\Omega = [0, N[$. The integer part function $X : [0, N[\to [\![0, N-1]\!]$ is defined by $X(\omega) = [\omega]$, the largest integer smaller than (or equal to) ω. This is an example of a discrete random variable defined on a non-countable probability space.

Indeed, let $([0, N[, \mathcal{B}([0, N[))$ be equipped with the uniform probability defined by

$$\mathbb{P}([a, b]) = \mathbb{P}([a, b[) = \frac{b - a}{N}, \quad 0 \leq a \leq b < N.$$

The distribution of X is given by

$$\mathbb{P}(X = n) = \mathbb{P}(\{\omega \in [0, N[: [\omega] = n\}) = \mathbb{P}_X([n, n+1[) = \frac{1}{N},$$

a uniform distribution on the finite set $[\![0, N-1]\!]$, while $[0, N[$ is not countable. ◁

Considering integration with respect to the counting measure yields the following formulas for the expected value of a discrete variable X. For any Borel function $h\colon \mathbb{R} \to \mathbb{R}$,

$$\mathbb{E}\,[h(X)] = \sum_{i \in I} \mathbb{P}(X = x_i) h(x_i).$$

Some cases are of particular interest; the k-th order moment of X, for $k \in \mathbb{N}^*$, when defined, is

$$\mathbb{E}\,(X^k) = \sum_{i \in I} \mathbb{P}(X = x_i) x_i^k,$$

with expectation $\mathbb{E}\,X = \sum_{i \in I} \mathbb{P}(X = x_i) x_i$, and—thanks to König-Huighens formula—variance

$$\mathrm{Var}\,X = \sum_{i \in I} (x_i - \mathbb{E}\,X)^2 \mathbb{P}(X = x_i)$$

$$= \sum_{i \in I} \mathbb{P}(X = x_i) x_i^2 - \Big[\sum_{i \in I} \mathbb{P}(X = x_i) x_i \Big]^2.$$

Expectation may be regarded as the barycenter of the system of masses $p_i = \mathbb{P}(X = x_i)$ associated with points x_i, and variance as the inertia moment of the system with respect to its barycenter.

▷ *Example 2.38 (Indicator Function)* The indicator function—see Definition 2.1—of any event A is a random variable. We compute

$$\mathbb{E}\,\mathbb{1}_A = 1 \times \mathbb{P}(\mathbb{1}_A = 1) + 0 \times \mathbb{P}(\mathbb{1}_A = 0) = \mathbb{P}(A).$$

Moreover $\mathbb{1}_A^2 = \mathbb{1}_A$, thus $\mathbb{E}\,(\mathbb{1}_A^2) = \mathbb{P}(A)$, from which we deduce that $\mathrm{Var}\,\mathbb{1}_A = \mathbb{P}(A)[1 - \mathbb{P}(A)]$.   ◁

▷ *Example 2.39 (A Non-finite Expected Value)* Let X be a random variable taking positive integer values, with $\mathbb{P}(X = n) = 6/\pi n^2$ for $n \geq 1$. Since the harmonic series is divergent, the expected value of this variable is not finite.   ◁

The following formula can sometimes be a useful tool for computing expected values. Its extension to general nonnegative variables and other moments will be presented in Exercise 2.6.

Proposition 2.40 *If X is a random variable taking nonnegative integer values, then*

$$\mathbb{E}\, X = \sum_{n \geq 0} \mathbb{P}(X > n).$$

Proof We have

$$X = \sum_{n \geq 0} n \mathbb{1}_{(X=n)} = \sum_{n \geq 1} n[\mathbb{1}_{(X>n-1)} - \mathbb{1}_{(X>n)}] = \sum_{n \geq 0} \mathbb{1}_{(X>n)},$$

and, using the Lebesgue monotone convergence theorem,

$$\mathbb{E}\, X = \mathbb{E}\left[\sum_{n \geq 0} \mathbb{1}_{(X>n)}\right] = \sum_{n \geq 0} \mathbb{P}(X > n),$$

that is the desired result.   □

2.3.2 Classical Discrete Distributions

We present in this section some classical discrete distributions. These are characterized by the probability of the values taken by the random variables.

Dirac Distribution

A random variable X has a Dirac distribution if $\mathbb{P}(X = C) = 1$, where C is a fixed real number; we will write $X \sim \delta_C$.

A Dirac (distributed) random variable X is a.s. equal to C—and may simply be denoted by C. Its expected value is C and its variance is zero. Both the variable and its distribution are said to be degenerated.

▷ *Example 2.41 (Indicator Functions)* The indicator function of the empty set is the null random variable. The indicator function of Ω is the constant random variable equal to 1.   ◁

Any discrete random variable X, taking values $\{x_n, n \in \mathbb{N}\}$, can be written

$$X = \sum_{n \geq 0} x_n \mathbb{1}_{A_n}, \quad x_n \in \mathbb{R}, \ A_n \in \mathcal{F},$$

where $A_n = \{\omega \in \Omega : X(\omega) = x_n\}$, with $A_n \cap A_m = \emptyset$ for $n \neq m$. Its distribution is a linear combination of Dirac distributions, that is $\mathbb{P}_X = \sum_{n \geq 0} \mathbb{P}(A_n) \delta_{x_n}$. Its distribution function can be written

$$F_X(x) = \sum_{n \geq 0} \mathbb{P}(A_n) \mathbb{1}_{[x_n, +\infty[}(x).$$

Uniform Distribution

A discrete random variable X has a uniform distribution over $\{x_1, \ldots, x_N\}$ if

$$\mathbb{P}(X = x_i) = \frac{1}{N}, \quad 1 \leq i \leq N.$$

We will write $X \sim \mathcal{U}(x_1, \ldots, x_N)$. We compute

$$\mathbb{E}\, X = \frac{1}{N} \sum_{i=1}^{N} x_i \quad \text{and} \quad \mathbb{V}\text{ar}\, X = \frac{1}{N} \sum_{i=1}^{N} x_i^2 - \frac{1}{N^2} \left(\sum_{i=1}^{N} x_i \right)^2.$$

If $\{x_1, \ldots, x_N\} = \{1, \ldots, N\}$, then

$$\mathbb{P}(X = i) = \frac{1}{N}, \quad \mathbb{E}\, X = \frac{1}{N} \times \frac{N(N+1)}{2} = \frac{N+1}{2}, \quad \text{and} \quad \mathbb{V}\text{ar}\, X = \frac{N^2 - 1}{12}.$$

▷ *Example 2.42 (Rolling a Die)* The random variable X equal to the result of rolling one fair die has a uniform distribution over $[\![1, 6]\!]$. We compute $\mathbb{E}\, X = 7/2 = 3.5$ and $\mathbb{V}\text{ar}\, X = 35/12 \stackrel{\sim}{=} 2.9$. ◁

Bernoulli Distribution

A random variable X has a Bernoulli distribution with parameter $p \in [0, 1]$ if it takes only two values, 0 and 1 unless otherwise stated, with

$$\mathbb{P}(X = 1) = p = 1 - \mathbb{P}(X = 0).$$

We will write $X \sim \mathcal{B}(p)$. For $p = 0$ or 1, it reduces to Dirac distributions. We compute easily $\mathbb{E}\, X = p$ and $\mathbb{V}\text{ar}\, X = p(1 - p)$.

The Bernoulli distribution is the distribution of any experiment with two outcomes, for instance, of any indicator function. The event $(X = 1)$ is usually referred to as "success" and the event $(X = 0)$ as "failure".

Binomial Distribution

A random variable X has a binomial distribution with parameters $n \in \mathbb{N}^*$ and $p \in [0, 1]$, if it takes the integer values $k \in [\![1, n]\!]$, with

$$\mathbb{P}(X = k) = \binom{n}{k} p^k (1 - p)^{n-k}, \quad 0 \leq k \leq n.$$

We will write $X \sim \mathcal{B}(n, p)$. For $n = 1$ it reduces to the Bernoulli distribution $\mathcal{B}(p)$. See Fig. 2.2 for an illustration.

We compute

$$\mathbb{E}\,X = \sum_{k=0}^{n} k \binom{n}{k} p^k (1-p)^{n-k} = np \sum_{k=1}^{n} \binom{n-1}{k-1} p^{k-1} (1-p)^{n-k}$$

$$= np \sum_{k=0}^{n-1} \binom{n-1}{k} p^k (1-p)^{n-k} = np.$$

In the same way, we obtain $\mathrm{Var}\,X = \mathbb{E}\,(X^2) - (\mathbb{E}\,X)^2 = np(1-p)$.

We will see in Chap. 3 that the sum of n independent random variables with the same Bernoulli distribution $\mathcal{B}(p)$ has a binomial distribution $\mathcal{B}(n, p)$. Nevertheless, all binomial variables are not issued from sums of binomial variables, as will be shown in Example 3.51.

Hyper-Geometric Distribution

A random variable X has a hyper-geometric distribution with positive integer parameters N, M and n, where $M < N$ and $n \leq M$, if it takes the integer values $k \in [\![0, M \wedge n]\!]$, with

$$\mathbb{P}(X = k) = \frac{\binom{M}{k}\binom{N-M}{n-k}}{\binom{N}{n}}, \quad 0 \leq k \leq M \wedge n.$$

We will write $X \sim \mathcal{H}(N, M, n)$. See Fig. 2.2 for an illustration.

We compute $\mathbb{E}\,X = nM/N$ and

$$\mathrm{Var}\,X = \frac{n}{N-1} \frac{M}{N} \left(1 - \frac{M}{N}\right)(N - n).$$

▷ *Example 2.43 (Urn Problem)* An urn contains N bowls among which M are white; n bowls are drawn at random. Let X be the random variable equal to the number of drawn white bowls. If the drawing is done:

• with replacement, X has a binomial distribution $\mathcal{B}(n, M/N)$. Indeed, the probability space is $(\Omega, \mathcal{P}(\Omega), \mathbb{P})$, where Ω is the set of all functions from $[\![1, n]\!]$ into $[\![1, N]\!]$ (so $|\Omega| = N^n$), and $\mathbb{P}$ is uniform. Suppose that the white bowls are numbered from 1 to M; the event $(X = k)$ is the set of all applications f of Ω such that $f(i) \leq M$ for k elements i. There exist $\binom{n}{k}$ possible choices for the k elements, M^k functions sending them into $[\![1, M]\!]$ and $(M - N)^{n-k}$ functions sending the remaining $n - k$ elements into $[\![M + 1, N]\!]$. Therefore

$$\mathbb{P}(X = k) = \binom{n}{k} \frac{M^k (M - N)^{n-k}}{N^n}, \quad \text{for } k \in [\![1, n]\!].$$

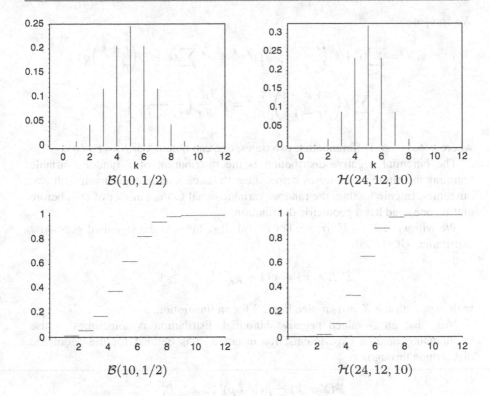

Fig. 2.2 Binomial and hyper-geometric distributions (top) and distribution functions (bottom)

- without replacement, X has a hyper-geometric distribution $\mathcal{H}(N, M, n)$. Indeed, Ω is then the set of all possible choices of n bowls among N without replacement, $|\Omega| = \binom{N}{n}$ and $\mathbb{P}$ is uniform. The event $(X = k)$ corresponds to the choice of k bowls among M and of $n - k$ bowls among $N - M$. ◁

Geometric and Negative Binomial Distributions

A random variable X has a negative binomial (or Pascal) distribution with parameters $r \in \mathbb{N}^*$ and $p \in [0, 1]$ if it takes all integer values $n \geq r$, with

$$\mathbb{P}(X = r + k) = \binom{r + k - 1}{r - 1} p^r (1 - p)^k, \quad k \geq 0.$$

We compute

$$\mathbb{E}\,X = \sum_{k\geq 0}(r+k)\binom{r+k-1}{r-1}p^r q^k = p^r\sum_{k\geq 0}(k+1)\binom{r+k}{r-1}q^k$$

$$= p^r\sum_{k\geq 1}k\binom{r+k-1}{r-1}q^{k-1} = p^r\frac{d}{dq}\left(\frac{1}{(1-q)^r}\right) = \frac{r}{p},$$

where $q = 1 - p$. Differentiating two times, one can obtain $\mathbb{V}\text{ar}\,X = rq/p^2$.

The binomial negative distribution is the distribution of a random variable counting the number of failures before the r-th success in an experiment with two outcomes. In coin tossing, the random variable equal to the number of tails before getting one head has a geometric distribution.

We will write $X \sim \mathcal{B}_-(r, p)$. For $r = 1$, it reduces to the so-called geometric distribution $\mathcal{G}(p)$, with

$$\mathbb{P}(X = n) = p(1-p)^{n-1}, \quad n \geq 1,$$

with expectation $\mathbb{E}\,X = 1/p$. See Fig. 2.3 for an illustration.

Note that an $\mathbb{N}$-valued negative binomial distribution is sometimes of use in modeling random experiments; for instance, it is defined for the geometric distribution through

$$\mathbb{P}(X = k) = p(1-p)^k, \quad k \geq 0,$$

so that $\mathbb{E}\,X = (1-p)/p$.

▷ *Example 2.44 (Continuation of Example 2.43)* Suppose n bowls are drawn at random with replacement. Let X_r be the random variable equal to the number of drawings necessary for obtaining r white bowls. Its distribution is the binomial negative distribution $\mathcal{B}_-(r, M/N)$.

For one drawing, the probability space is $(\{0, 1\}, \mathcal{P}(\{0, 1\}), \mathbb{P})$, in which $\mathbb{P}(\text{"the bowl is white"}) = M/N$.

We can write $(X_r = r + k)=$ "the $(r + k)$-th drawn bowl is white and there are $r - 1$ white bowls among the first $(r + k - 1)$-th drawn bowls".

Since the number of necessary drawings is not known in advance, the infinite product space $(\{0, 1\}^{\mathbb{N}}, \mathcal{P}(\{0, 1\})^{\otimes\mathbb{N}}, \mathbb{P}^{\otimes\mathbb{N}})$ has to be considered to investigate X_r. This space will be rigorously defined in Chap. 4.   ◁

Poisson Distribution

A random variable X has a Poisson distribution with parameter $\lambda \in \mathbb{R}_+^*$, if it takes all integer values, with

$$\mathbb{P}(X = k) = e^{-\lambda}\frac{\lambda^k}{k!}, \quad k \in \mathbb{N}.$$

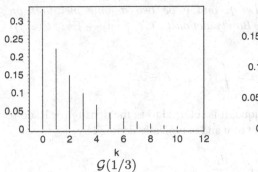
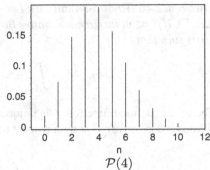

Fig. 2.3 Geometric and Poisson distribution functions

We will write $X \sim \mathcal{P}(\lambda)$. It is also characterized by the recursive relation $\mathbb{P}(X = k) = \lambda \mathbb{P}(X = k - 1)/k$, with $\mathbb{P}(X = 0) = e^{-\lambda}$. See Fig. 2.3 for an illustration.

We compute

$$\mathbb{E}\,X = \sum_{n \geq 1} \frac{\lambda^n e^{-\lambda}}{(n-1)!} = e^{-\lambda}\lambda \sum_{n \geq 0} \frac{\lambda^n}{n!} = \lambda,$$

and, in the same way, $\mathrm{Var}\,X = \lambda$.

The Poisson (distributed) random variables typically count the occurrences of rare events, such as accidents, fabrication errors, viral infections, …

Further, Zeta distributions are studied in Exercise 2.4, the compound Poisson distribution in Chap. 4, etc.

2.4 Continuous Random Variables

The definition of continuous random variables and of probability densities is based on the notion of absolute continuity for measures.

2.4.1 Absolute Continuity of Measures

Most results are here presented without proofs. We refer the interested reader to a book on measure theory for details.

Definition 2.45 Let μ and ν be two nonnegative measures on a measure space $(\Omega, \mathcal{F})$. The measure μ is said to be absolutely continuous with respect to (or dominated by) ν if $\mu(A) = 0$ for all $A \in \mathcal{F}$ such that $\nu(A) = 0$.

We will write $\mu \ll \nu$.

Theorem 2.46 (Radon-Nikodym) *Let μ and ν be two σ-finite measures on $(\Omega, \mathcal{F})$. If $\mu \ll \nu$, then a nonnegative Borel function $f : (\Omega, \mathcal{F}, \nu) \to (\mathbb{R}_+, \mathcal{B}(\mathbb{R}_+))$ exists such that*

$$\mu(A) = \int_A d\mu = \int_A f d\nu, \quad A \in \mathcal{F}.$$

The ν-a.e. defined function f is unique. It is referred to as the density (or Radon-Nikodym derivative) of μ with respect to ν and is typically denoted by

$$f = \frac{d\mu}{d\nu}.$$

If h is a μ-integrable function, then $\int_\Omega h d\mu = \int_\Omega h f d\nu$.
    Conversely, the next result holds true.

Theorem 2.47 *Let $f : \Omega \to \mathbb{R}_+$ be a nonnegative Borel function and let ν be a σ-finite measure on $(\Omega, \mathcal{F})$. A measure μ is defined on $(\Omega, \mathcal{F})$ by setting $\mu(A) = \int_A f d\nu$ for all $A \in \mathcal{F}$. The measure μ is absolutely continuous with respect to ν and its density is f.*

Moreover, if $\int_\mathbb{R} f(x) \, dx = 1$, then μ is a probability.

Proof We prove the theorem for a function with integral equal to 1.
    Let F be the function defined by $F(x) = \int_{-\infty}^x f(t) \, dt$. It is non decreasing, continuous from the right, and satisfies $\lim_{x \to -\infty} F(x) = 0$ and $\lim_{x \to +\infty} F(x) = \int_\mathbb{R} f(x) \, dx = 1$. Hence, this is the distribution function of a unique probability μ on $\mathbb{R}$. By definition, the density of μ is f.   □

▷ *Example 2.48 (Lebesgue-Stieltjes Measure)* Let f be any nonnegative integrable Borel function. The mass function F defined by $F(x) = \int_{-\infty}^x f(t) \, dt$ is associated with the Lebesgue-Stieltjes measure λ_F defined by $\lambda_F(]a, b]) = \int_a^b f(t) \, dt$.   ◁

Contrary to distribution functions, no one-to-one correspondence exists between densities and distributions absolutely continuous with respect to the Lebesgue measure on $\mathbb{R}$. Indeed, any two λ-a.e. equal densities give the same distribution.

Definition 2.49 Two measures μ and ν defined on a measurable space $(\Omega, \mathcal{F})$ are said to be singular if some $A \in \mathcal{F}$ exists such that $\mu(A) = 0$ and $\nu(\Omega \setminus A) = 0$.

Theorem 2.50 (Lebesgue Decomposition) *Let μ and ν be two measures defined on a measurable space $(\Omega, \mathcal{F})$. There exists a unique decomposition $\mu = \mu_1 + \mu_2$ such that μ_1 and μ_2 are singular and $\mu_1 \ll \nu$.*

Theorem 2.51 (Criterion of Equality of Measures) *Let μ and v be two finite measures defined on $\mathbb{R}^d$. Then $\mu = v$ if, for all bounded Borel function $h \colon \mathbb{R}^d \to \mathbb{R}$, with $d \geq 1$,*

$$\int_{\mathbb{R}^d} h\, d\mu = \int_{\mathbb{R}^d} h\, dv.$$

Note that instead of bounded Borel functions, one may consider continuous functions with bounded support.

Proof By Theorem 2.6, the indicator function of any $B \in \mathcal{B}(\mathbb{R}^d)$ is known to be the non-decreasing limit of a sequence (h_n) of bounded Borel functions. We have

$$\int_{\mathbb{R}^d} h_n\, d\mu_i \longrightarrow \int_B d\mu_i, \quad i = 1, 2.$$

Therefore, $\mu_1(B) = \mu_2(B)$, and the conclusion follows. $\qquad\square$

The following result, an immediate consequence of the change of variables theorem and of the criterion of equality of measures, is a useful tool for identifying distributions.

Proposition 2.52 *Let μ be a measure defined on an open subset D_1 of $\mathbb{R}^d$. Let $\psi \colon D_1 \to D_2$ be a diffeomorphism, where D_2 is an open subset of $\mathbb{R}^d$ too. Assume that $\mu \ll \lambda$, with density f. Then the image measure μ_ψ of μ by ψ is absolutely continuous with respect to the Lebesgue measure λ too, with density*

$$f_{\mu_\psi}(x) = |J_{\psi^{-1}}(x)| f \circ \psi^{-1}(x),$$

where $J_{\psi^{-1}}$ denotes the Jacobian of ψ^{-1}.

Finally, the notion of entropy defined in Chap. 1 for discrete distributions extends as follows.

Definition 2.53 Let P be a probability on $(\mathbb{R}, \mathcal{B}(\mathbb{R}))$. Assume that P is absolutely continuous with respect to the Lebesgue measure, with density f. The entropy of P is defined as

$$\mathbb{S}(P) = -\int_{\mathbb{R}} f(x) \log f(x)\, dx.$$

Further, the Kullback-Leibler information (or relative entropy) of a probability P with respect to another Q is

$$\mathbb{K}(P|Q) = \int_{\mathbb{R}} \log\left(\frac{dP}{dQ}\right) dP,$$

when $P \ll Q$. These quantities are not necessarily well-defined.

2.4.2 Densities

Absolute continuity and densities are linked as follows for distributions of random variables.

Definition 2.54 A random variable is said to be:

- continuous if its distribution function is continuous.
- absolutely continuous (or with density) if its distribution is absolutely continuous with respect to the Lebesgue measure λ on $\mathbb{R}$.

A random variable X is continuous if and only if $\mathbb{P}(X = x) = 0$ for all $x \in \mathbb{R}$. It is absolutely continuous if and only if a nonnegative Borel function $f_X : \mathbb{R} \to \mathbb{R}_+$ exists such that $d\mathbb{P}_X(x) = f_X(x)dx$, in other words—thanks to the criterion of equality of measures and Radon-Nikodym theorem—if

$$\mathbb{E}\left[h(X)\right] = \int_{\mathbb{R}} h(x) f_X(x) \, dx$$

for all bounded Borel functions h. In particular, the moments of X are computed through

$$\mathbb{E}\left(X^k\right) = \int_{\mathbb{R}} x^k f_X(x) \, dx.$$

The function f_X is called the (probability) density (function) of the random variable X; it is a.s. unique. We will also write f when no confusion may arise. The variable is said to be (continuous) with density f.

An heuristic interpretation of the density is the following.

$$\mathbb{P}(X \in \Delta x) = f_X(x)\Delta x + o(\Delta x), \quad \Delta x \to 0^+,$$

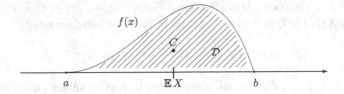

Fig. 2.4 Physical interpretation of the expectation—Example 2.55

where Δx is the set $[x, x + \Delta x]$, and $(X \in \Delta x) = (x < X \le x + \Delta x)$, or equivalently

$$f_X(x) = \lim_{\Delta x \to 0^+} \frac{1}{\Delta x} \mathbb{P}(X \in \Delta x).$$

▷ *Example 2.55 (Physical Interpretation of the Expectation)* Let X be a random variable with density f supported within $[a, b]$. Then

$$\mathbb{E}\, X = \int_a^b x f(x) dx = \int_a^b x dx \int_0^{f(x)} dy = \iint_{\mathcal{D}} x dx dy,$$

that is the abscissa of the mass center C of a thin plate $\mathcal{D}$ limited by the curve of f and the segment $[a, b]$, with elementary area mass equal to 1; see Fig. 2.4.   ◁

Properties of Densities of Random Variables

1. For all $B \in \mathcal{B}(\mathbb{R})$, we have $\mathbb{P}(X \in B) = \mathbb{E}\, \mathbb{1}_{(X \in B)} = \int_B f(x)\, dx$, with the particular case $F_X(x) = \int_{-\infty}^x f(t)\, dt$.
2. If the distribution function F_X of the random variable X is continuously differentiable with derivative f, then X is absolutely continuous with density f.
3. Densities are λ-a.e. nonnegative functions. Indeed, if $f(x) < 0$ on a non-null set $B \in \mathcal{B}(\mathbb{R})$, then $\mathbb{P}(X \in B) = \int_B f(x)\, dx < 0$, which is not possible since $\mathbb{P}$ is a nonnegative measure.
4. The integral of a density is equal to 1; indeed,

$$\int_{\mathbb{R}} f(x)\, dx = \int_{\mathbb{R}} d\mathbb{P}_X(x) = \mathbb{P}(X \in \mathbb{R}) = \mathbb{P}(\Omega) = 1.$$

Definition 2.56 Let $X_1, \dots, X_n$ and Z be random variables. If Z has the same distribution as X_i with probability p_i, that is $\mathbb{P}(Z = X_i) = p_i$, the distribution of Z is called a mixture of the n distributions.

A mixture of distribution is studied in Exercise 2.5.

▷ *Example 2.57 (Mixture of Distributions)* For $n = 2$, we get $\mathbb{P}(Z = X_1) = p = 1 - \mathbb{P}(Z = X_2)$. The distribution functions of the three variables satisfy

$$F_Z(z) = pF_{X_1}(z) + (1 - p)F_{X_2}(z), \quad z \in \mathbb{R}.$$

If both X_1 and X_2 are discrete variables taking respective values $\{x_i : i \in I\}$ and $\{y_j : j \in J\}$, then Z is discrete, with distribution given by $\mathbb{P}(Z = z) = p\mathbb{P}(X_1 = z) + (1 - p)\mathbb{P}(X_2 = z)$, for $z \in \{x_i, i \in I\} \cup \{y_j, j \in J\}$.

If both X_1 and X_2 have respective densities f_{X_1} and f_{X_2}, then Z is has the density $f_Z(z) = pf_{X_1}(z) + (1 - p)f_{X_2}(z)$.

If X_1 is discrete and X_2 is continuous, the distribution function of X_1 is a step function, and the distribution function of X_2 is continuous. Then Z is neither discrete nor continuous.

In all above cases, $\mathbb{E}\, Z = p\mathbb{E}\, X_1 + (1 - p)\mathbb{E}\, X_2$ holds true.   ◁

More generally, every distribution function F can be proven to be a convex combination of three distribution functions, that is $F = a_1 F_1 + a_2 F_2 + a_3 F_3$, where F_1 is a purely atomic function, F_2 is continuous but not differentiable, and F_3 is differentiable. This is an interpretation of Lebesgue's and Jordan's decomposition theorems.

2.4.3   Classical Distributions with Densities

We present in this section some classical distributions with densities. We will also present the Pareto distribution in Exercise 2.4, the symmetric exponential (or Laplace) in Exercise 3.6, the hypo-exponential in Example 3.47, . . .

Uniform Distribution
A random variable X has a uniform distribution with real parameters $a < b$, if its density is

$$\frac{1}{b - a}\mathbb{1}_{[a,b]}(x).$$

We will write $X \sim \mathcal{U}(a, b)$.

▷ *Example 2.58 (Uniform Distribution on $[0, 1]$)* Since the density of a random variable $X \sim \mathcal{U}(0, 1)$ is $f_X(x) = \mathbb{1}_{[0,1]}(x)$, its distribution function is

$$F_X(x) = \begin{cases} 0 & \text{if } x < 0, \\ x & \text{if } 0 \le x \le 1, \\ 1 & \text{if } x > 1; \end{cases}$$

Fig. 2.5 Uniform
distribution $\mathcal{U}(0, 1)$

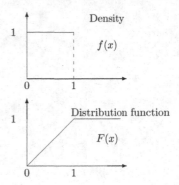

see Fig. 2.5. We compute

$$\mathbb{E}\, X = \int_{\mathbb{R}} x \mathbb{1}_{[0,1]}(x)\, dx = \int_0^1 x\, dx = \frac{1}{2}.$$

In the same way, we compute $\mathbb{E}\, X^n = 1/(n+1)$ for all $n \geq 1$, which induces, in particular, that $\mathbb{V}\mathrm{ar}\, X = 1/12$. ◁

Gaussian Distribution

A random variable X has a Gaussian (or normal) distribution with parameters $m \in \mathbb{R}$ and $\sigma^2 \in \mathbb{R}_+$ if its density on $\mathbb{R}$ is

$$\frac{1}{\sqrt{2\pi}\sigma} e^{-(x-m)^2/2\sigma^2}.$$

We will write $X \sim \mathcal{N}(m, \sigma^2)$. If $\sigma = 0$, it reduces to the Dirac distribution δ_m. If $m = 0$ and $\sigma^2 = 1$, the distribution is said to be standard normal. Figure 2.6 shows different examples of the celebrated bell curves.

Let $Z \sim \mathcal{N}(0, 1)$. The function $f_Z(x) = \frac{1}{\sqrt{2\pi}} e^{-x^2/2}$ is indeed a probability density, because, thanks to a polar change of variables,

$$\left(\int_{\mathbb{R}} e^{-x^2/2}\, dx\right)^2 = \iint_{\mathbb{R}^2} e^{-(x^2+y^2)/2}\, dxdy = \int_0^{2\pi} \int_{\mathbb{R}_+} e^{-r^2/2} r\, dr d\theta$$

$$= \int_0^{2\pi} \int_{\mathbb{R}_+} e^{-u}\, du d\theta = \int_0^{2\pi} d\theta = 2\pi.$$

We compute $\mathbb{E}\,(Z^{2n+1}) = \frac{1}{\sqrt{2\pi}} \int_{\mathbb{R}} x^{2n+1} e^{-x^2/2}\, dx = 0$ for all $n \in \mathbb{N}$, as the integral on $\mathbb{R}$ of an odd function, and, through integration by parts on $\mathbb{R}$ of an even

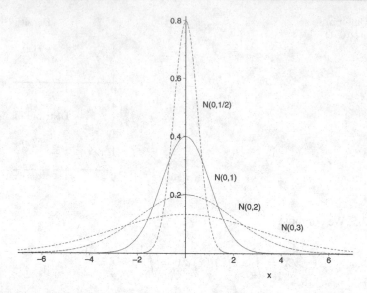

Fig. 2.6 Densities of various Gaussian distributions

function,

$$\mathbb{E}\,(Z^2) = 2\frac{1}{\sqrt{2\pi}} \int_{\mathbb{R}^+} x e^{-x^2/2}\,dx = \frac{1}{\sqrt{2\pi}} \left[-2x e^{-x^2/2}\right]_0^{+\infty} + 2\int_{\mathbb{R}^+} f_Z(x)\,dx = 1.$$

Similarly, $\mathbb{E}\,(Z^{2n}) = (2n-1)\mathbb{E}\,(Z^{2n-2})$ for all $n \geq 2$. In particular, the expected value of $Z \sim \mathcal{N}(0, 1)$ is 0, its variance is 1.

Example 2.59 below shows that if $X \sim \mathcal{N}(m, \sigma^2)$, then $(X - m)/\sigma = Z \sim \mathcal{N}(0, 1)$, and hence $\mathbb{E}\,X = \mathbb{E}\,Z + m = m$ and $\mathbb{V}\mathrm{ar}\,X = \sigma^2 \mathbb{V}\mathrm{ar}\,Z = \sigma^2$.

Note that $\mathbb{P}(|X - m| \leq 3\sigma) \stackrel{\sim}{=} 0.997$ and $\mathbb{P}(|X - m| \leq 1.96\sigma) \stackrel{\sim}{=} 0.95$, two results of use in statistics.

Gamma, Exponential, Chi-Squared, Erlang Distributions

A random variable X has a gamma distribution with parameters $a \in \mathbb{R}_+^*$ and $b \in \mathbb{R}_+^*$ if its density is

$$\frac{b^a}{\Gamma(a)}\, e^{-bx} x^{a-1} \mathbb{1}_{\mathbb{R}_+}(x),$$

where

$$\Gamma(a) = \int_0^{+\infty} e^{-x} x^{a-1}\,dx.$$

We will write $X \sim \gamma(a, b)$. The expected value of X is a/b and its variance is a/b^2. Note that $\Gamma(a+1) = a\Gamma(a)$ for all $a \in \mathbb{R}_+^*$, with $\Gamma(n) = (n-1)!$ for $n \in \mathbb{N}^*$, and that $\Gamma(1/2) = \sqrt{2\pi}$.

A distribution $\gamma(1, \lambda)$ is called an exponential distribution, usually denoted by $\mathcal{E}(\lambda)$. If $X \sim \mathcal{E}(\lambda)$, its density is

$$\lambda e^{-\lambda x} \mathbb{1}_{\mathbb{R}_+}(x),$$

and we compute for all $n \in \mathbb{N}^*$, through integration by parts,

$$\mathbb{E}(X^n) = \int_{\mathbb{R}} \lambda x^n e^{-\lambda x} dx = \left[-x^n e^{-\lambda x}\right]_0^{+\infty} + \frac{n}{\lambda} \int_{\mathbb{R}} \lambda x^{n-1} e^{-\lambda x} dx = \frac{n}{\lambda}\mathbb{E}(X^{n-1}),$$

and hence $\mathbb{E}(X^n) = n!/\lambda^n$. In particular, $\mathbb{E}X = 1/\lambda$ and $\operatorname{Var}X = 1/\lambda^2$.

Exponential distributions are widely used in reliability. A system, for example, constituted of electronic components, whose lifetime is modelled by an exponential distribution, is referred to as memoryless; see Exercise 2.11. Exponential distributions are also well-known to model accurately lifetimes of radio-active substances.

A distribution $\gamma(n/2, 1/2)$ for $n \in \mathbb{N}^*$ is called a chi-squared distribution with n degrees of freedom, and usually denoted by $\chi^2(n)$. Its density is

$$\frac{1}{2^{n/2}\Gamma(n/2)} x^{\frac{n}{2}} e^{-x/2} \mathbb{1}_{\mathbb{R}_+}(x).$$

It is the distribution of the sum of n squares of standard Gaussian variables; see Chap. 3 for details. In particular, the chi-squared distribution with 1 degree of freedom, $\chi^2(1)$, with density $e^{-x/2}/\sqrt{2\pi x}$ on $\mathbb{R}_+$, is the distribution of the square of a random variable with standard normal distribution; see Example 2.60 below.

A distribution $\gamma(n, \lambda)$, for $n \in \mathbb{N}^*$, is called an Erlang distribution, usually denoted by $\mathcal{E}(n, \lambda)$. The Erlang distribution $\mathcal{E}(k, 1/2)$ is a chi-squared distribution $\chi^2(2k)$ and the distribution $\mathcal{E}(1, \lambda)$ is an exponential distribution $\mathcal{E}(\lambda)$. General Erlang distributions model pertinently inter-arrival and service times of queuing problems in telecommunication systems.

Figure 2.7 shows different examples of chi-squared, exponential, and Erlang distributions.

Log-Normal Distribution

A random variable X has a log-normal distribution with parameters $m \in \mathbb{R}_+$ and $\sigma^2 \in \mathbb{R}_+^*$ if its density is

$$\frac{1}{\sigma x \sqrt{2\pi}} e^{-(\log x - m)^2/2\sigma^2} \mathbb{1}_{\mathbb{R}_+}(x).$$

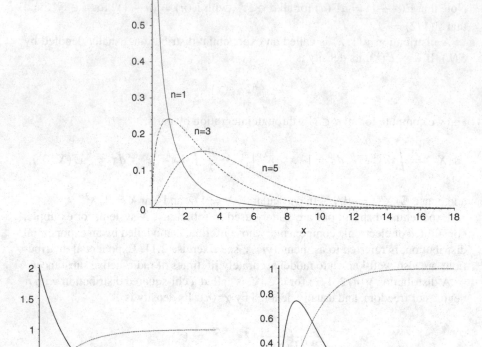

Fig. 2.7 Densities of various gamma distributions: chi-squared (top), exponential (bottom left), and Erlang (bottom right) distributions

We will write $X \sim \mathcal{LN}(m, \sigma^2)$. We compute

$$\mathbb{E}\,X = e^{m+\sigma^2/2} \quad \text{and} \quad \mathbb{V}\text{ar}\,X = e^{2(m+\sigma^2)} - e^{2m+\sigma^2}.$$

Figure 2.8 shows an example of a log-normal distribution.

If X has a log-normal distribution, then $Y = \log X$ has a normal distribution with the same parameters, hence its name. In reliability, the log-normal distribution accurately models repairing times.

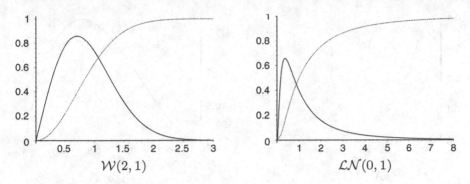

Fig. 2.8 Densities and distribution functions of Weibull and log-normal distributions

Weibull Distribution

A random variable X has a Weibull distribution with parameters $\alpha \in \mathbb{R}_+^*$ and $\lambda \in \mathbb{R}_+^*$ if its density is

$$\alpha\lambda(\lambda x)^{\alpha-1}e^{-(\lambda x)^\alpha}\mathbb{1}_{\mathbb{R}_+}(x).$$

We will write $X \sim \mathcal{W}(\alpha, \lambda)$. We compute

$$\mathbb{E}\,X = \frac{1}{\lambda}\Gamma(1+1/\alpha) \quad \text{and} \quad \mathrm{Var}\,X = \frac{1}{\lambda^2}[\Gamma(1+2/\alpha)^2 - \Gamma(1+1/\alpha)^2].$$

Figure 2.8 shows an example of a Weibull distribution.

Note that $X \sim \mathcal{W}(\alpha, \lambda)$ if and only if $X^\alpha \sim \mathcal{E}(\lambda^\alpha)$. The Weibull distributions $\mathcal{W}(2, 1/\sqrt{2}\,a)$ are called Rayleigh distributions. In reliability, Weibull distributions model lifetimes of mechanical systems. Rayleigh distributions model propagation of radio-electric waves.

Inverse-Gaussian Distribution

A random variable X has an inverse-Gaussian distribution with parameters $\lambda \in \mathbb{R}_+^*$ and $\mu \in \mathbb{R}_+^*$ if its density is

$$\frac{\sqrt{\lambda}}{\sqrt{2\pi x^3}}e^{-\lambda(x-\mu)^2/2\mu^2 x}\mathbb{1}_{\mathbb{R}_+^*}(x).$$

We will write $X \sim \mathcal{N}^{-1}(\mu, \lambda)$. We compute $\mathbb{E}\,X = \mu$ and $\mathrm{Var}\,X = \mu^3/\lambda$. Figure 2.9 shows an example of an inverse-Gaussian distribution.

Inverse-Gaussian distributions model first-passage times of the Brownian motion, and also repairing times in reliability.

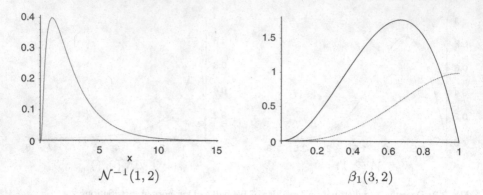

Fig. 2.9 Densities and distribution functions of inverse-Gaussian and beta distributions

Beta Distribution

A random variable X has a beta distribution (of the first kind) with parameters $p \in \mathbb{R}^*_+$ and $q \in \mathbb{R}^*_+$ if its density is

$$f(x) = \frac{1}{\beta(p,q)} x^{p-1}(1-x)^{q-1} \mathbb{1}_{[0,1]}(x),$$

where

$$\beta(p,q) = \int_0^1 x^{p-1}(1-x)^{q-1}\, dx = \frac{\Gamma(p)\Gamma(q)}{\Gamma(p+q)}.$$

We will write $X \sim \beta_1(p,q)$. The $\beta_1(1,1)$ distribution is the uniform distribution on $[0,1]$. We compute $\mathbb{E}\,X = p/(p+q)$ and $\mathrm{Var}\,X = pq/(p+q)^2(1+p+q)$. Figure 2.9 shows an example of a beta distribution.

Fisher Distribution

A random variable X has a Fisher distribution with parameters $n \in \mathbb{N}^*$ and $m \in \mathbb{N}^*$, if its density on $\mathbb{R}$ is

$$f(x) = \frac{n^{n/2} m^{m/2}}{\beta(n/2, m/2)} \frac{x^{n/2-1}}{(m+nx)^{(n+m)/2}}.$$

We will write $X \sim \mathcal{F}(n,m)$. If $m > 4$, then its expectation and variance are finite, and we compute $\mathbb{E}\,X = m/(m-2)$ and $\mathrm{Var}\,X = 2m^2(m+n-2)/n(m-4)(m-2)^2$. Figure 2.10 shows an example of a Fisher distribution.

Student and Cauchy Distributions

A random variable X has a Student distribution (or t-distribution) with parameter $n \in \mathbb{N}^*$ if its density on $\mathbb{R}$ is

$$f(x) = \frac{n^{-1/2}}{\beta(1/2, n/2)}\left(1 + \frac{x^2}{n}\right)^{-(n+1)/2}.$$

We will write $X \sim \tau(n)$. If $n > 2$, then $\mathbb{E}\, X = 0$ and $\operatorname{Var} X = n/(n-2)$.

A random variable X has a Cauchy distribution $\mathcal{C}(\alpha)$ with parameter $\alpha > 0$ if its density on $\mathbb{R}$ is

$$f(x) = \frac{1}{\pi}\frac{\alpha^2}{\alpha^2 + x^2}.$$

This is a typical example of a distribution with median zero and infinite expected value. Figure 2.10 shows an example of a Cauchy distribution.

The Student distribution with parameter $n = 1$ is a Cauchy distribution with parameter 1.

Chi-squared, Fisher, and Student distributions play a particular important part in inferential statistics. Alternative definitions of these distributions can be given in terms of combinations of Gaussian random variables; see Proposition 3.59 in Chap. 3.

Pareto distributions will be defined and studied in Exercise 2.4. They originate from the description of the distribution of wealth in a society by Vilfredo Pareto in the early twentieth century. Translated exponential distributions will be considered in Exercise 2.11 and symmetric exponential distributions in Exercises 2.13 and 3.6.

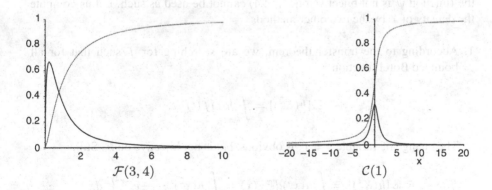

Fig. 2.10 Densities and distribution functions of Fisher and Cauchy distributions

2.4.4   Determination of Distributions

Different methods are of us in order to infer the density of a function $Y = \psi(X)$ from the density of the random variable X.

Let X have the density $f(x)\mathbb{1}_I(x)$ on an open set $I \subset \mathbb{R}$ and let $\psi : I \to I'$ be a diffeomorphism, with $I' \subset \mathbb{R}$. Then $\mathbb{P}_Y$ is the image measure of $\mathbb{P}_X$ by ψ. Therefore, the change of variables theorem, precisely (2.2) p. 42, gives

$$f_Y(y) = (\psi^{-1})'(y) f_X(\psi^{-1}(y))\mathbb{1}_{I'}(y). \tag{2.4}$$

▷ *Example 2.59 (Density of the Linear Transform of a Variable)* Let X be any random variable with density f on $]\alpha, \beta[\subset \overline{\mathbb{R}}$. Set $Y = aX + b$, where $a \in \mathbb{R}_+^*$ and $b \in \mathbb{R}$. Using (2.4), we get directly

$$f_Y(y) = \mathbb{1}_{[a\alpha+b, a\beta+b]}(y)\frac{1}{a} f_X\left(\frac{y-b}{a}\right).$$

For instance, if $X \sim \mathcal{N}(0, 1)$, then $Y \sim \mathcal{N}(m, \sigma^2)$, by setting $a = \sigma$ and $b = m$. ◁

Even if ψ is not a bijection, (2.4) may still be used, by cutting the interval I into parts; see (3.9) in Chap. 33. An alternative to remembering the cumbersome formula (2.4) is given by computing $\mathbb{E}[h(Y)] = \int_{\mathbb{R}} h(y) f_Y(y) dy$ for any bounded Borel function h. Another possibility is to determine first the distribution function of Y from the distribution function of X and then to take its derivative.

▷ *Example 2.60 (Density of the Square of a Normal Variable)* Let X be any random variable with standard normal distribution. Let $Y = X^2 = \psi(X)$. Since the function ψ is not bijective on $\mathbb{R}$, (2.4) cannot be used as such. Let us compute the density of Y by the two other methods.

1. According to the transfer theorem, we are searching for f such that for all bounded Borel function h,

$$\mathbb{E}[h(X^2)] = \int_{\mathbb{R}} h(t) f(t)\, dt,$$

The change of variable $u = t^2$ is obvious, but is not bijective on $\mathbb{R}$. Still,

$$\mathbb{E}[h(X^2)] = \int_{\mathbb{R}} h(x^2) d\mathbb{P}_X(x) = \int_{\mathbb{R}} h(x^2)\frac{1}{\sqrt{2\pi}} e^{-x^2/2}\, dx$$

$$= 2\int_{\mathbb{R}_+} h(x^2)\frac{1}{\sqrt{2\pi}} e^{-x^2/2}\, dx,$$

Fig. 2.11 Determining the
distribution function of
X^2—Example 2.60

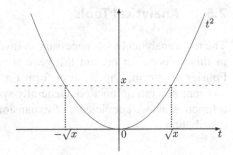

and $u = t^2$ (with $t = \sqrt{u}$ and $dt = du/2\sqrt{u}$) defines a diffeomorphism on $\mathbb{R}_+^*$.
Therefore,

$$\mathbb{E}[h(X^2)] = 2 \int_{\mathbb{R}_+} h(u)e^{-u/2}\frac{1}{\sqrt{2\pi}}\frac{du}{2\sqrt{u}},$$

which says that $d\mathbb{P}_{X^2}(x) = \mathbb{1}_{\mathbb{R}_+}(x)\frac{1}{\sqrt{2\pi x}}e^{-x/2}\,dx$.

2. The distribution function of X^2 is null for $x \in \mathbb{R}_-$, and is, for $x \in \mathbb{R}_+^*$,

$$F_{X^2}(x) = \mathbb{P}(X^2 \leq x) = \mathbb{P}(-\sqrt{x} \leq X \leq \sqrt{x}) = F_X(\sqrt{x}) - F_X(-\sqrt{x})$$

(see Fig. 2.11). Taking the derivative on both sides yields

$$f_{X^2}(x) = \frac{1}{2\sqrt{x}}[f_X(\sqrt{x}) + f_X(-\sqrt{x})]\mathbb{1}_{\mathbb{R}_+}(x) = \frac{1}{\sqrt{2\pi x}}e^{-x/2}\mathbb{1}_{\mathbb{R}_+^*}(x),$$

that defines the $\chi^2(1)$-distribution.   ◁

▷ *Example 2.61 (The $\sigma^2\chi^2(n)$-Distribution)* Let $Y \sim \chi^2(n)$, with density is

$$f_Y(y) = \frac{1}{\Gamma(n/2)2^{n/2}}e^{-y/2}y^{-1+n/2}\mathbb{1}_{\mathbb{R}_+}(y).$$

Using (2.4) with the change of variable $z = \sigma^2 y$ yields the density of $Z = \sigma^2 Y$,

$$f(z) = \frac{1}{\Gamma(n/2)2^{n/2}\sigma^{n/2}}e^{-z/2\sigma^2}z^{-1+n/2}\mathbb{1}_{\mathbb{R}_+}(z).$$

In other words, Z has a $\gamma(n/2, 1/2\sigma^2)$-distribution, sometimes denoted by
$\sigma^2\chi^2(n)$, with expected value $n\sigma^2$ and variance $2n\sigma^4$.   ◁

2.5   Analytical Tools

The main analytical tools necessary in investigating random variables are presented
in this section. We present the generating function for integer valued variables,
Fourier transform, Laplace transform for nonnegative variables and the so-called
moment generating function. Generally speaking, a moment generating function is
a function whose coefficients of expansion as a power series yields moments of a
distribution.

2.5.1   Generating Functions

The generating function which yields factorial moments is mainly used for random
variables taking integer values.

Definition 2.62 Let X be a random variable taking values in $\mathbb{N}$. The (probability)
generating function of X is defined at least for $t \in [-1, 1]$ by

$$g_X(t) = \mathbb{E}\,(t^X) = \sum_{n \geq 0} \mathbb{P}(X = n)t^n.$$

The generating function g_X is a power series whose convergence radius is at least
equal to 1.

▷ *Example 2.63 (Generating Function of the Negative Binomial Distribution)* Let
$X \sim \mathcal{B}_-(r, p)$. The generating function of X is

$$g(t) = \sum_{k \geq 0} \mathbb{P}(X_r = r + k)t^{r+k} = p^r t^r \sum_{k \geq 0} \binom{r + k - 1}{r - 1} t^k q^k = \frac{p^r t^r}{(1 - qt)^r},$$

defined for $|t| < 1/q$, where $q = 1 - p$.   ◁

Properties of Generating Functions
1. $g_X(0) = \mathbb{P}(X = 0)$ and $g_X(1) = 1$. Moreover, $g_X(t) \in [0, 1]$ for $t \in [0, 1]$.
2. g_X belongs to $\mathcal{C}^\infty(]-1, 1[)$, with

$$g_X^{(n)}(0) = n!\,\mathbb{P}(X = n), \quad n \geq 0,$$

and hence g_X characterizes the distribution of X.

Determining the generating function often constitutes a valuable alternative
method for computing the moments of a random variable.

Proposition 2.64 *Let X be a random variable taking integer values. Let g_X denote its generating function.*

1. $\mathbb{E}\,X = \lim_{t \to 1^-} g_X'(t)$.
2. If the expected value of X^2 is finite, then

$$\mathbb{V}\text{ar}\,X = g_X'(1)[1 - g_X'(1)] + \lim_{t \to 1^-} g_X''(t).$$

Proof

1. We have

$$g_X'(t) = \mathbb{E}\,(Xt^{X-1}) = \sum_{n \geq 1} n\mathbb{P}(X = n)t^{n-1},$$

and g_X' converges to a (possibly non-finite) limit, when $t \to 1^-$; hence, by continuity, $X(\omega)t^{X(\omega)-1}$ converges to $X(\omega)$ for all ω, so the first equality holds.
2. In the same way, $g_X''(t) = \mathbb{E}\,[X(X-1)t^{X-2}]$, so

$$\lim_{t \to 1^-} g_X''(s) = \mathbb{E}\,[X(X-1)] = \mathbb{E}\,(X^2) - \mathbb{E}\,X,$$

and the second equality follows. $\qquad\qquad\qquad\qquad\qquad\qquad\qquad\qquad\square$

▷ *Example 2.65 (Moments of a Poisson Distribution)* Let $X \sim \mathcal{P}(\lambda)$. We have

$$g_X(t) = \sum_{n \geq 0} t^n\,\mathbb{P}(X = n) = e^{-\lambda}\sum_{n \geq 0} \frac{(t\lambda)^n}{n!} = e^{\lambda(t-1)}.$$

This function is two times differentiable on $\mathbb{R}$, with $g_X'(t) = \lambda g_X(t)$ and $g_X''(t) = \lambda^2 g_X(t)$. Hence, using Proposition 2.64, we obtain $\mathbb{E}\,X = g_X'(1) = \lambda g_X(1) = \lambda$, and $\mathbb{V}\text{ar}\,X = g_X''(1) + g_X'(1) - [g_X'(1)]^2 = \lambda^2 g_X(1) + \lambda - \lambda^2 = \lambda$. ◁

Note that if $\mathbb{E}\,X^{n-1}$ is finite and $g_X^{(n-1)}$ is differentiable from the left at 1, then the factorial moments of X are given by

$$\mathbb{E}\,[X(X-1)\ldots(X-n+1)] = \lim_{t \to 1^-} g_X^{(n)}(t).$$

2.5.2 Fourier Transform and Characteristic Functions

Different definitions of Fourier transforms exist, all equivalent up to a multiplicative constant. The next is the usual one in probability theory.

Definition 2.66 The Fourier transform is defined:

- for a function $f \in L^1(\mathbb{R}^d)$, as the function $\widehat{f}$ defined on $\mathbb{R}^d$ by

$$\widehat{f}(t) = \int_{\mathbb{R}^d} e^{i<t,x>} f(x)\, dx, \quad t \in \mathbb{R}^d,$$

where $< t, x > = \sum_{i=1}^d t_i x_i$.
- for a finite measure μ on $(\mathbb{R}^d, \mathcal{B}(\mathbb{R}^d))$, as the function $\widehat{\mu}$ defined on $\mathbb{R}^d$ by

$$\widehat{\mu}(t) = \int_{\mathbb{R}^d} e^{i<t,x>} d\mu(x), \quad t \in \mathbb{R}^d.$$

If μ is absolutely continuous with respect to the Lebesgue measure on $\mathbb{R}^d$, then its Fourier transform is the Fourier transform of its density.

The Fourier transform is indeed defined on $\mathbb{R}^d$ because $|e^{i<t,u>}| = 1$ for all $t \in \mathbb{R}^d$ and is bounded by 1. Moreover, $\widehat{f}$ is uniformly continuous since (for $d = 1$)

$$|\widehat{f}(t+h) - \widehat{f}(t)| \le \int_{\mathbb{R}} |e^{ihx} - 1| f(x)\, dx,$$

and using the dominated convergence theorem. Moreover, it is continuously differentiable on $\mathbb{R}^d$. Still, it is not integrable in general.

We state without proofs the next two results.

Theorem 2.67 (Fourier's Inversion Formula) *If both f and $\widehat{f}$ are integrable, then f is continuous and*

$$f(x) = \frac{1}{(2\pi)^d} \int_{\mathbb{R}^d} \widehat{f}(t) e^{-i<t,x>}\, dt, \quad \lambda - a.e..$$

The measure of any Borel set can also be obtained using Fourier transform.

Proposition 2.68 *Let μ be a measure on $(\mathbb{R}^d, \mathcal{B}(\mathbb{R}^d))$. Let $\widehat{\mu}$ denote its Fourier transform. Then*

$$\mu(]x_1, y_1] \times \cdots \times]x_d, y_d]) =$$

$$= \frac{1}{(2\pi)^d} \lim_{t \to +\infty} \int_{-t}^t \cdots \int_{-t}^t \prod_{k=1}^d \frac{e^{-iu_k x_k} - e^{-iu_k y_k}}{iu_k} \widehat{\mu}(u_1, \ldots, u_d)\, du_1 \ldots du_d.$$

The Fourier transform is used for real random variables as follows.

Definition 2.69 Let X be a real random variable. The characteristic function of X is the Fourier transform φ_X of its distribution $\mathbb{P}_X$, that is

$$\varphi_X(t) = \mathbb{E}\left(e^{itX}\right) = \int_{\mathbb{R}} e^{itx} d\mathbb{P}_X(x), \quad t \in \mathbb{R}.$$

For an integer valued variable, we have seen in the preceding section that the generating function is $g_X(s) = \mathbb{E}(s^X)$. Setting $s = e^{it}$ in the above definition shows that characteristic functions constitute an extension of generating functions to general real random variables.

Many properties of characteristic functions derive from the properties of Fourier transforms of measures.

Properties of Characteristic Functions
1. $\varphi_X(0) = 1$ and $|\varphi_X(t)| \leq 1$ for all $t \in \mathbb{R}$.
2. The characteristic function characterizes the distribution: two variables with the same characteristic function have the same distribution.
3. If X has a density f_X, then φ_X is the Fourier transform of f_X.
4. $\varphi_{-X}(t) = \varphi_X(-t) = \overline{\varphi_X(t)}$ so φ_X takes values in $\mathbb{R}$ if and only if X is symmetric, and then φ_X is an even function.
5. $\varphi_{aX+b}(t) = e^{itb}\varphi_X(at)$ for all $a \in \mathbb{R}^*$ and $b \in \mathbb{R}$.

▷ *Example 2.70 (Characteristic Functions of Binomial Distributions)* If X has the binomial distribution $\mathcal{B}(n, p)$, then

$$\varphi_X(t) = \sum_{k=0}^{n} e^{itk}\binom{n}{k} p^k (1-p)^{n-k} = [pe^{it} + (1-p)]^n.$$

The generating function of X follows directly as $g_X(s) = [ps + (1-p)]^n$. ◁

▷ *Example 2.71 (Characteristic Functions of Normal Distributions)* If $X \sim \mathcal{N}(0, 1)$, then

$$\varphi_X(t) = \frac{1}{\sqrt{2\pi}} \int_{\mathbb{R}} e^{itx} e^{-x^2/2} dx.$$

Differentiating with respect to t gives

$$\varphi_X'(t) \stackrel{(1)}{=} \frac{1}{\sqrt{2\pi}} i \int_{\mathbb{R}} x e^{itx} e^{-x^2/2} dx$$

$$\stackrel{(2)}{=} -\frac{1}{\sqrt{2\pi}} t \int_{\mathbb{R}} e^{itx} e^{-x^2/2} dx = -t\varphi(t).$$

(1) by Proposition 2.19 and (2) by integrating by parts. Therefore, $\varphi' = -t\varphi$ with $\varphi(0) = 1$, a differential equation whose solution is $\varphi_X(t) = e^{-t^2/2}$ on $\mathbb{R}$.

If $Y \sim \mathcal{N}(m, \sigma^2)$, we compute for all $t \in \mathbb{R}$,

$$\varphi_Y(t) = \frac{1}{\sqrt{2\pi}\sigma} \int_{\mathbb{R}} e^{ity} e^{-(y-m)^2/2\sigma^2} \, dx \stackrel{(1)}{=} \frac{1}{\sqrt{2\pi}} e^{itm} \int_{\mathbb{R}} e^{it\sigma x} e^{-x^2/2} \, dx$$

$$= e^{itm} \varphi_X(\sigma t) = e^{itm} e^{-\sigma^2 t^2/2}.$$

(1) by the change of variable $x = (y - m)/\sigma$. ◁

Fourier's inversion formula reads as follows for densities.

Proposition 2.72 *If φ_X is integrable, then X has a density with respect to the Lebesgue measure, given by*

$$f_X(x) = \frac{1}{2\pi} \int_{\mathbb{R}} e^{-itx} \varphi_X(t) \, dt, \quad \lambda - a.e..$$

The moments of an integer valued random variable can be deduced from its generating function. Similarly, up to some existence conditions, the moments of a general random variable can be deduced from its characteristic function.

Theorem 2.73 *Let X be a random variable and $n \geq 1$.*

1. *If $\mathbb{E}(|X|^n) < +\infty$, then $\varphi_X \in C^n(\mathbb{R})$, and we have*

$$\varphi_X^{(m)}(t) = i^m \mathbb{E}(X^m e^{itX}).$$

In particular, $\varphi_X^{(m)}(0) = i^m \mathbb{E}(X^m)$ for $1 \leq m \leq n$.
2. *If φ_X is indefinitely differentiable at 0, then $\mathbb{E}(|X|^n) < +\infty$ for all integers $n > 0$.*

Proof

1. Proof by induction. The equality holds true for $n = 1$. Assume that $\mathbb{E}(|X|^{n+1}) < +\infty$. We suppose that

$$\varphi_X^{(n)}(t) = i^n \int_{\mathbb{R}} x^n e^{itx} d\mathbb{P}_X(x).$$

If $h \neq 0$, then by applying Theorem 2.19 yields

$$\frac{1}{h} \left[\varphi_X^{(n)}(t + h) - \varphi_X^{(n)}(t) \right] = \int_{\mathbb{R}} \frac{i^n x^n}{h} [e^{ix(t+h)} - e^{itx}] d\mathbb{P}_X(x).$$

Let us set $g_h(x) = x^n[e^{itx(t+h)} - e^{itx}]/h$ and $G(x) = x^{n+1}t$. We have

$$|g_h(x)| \leq \frac{1}{|h|}|x^n||e^{ixh} - 1|,$$

but $|e^{ixh} - 1| \leq |xh|$, so that $|g_h| \leq |G|$. Hence, the dominated convergence theorem applies for h tending to zero and yields

$$\varphi_X^{(n+1)}(t) = \int_{\mathbb{R}} i^{n+1}tx^n e^{itx} d\mathbb{P}_X(x) = \mathbb{E}(X^n e^{itX}).$$

2. The proof is omitted.   □

Moreover, Point 1 of Theorem 2.73 and Taylor-Young formula jointly yield

$$\varphi_X(t) = \sum_{k=0}^{n} \frac{(it)^k}{k!} \mathbb{E}(X^k) + o(t^n).$$

On the contrary, φ_X can be n-times continuously differentiable on $\mathbb{R}$ without $\mathbb{E}(X^m)$ being finite for $1 \leq m \leq n$.

▷ Example 2.74 (Moments of Normal Distributions) If $X \sim \mathcal{N}(0, \sigma^2)$, its charac teristic function is $\varphi_X(t) = e^{-\sigma^2 t^2/2}$, and we compute by induction $\varphi_X^{(2n+1)}(0) = 0$ and

$$\varphi_X^{(2n)}(0) = (-1)^n \sigma^{2n}(2n-1)(2n-3)\ldots 3.1, \quad n \geq 0.$$

Thanks to Theorem 2.73, $\mathbb{E}(X^{2n+1}) = 0$ and $\mathbb{E}(X^{2n}) = \sigma^{2n}(2n)!/n!\, 2^n$.   ◁

2.5.3 Laplace Transform

Similarly to Fourier transforms, Laplace transforms characterize distributions. Contrary to Fourier transform, on the one hand, no simple inversion formula exists, but the Laplace transforms of usual distributions are given in tables.

Definition 2.75 Let X be a nonnegative random variable. The Laplace transform ψ_X of X (or of its distribution) is defined by

$$\psi_X(t) = \mathbb{E}(e^{-tX}), \quad t \in \mathbb{R}_+.$$

The Laplace transform is mainly used for nonnegative random variables. The definition carries over to general variables, but the support of ψ_X would then possibly be reduced to $\{0\}$.

Properties of Laplace Transforms
1. $\psi_X(0) = 1$.
2. $\psi_X(t) = \varphi_X(it)$ for all $t \in \mathbb{R}_+$.
3. For all $a \in \mathbb{R}_+^*$ and $b \in \mathbb{R}$, $\psi_{aX}(t) = e^{-bt}\psi_X(at)$ for all $t \in \mathbb{R}_+$.
4. If X is integer valued, then $\psi_X(t) = g_X(e^{-t})$ for all $t \in \mathbb{R}_+$.

We state the next result without proof.

Proposition 2.76 *A function $\psi : \mathbb{R}_+ \to \mathbb{R}$ is the Laplace transform of a nonnegative random variable if and only if $\psi(0) = 1$ and ψ is indefinitely differentiable, with $(-1)^n\psi^{(n)}(t) \geq 0$ for all n.*

An indefinitely differentiable function $\psi : \mathbb{R}_+ \to \mathbb{R}$ such that $(-1)^n\psi^{(n)}(t)$ is nonnegative for all n is said to be completely monotonous.

▷ *Example 2.77 (Laplace Transform of the Exponential Distribution)* If $X \sim \mathcal{E}(\lambda)$, then its Laplace transform is the completely monotonous function $t \to \lambda/(\lambda + t)$.
◁

Moments of random variables can also be deduced from Laplace transforms, through the formula $\mathbb{E}(X^n) = (-1)^n\psi_X^{(n)}(0)$ that can be proven by induction from the definition.

▷ *Example 2.78 (Cumulant Generating Functions)* The logarithm of the Laplace transform of a distribution P, say $\log \psi_P$, is sometimes referred to as the cumulant generating function. For Poisson distribution $\mathcal{P}(\lambda)$ and Gamma distribution $\gamma(a, \lambda)$, they are linked through

$$\log \psi_\gamma(t) = a(\log \psi_\mathcal{P})^{-1}(t), \quad t \in \mathbb{R}.$$

Indeed, on the one hand, if $X \sim \gamma(a, \lambda)$, then

$$\psi_X(t) = \frac{\lambda^a}{\Gamma(a)} \int_0^{+\infty} e^{-(\lambda+t)x} x^{a-1}\, dx.$$

For $t < b$, using the change of variable $u = (\lambda + t)x$ and the properties of the $\gamma(a, 1)$ distribution, this becomes

$$\psi_X(t) = \frac{\lambda^a}{\Gamma(a)} \int_0^{+\infty} e^{-u} \frac{u^{a-1}}{(\lambda+t)^a}\, du = \left(\frac{\lambda}{\lambda + t}\right)^a.$$

On the other hand, if $X \sim \mathcal{P}(\lambda)$, then

$$\psi_X(t) = \sum_{n \geq 0} e^{-\lambda} \frac{\lambda^n e^{-nt}}{n!} = e^{-\lambda} \sum_{n \geq 0} \frac{(\lambda e^{-t})^n}{n!} = e^{\lambda(e^{-t}-1)},$$

which induces $(\log \psi_{\mathcal{P}})^{-1}(t) = \log[\lambda/(\lambda + t)]$.

Binomial and negative binomial distributions on the one hand, and Gaussian and inverse-Gaussian distributions on the other hand, are similarly linked through their cumulant generating functions. ◁

2.5.4 Moment Generating Functions and Cramér Transform

The so-called moment generating functions yield the ordinary moments of random variables.

Definition 2.79 The moment generating function M_X of a random variable X (or of its distribution) is defined by

$$M_X(t) = \mathbb{E}(e^{tX}),$$

for all $t \in \mathbb{R}$ such that this quantity is finite.

The moment generating function is defined on an interval I_X of $\mathbb{R}$, possibly reduced to $\{0\}$. This is a convex function, which is indefinitely differentiable on $I_X \neq \{0\}$.

Its name comes from the following result proven by induction using the definition.

Proposition 2.80 *If some $t_0 > 0$ exists such that M_X is defined on $[-t_0, t_0]$, then:*

1. *X has finite moments of all orders, with $\mathbb{E}(X^n) = M_X^{(n)}(0)$;*
2. *$M_X^{(n)}(1) = \mathbb{E}(X^n e^X)$;*
3. *$M_X(t) = \sum_{n \geq 0} \frac{t^n}{n!} \mathbb{E}(X^n)$.*

▷ *Example 2.81 (Moment Generating Function of Normal Distributions)* Let X have a standard $\mathcal{N}(0, 1)$ distribution. We have

$$M_X(t) = \frac{1}{\sqrt{2\pi}} \int_{\mathbb{R}} e^{tx} e^{-x^2/2} \, dx = e^{t^2/2}.$$

If $Y = m + \sigma X$, that is $Y \sim \mathcal{N}(m, \sigma^2)$, then

$$M_Y(t) = \mathbb{E}\left(e^{mt} e^{t\sigma X}\right) = e^{mt} M_X(t\sigma) = e^{mt + \sigma^2 t^2 / 2}, \quad t \in \mathbb{R}.$$

$\triangleleft$

$\triangleright$ *Example 2.82 (Expectation of Binomial Distributions)* Let $X \sim \mathcal{B}(n, p)$. We compute

$$M_X(t) = \sum_{k=0}^{n} e^{tk} \binom{n}{k} p^k (1-p)^{n-k} = (1 - p + pe^t)^n,$$

and hence $M'(t) = npe^t(1 - p + pe^t)^{n-1}$, so that $\mathbb{E}\, X = np$. $\triangleleft$

The moment problem consists in identifying conditions under which a distribution is uniquely defined by its moments. No necessary and sufficient condition is known, but various sufficient conditions do exist, among which the simplest is $I_X \neq \{0\}$.

Cramér transform is defined using the moment generating function.

Definition 2.83 Let X be a non-constant random variable such that M_X is defined on $I_X \neq \{0\}$. The Cramér transform $h_X : I_X \to \mathbb{R}$ of X is defined by

$$h_X(t) = \sup_{u \in I_X} [ut - \log M_X(u)].$$

$\triangleright$ *Example 2.84 (Cramér Transform of the Standard Normal Distribution)* Let $X \sim \mathcal{N}(0, 1)$. Using Example 2.81, we obtain $M_X(u) = e^{u^2/2}$ for all $u \in \mathbb{R}$, so that $h_X(t) = \sup_{u \in \mathbb{R}}(ut - u^2/2) = t^2/2$ for all $t \in \mathbb{R}$. See Fig. 2.12 for $t = 3$.

$\triangleleft$

Fig. 2.12 Cramér transform of distribution $\mathcal{N}(0, 1)$ at $t = 3$

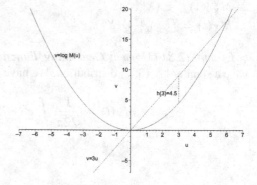

Properties of Cramér Transform
1. h_X is convex, since it is the sup of affine functions.
2. h_X is nonnegative, because $h_X(t) \geq ut - \log M_X(u)$ for all u, so, in particular, for $u = 0$.
3. The minimum value of h_X is 0, obtained at $t = \mathbb{E} X$. Indeed, by Jensen's inequality, $M_X(u) \geq e^{u \mathbb{E} X}$, so $u \mathbb{E} X - \log M_X(u) \leq 0$ for all u; hence $h_X(\mathbb{E} X) = 0$ and $h_X(t) \geq h_X(\mathbb{E} X)$.

The following inequalities bound the distance between a random variable and its expected value. They belong to the collection of large deviation and concentration inequalities.

Theorem 2.85 (Chernoff) *Let X be a random variable such that $I_X = \mathbb{R}$. Then, for all $(a, b) \in \mathbb{R}^2$,*

$$\mathbb{P}(X \geq a) \leq \exp(-h_X(a)) \quad and \quad \mathbb{P}(X \leq b) \leq \exp(-h_X(b)).$$

Clearly, these inequalities are mainly interesting for $a \geq \mathbb{E} X$ and $b \leq \mathbb{E} X$.

Proof We have $(X > a) = (e^{tX} > e^{ta})$ for all $t \in \mathbb{R}_+^*$, and hence by Markov's inequality,

$$\mathbb{P}(X > a) = \mathbb{P}(e^{tX} > e^{ta}) \leq \frac{M_X(t)}{e^{ta}} = e^{-at + \log M_X(t)}, \quad t > 0,$$

from which the first inequality derives. The second can be proven similarly. □

2.6 Reliability and Survival Analysis

The goal is here to investigate the stochastic behavior of systems with failures (break-down or illness) by observing these along time. The term system is very large. Reliability is generally related to technological or economic systems, while survival analysis is used for biological systems.

▷ *Example 2.86 (Some Typical Systems)*

1. the functioning time of a lamp with one or several bulbs;
2. health of individuals—failure means illness or death;
3. careers of individuals—failure means unemployment;
4. fatigue models—materials (pieces of chain, wire, or cable) subject to stress up to break. ◁

In order to make the investigation easier, we suppose that the system starts running at time $t = 0$. We study the nonnegative random variable T equal to the lifetime of the system, with continuous distribution function F.

Definition 2.87 The function $R : \mathbb{R}_+ \to [0, 1]$, defined by

$$R(t) = P(T > t),$$

is called the reliability or survival function of the system.

The reliability $R(t)$ is the probability for the system of surviving (functioning) free of failure at least until time t, with $R(0) = 1$. Since $R(t) = 1 - F(t)$, its main properties derive easily from the properties of the distribution function F of the lifetime T. In particular, the reliability is a decreasing function converging to zero at infinity.

Definition 2.88 The function h defined on $\mathbb{R}_+$ by

$$h(t) = \lim_{\Delta t \to 0^+} \frac{P(T \in \Delta t \mid T > t)}{\Delta t}$$

is called either the (instantaneous) hazard rate (function), failure rate, or repairing rate, depending on the studied type of phenomenon.

The failure rate $h(t)$ is the probability that the system which has functioned without failure up to time t will fail (or die) in the small period of time $[t, t + \Delta t]$.

Note that h is a nonnegative function, and that $\int_0^{+\infty} h(u)du = +\infty$. If T has a density f, then $h(t) = f(t)/R(t)$ and the above definition yields the reliability through the relation

$$R'(t) + h(t)R(t) = 0,$$

obtained by definition of the hazard rate. If the reliability at time $t = 0$ is equal to 1, the solution of this equation is

$$R(t) = \exp\left(-\int_0^t h(u)du\right).$$

The mean time to failure is $\mathbb{E}\,T$. It is a pertinent index for comparing reliability of different systems or components.

▷ *Example 2.89 (Systems Without Memory)* If an electrical device has the constant failure rate $\lambda > 0$, its life time T has an exponential distribution $\mathcal{E}(\lambda)$. Hence, its mean time to failure is $\mathbb{E}\,T = 1/\lambda$ and its reliability is $R(t) = \exp(-\lambda t)$. The probability that it functions at time t_2 given that it had functioned without failure up to time t_1 is

$$\mathbb{P}(T > t_2 \mid T > t_1) = \frac{\mathbb{P}(T > t_2)}{\mathbb{P}(T > t_1)} = \exp[-\lambda(t_1 - t_2)], \quad t_2 > t_1 > 0,$$

by lack of memory of the exponential distribution; see Exercise 2.11. ◁

2.7 Exercises and Complements

▽ **Exercise 2.1 (Images of σ-Algebras and Probabilities)**

1. Show that the image of a σ-algebra by a function is a σ-algebra if and only if the function is bijective.
2. Let $X : (\Omega, \mathcal{F}, \mathbb{P}) \to (\Omega', \mathcal{F}')$ be a random variable. Show that $\mathcal{F}'_X = \{A' \in \mathcal{P}(\Omega') : X^{-1}(A') \in \mathcal{F}\}$ is a σ-algebra on Ω' and that $\mathbb{P}_X$ is a probability on $(\Omega', \mathcal{F}'_X)$.

Solution
1. Let $f : (\Omega, \mathcal{F}) \to \Omega'$. Clearly, $f(\Omega) \in f(\mathcal{F})$. We have

$$\cup_{n\geq 0} f(A_n) = f(\cup_{n\geq 0} A_n) \in f(\mathcal{F}),$$

so Points i. and iii. of Definition 1.8 are fulfilled. Finally, we have $\overline{f(A)} = \{\omega' \in \Omega' : \nexists \omega \in A, \ \omega' = f(\omega)\}$. If f is bijective, then

$$\exists \omega \in \overline{A} \text{ such that } \omega' = f(\omega) \quad \Leftrightarrow \quad \nexists \omega \subset A, \text{ such that } \omega' = f(\omega).$$

Hence Point ii. of Definition 1.8 is fulfilled, and the image of $\mathcal{F}$ is indeed a σ-algebra. On the contrary, if f is not bijective, only the direct implication holds.
2. We have $X^{-1}(\Omega') = \Omega$, so $\Omega' \in \mathcal{F}'_X$. If $A' \in \mathcal{F}'_X$, then $X^{-1}(\overline{A'}) = \overline{X^{-1}(A')} \in \mathcal{F}$, so $\overline{A'} \in \mathcal{F}'_X$. If (A'_n) is a sequence of elements of $\mathcal{F}'_X$, then (see Sec. 2.1.1)

$$X^{-1}(\cup_{n\geq 0} A'_n) = \cup_{n\geq 0} X^{-1}(A'_n) \in \mathcal{F}.$$

Hence $\cup_{n\geq 0} A'_n \in \mathcal{F}'_X$, and $\mathcal{F}'_X$ is indeed a σ-algebra.
   Finally, if (A'_n) is a sequence of pairwise disjoint elements of $\mathcal{F}'_X$, then

$$\mathbb{P}_X(\cup_{n\geq 0} A'_n) = \mathbb{P}[X^{-1}(\cup_{n\geq 0} A'_n)] = \sum_{n\geq 0} \mathbb{P}[X^{-1}(A'_n)] = \sum_{n\geq 0} \mathbb{P}_X(A'_n),$$

so $\mathbb{P}_X = \mathbb{P} \circ X^{-1} : \mathcal{F}'_X \to [0, 1]$ is a probability on $\mathcal{F}'_X$. △

∇ Exercise 2.2 (Borel Functions)

1. Show that every non-decreasing function $f : \mathbb{R} \to \mathbb{R}$ is a Borel function.
2. Same question for a continuous function $f : \mathbb{R}^d \to \mathbb{R}$.
3. Show that a step function $f = \sum_{i=1}^{n} \alpha_i \mathbb{1}_{A_i}$ defined on a measure space $(\Omega, \mathcal{F})$ is a Borel function if and only if $A_i \in \mathcal{F}$ for all $i \in [\![1, n]\!]$.
4. Let $(\Omega, \mathcal{F})$ be any finite measure space and let $f : \Omega \to \mathbb{R}$ be a Borel function. Show that $f(\Omega)$ is finite and then that f is a step function.

Solution
The σ-algebra $\mathcal{B}(\mathbb{R})$ is known by Theorem 1.13 to be generated by all intervals $]a, +\infty[$ for $a \in \mathbb{R}$. Hence, it is enough to prove that $I_a = f^{-1}(]a, +\infty[)$ is an interval.

1. The function f is a Borel function if for every z such that $x < z < y$ with $x \in I_a$ and $y \in I_a$, we have $z \in I_a$ too. Since f is non decreasing and $f(x) > a$, we have $f(z) \geq f(x) > a$, so indeed $z \in I_a$.
2. The inverse image of an open interval by a continuous function is an open set, and, by definition, $\mathcal{B}(\mathbb{R})$ is the σ-algebra defined by the open sets, so the result follows.
3. Since $A_i = f^{-1}(\{\alpha_i\})$, if f is a Borel function, then $A_i \in \mathcal{F}$ for all i. Conversely, if $A_i \in \mathcal{F}$, then $\mathbb{1}_{A_i}$ is a Borel function and f is a sum of Borel functions.
4. We know that $|f(\Omega)| \leq |\Omega| < +\infty$. Moreover, since f is $\mathcal{F}$-measurable, $f^{-1}(y) = \{\omega \in \Omega : f(\omega) = y\}$ belongs to $\mathcal{F}$ for all $y \in \mathbb{R}$, so the range of the pseudo-inverse f^{-1} is finite. This defines an injection from $f(\Omega)$ into a finite set. Hence, $f(\Omega)$ is also finite and f can be written

$$f = \sum_{y \in f(\Omega)} y \mathbb{1}_{f^{-1}(y)}.$$

Therefore, the set of functions measurable for a finite σ-algebra is the set of measurable step functions. △

∇ Exercise 2.3 (Matching Problem—Continuation of Exercise 1.3)
Let S_n denote the number of matches. Determine its distribution and compute the limit of $\mathbb{P}(S_n = k)$ when n tends to infinity.

Solution
We can write

$$(S_n = k) = \bigcup_{1 \leq i_1 < \cdots < i_k \leq n} \left[\left(\bigcap_{l=1}^{k} A_{i_l} \right) \bigcap \left(\bigcap_{\substack{j=1 \\ j \notin I_k}}^{n} \overline{A}_j \right) \right],$$

where A_i = "The i-th sportsman takes his own bag", for $i \in [\![1, n]\!]$ and $I_k = \{i_l : l \in [\![1, k]\!]\}$. This is an union of disjoint sets, and hence

$$\mathbb{P}(S_n = k) = \sum_{1 \le i_1 < \cdots < i_k \le n} \mathbb{P}\left(\bigcap_{l=1}^{k} A_{i_l}\right) \mathbb{P}\left(\bigcap_{\substack{j=1 \\ j \notin I_k}}^{n} \overline{A}_j \,\Big|\, \bigcap_{l=1}^{k} A_{i_l}\right).$$

We compute

$$\mathbb{P}\left(\bigcap_{\substack{j=1 \\ j \notin I_k}}^{n} \overline{A}_j \,\Big|\, \bigcap_{l=1}^{k} A_{i_l}\right) = \mathbb{P}\left(\overline{\bigcup_{j=1}^{n-k} A_j} \,\Big|\, \bigcap_{l=n-k+1}^{n} A_l\right) = \mathbb{P}\left(\overline{\bigcup_{j=1}^{n-k} A_j}\right) \overset{(1)}{=} \sum_{j=0}^{n-k} \frac{(-1)^j}{j!}.$$

(1) by Exercise 1.3. Therefore

$$P(S_n = k) = \binom{n}{k} \frac{(n-k)!}{n!} \sum_{j=0}^{n-k} \frac{(-1)^j}{j!} = \frac{1}{k!} \sum_{j=0}^{n-k} \frac{(-1)^j}{j!}.$$

Thus, $P(S_n = k)$ converges to $1/k!\,e$ when n tends to infinity, where e is Euler's number, the base of the natural logarithm. $\triangle$

∇ Exercise 2.4 (Zeta and Pareto Distributions)

1. The Riemann Zeta function is defined by

$$\zeta(\alpha) = \sum_{k \ge 1} \frac{1}{k^\alpha},$$

finite for $\alpha > 1$. Let X_α, for $\alpha \in \mathbb{N}^*$, be a random variable with distribution zeta given by

$$\mathbb{P}(X_\alpha = k) = \frac{K(\alpha)}{k^{\alpha+1}}, \quad k \in \mathbb{N}^*.$$

(a) Compute the normalization constant $K(\alpha)$ of this distribution.
(b) Compute the moments of X_α when finite.
2. Let $\alpha > 0$ and $a > 0$. Let $X_{a,\alpha}$ be a random variable with Pareto distribution, with density given by

$$f(x) = \frac{K(a, \alpha)}{x^{\alpha+1}} \mathbb{1}_{]a,+\infty[}(x).$$

(a) Compute the normalization constant $K(a, \alpha)$ of this distribution.

(b) Determine the distribution function of $X_{a,\alpha}$.
(c) Compute the moments of $X_{a,\alpha}$ when finite. Give its variance.
(d) Determine the distribution of $Y_\alpha = \log X_{1,\alpha}$.

Solution
1. (a) Necessarily, $\sum_{k\geq 1} \mathbb{P}(X_\alpha = k) = 1$, so $K(\alpha) = \zeta(\alpha+1)^{-1}$ for $\alpha > 1$.
   (b) For $\alpha > n$, we compute

$$\mathbb{E}(X_\alpha^n) = \sum_{k\geq 1} \frac{1}{\zeta(\alpha+1)k^{\alpha+1-n}} = \frac{\zeta(\alpha+1-n)}{\zeta(\alpha+1)}.$$

Zipf's distributions, which are truncated Zeta distributions, model the frequency of words in languages: the frequency of any word is inversely proportional to some power of its rank in the frequency table.

2. (a) From Theorem 2.47, we know that f is a probability density if it is nonnegative and integrable on $\mathbb{R}$, with integral equal to 1. We compute

$$\int_\mathbb{R} f(x)\,dx = \frac{K(a,\alpha)}{\alpha a^\alpha}.$$

Consequently, $K(a,\alpha) = \alpha a^\alpha$.
   (b) Clearly, for $x < a$, we have $F(x) = 0$ and, for $x \geq a$,

$$F(x) = \int_a^x f(t)\,dt = 1 - \left(\frac{a}{x}\right)^\alpha.$$

   (c) The n-th order moment of X is finite if $\alpha + 1 - n > 1$, or $n < \alpha$. Then $\mathbb{E}(X^n) = \alpha a^n/(\alpha - n)$. Therefore, for $\alpha > 2$, we get $\operatorname{Var} X_{a,\alpha} = \alpha a^2/(\alpha - 2)(\alpha - 1)^2$.
   (d) The change of variable $y = \log x$ is a diffeomorphism from $[1, +\infty[$ onto $\mathbb{R}_+$. The density of Y_α thus derives from formula (2.4) p. 68, specifically $f_{Y_\alpha}(y) = \alpha e^{-\alpha y} \mathbb{1}_{\mathbb{R}_+}(y)$, meaning that $Y_\alpha \sim \mathcal{E}(\alpha)$.

The Pareto distribution appears as the truncation on $]a, +\infty[$ of the power distribution with parameter α with density $\alpha/x^{\alpha+1}$ on $\mathbb{R}_+^*$. This distribution was used by Vilfredo Pareto in the twentieth century for modeling household incomes: roughly speaking, the number of households whose income is more than x decreases as $1/x^\alpha$, for some $\alpha > 1$. $\triangle$

∇ **Exercise 2.5 (Mixture of Distributions)** Let $F : \mathbb{R} \to [0, 1]$ be the function plotted on the left of Fig. 2.13.

1. Show that F is the distribution function of a distribution P on $(\mathbb{R}, \mathcal{B}(\mathbb{R}))$.
2. Find the mass points of P.

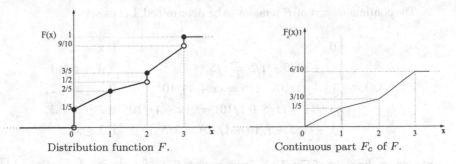

Distribution function F. Continuous part F_c of F.

Fig. 2.13 Decomposition of a mixed distribution function—Exercise 2.5

3. Write F as a convex combination of a continuous distribution function and of a step distribution function.
4. Compute the expected value, the second order moment, and then the variance of a random variable X with distribution P.

Solution

1. The function F is non decreasing and continuous from the right, null for $x \leq 0$ and equal to 1 for $x \geq 3$, thus a unique probability P exists whose distribution function is F; see Sec. 2.1.2.
2. Since

$$
F(x) = \begin{cases}
0 & \text{if } x < 0, \\
(1+x)/5 & \text{if } 0 \leq x < 1, \\
(3+x)/10 & \text{if } 1 \leq x < 2, \\
3x/10 & \text{if } 2 \leq x < 3, \\
1 & \text{if } 3 \leq x.
\end{cases}
$$

The mass points of P are $0, 2, 3$, identified by the discontinuities of F, with $P(\{0\}) = 1/5$, $P(\{2\}) = 3/5 - 1/2 = 1/10$ and $P(\{3\}) = 1 - 9/10 = 1/10$.

3. The step function given by

$$
F_d(x) = \begin{cases}
0 & \text{if } x < 0, \\
1/5 & \text{si } 0 \leq x < 2, \\
3/10 & \text{if } 2 \leq x < 3, \\
2/5 & \text{if } 3 \leq x
\end{cases}
$$

is non decreasing and continuous on the right. Moreover $F_1(x) = \frac{5}{2} F_d(x)$ converges to 0 when x tends to $-\infty$ and to 1 when x tends to $+\infty$; it is the distribution function of a discrete distribution.

The continuous part of F remains to be determined. Let us set

$$F_c(x) = \begin{cases} 0 & \text{if } x < 0, \\ (1+x)/5 - 1/5 = x/5 & \text{if } 0 \leq x < 1, \\ (3+x)/10 - 1/5 = (x+1)/10 & \text{if } 1 \leq x < 2, \\ 3x/10 - (1/5 + 1/10) = 3(x-1)/10 & \text{if } 2 \leq x < 3, \\ 1 - (1/5 + 1/10 + 1/10) = 3/5 & \text{if } 3 \leq x, \end{cases}$$

plotted on the right of Fig. 2.13. The function $F_2 = \frac{5}{3} F_c$ is non decreasing and continuous. It converges to 0 when x tends to $-\infty$ and to 1 when x tends to $+\infty$; it is the distribution function of a continuous distribution. Finally,

$$F = F_d + F_c = \frac{2}{5} F_1 + \frac{3}{5} F_2.$$

4. We compute

$$\mathbb{E}\, X = \int_{-\infty}^{0} x \times 0 \, dx + \int_{0}^{1} x \times \frac{1}{5} \, dx + \int_{1}^{2} x \times \frac{1}{10} \, dx$$

$$+ \int_{2}^{3} x \times \frac{3}{10} \, dx + \int_{3}^{+\infty} x \times 0 \, dx + 0 \times \frac{1}{5} + 2 \times \frac{1}{10} + 3 \times \frac{1}{10},$$

that is

$$\mathbb{E}\, X = \frac{1}{5} \left[\frac{x^2}{2} \right]_0^1 + \frac{1}{10} \left[\frac{x^2}{2} \right]_1^2 + \frac{3}{10} \left[\frac{x^2}{2} \right]_2^3 + \frac{2}{10} + \frac{3}{10} = \frac{3}{2}.$$

Similarly, $\mathbb{E}\,(X^2) = 7/2$, so that $\mathbb{V}\text{ar}\, X = 7/2 - (3/2)^2 = 5/4$.   △

∇ **Exercise 2.6 (An Alternative Method for Computing Moments)**   Let X be any nonnegative random variable.

1. Show that for every $p \geq 1$ such that $\mathbb{E}\,(X^p)$ is finite,

$$\mathbb{E}\,(X^p) = \int_{\mathbb{R}_+} p t^{p-1} \mathbb{P}(X > t) \, dt. \tag{2.5}$$

2. Deduce from 1 that

$$\sum_{n \geq 0} \mathbb{P}(X > n+1) \leq \mathbb{E}\, X \leq \sum_{n \geq 0} \mathbb{P}(X > n).$$

3. Show that

$$\sum_{n\geq 1} \mathbb{P}(X \geq n) \leq \mathbb{E}\,X \leq \sum_{n\geq 0} \mathbb{P}(X \geq n).$$

Solution

1. We know that $\mathbb{P}(X > t) = \int_t^{+\infty} d\mathbb{P}_X(x)$ by the transfer theorem. Hence

$$p \int_{\mathbb{R}_+} t^{p-1} \mathbb{P}(X > t)\,dt = \int_{\mathbb{R}_+} \int_t^{+\infty} p t^{p-1} d\mathbb{P}_X(x)\,dt$$

$$\overset{(1)}{=} \int_{\mathbb{R}_+} \int_0^x p t^{p-1}\,dt\,d\mathbb{P}_X(x) = \int_{\mathbb{R}_+} x^p d\mathbb{P}_X(x) = \mathbb{E}\,(X^p).$$

(1) by Fubini's theorem.

2. Suppose that X is a.s. finite; otherwise the result is obvious, since then all quantities are infinite. By 1., we have

$$\mathbb{E}\,X = \int_0^{+\infty} \mathbb{P}(X > t)\,dt = \sum_{n\geq 0} \int_n^{n+1} \mathbb{P}(X > t)\,dt.$$

Clearly,

$$\sum_{n\geq 0} \mathbb{P}(X > n+1) = \sum_{n\geq 0} \int_n^{n+1} \mathbb{P}(X > n+1)\,dt$$

$$\sum_{n\geq 0} \mathbb{P}(X > n) = \sum_{n\geq 0} \int_n^{n+1} \mathbb{P}(X > n)\,dt.$$

Since $\mathbb{P}(X > n+1) \leq \mathbb{P}(X > t) \leq \mathbb{P}(X > n)$ for all $n \leq t < n+1$, the result follows. For an integer valued variable, this amounts to Proposition 2.40.

3. We have

$$\sum_{n\geq 1} \mathbb{P}(X \geq n) = \sum_{n\geq 1} \mathbb{P}\Big[\bigcup_{i\geq n}(i \leq X < i+1)\Big] = \sum_{n\geq 1}\sum_{i\geq n} \mathbb{P}(i \leq X < i+1)$$

$$= \sum_{i\geq 1}\sum_{n=1}^{i} \mathbb{P}(i \leq X < i+1) = \sum_{i\geq 1} \mathbb{P}(i \leq X < i+1)i$$

or

$$\sum_{n\geq 1} \mathbb{P}(X \geq n) \leq \sum_{i\geq 1} \mathbb{E}\,[X\mathbb{1}_{(i\leq X<i+1)}] \leq \mathbb{E}\,X.$$

Finally, $\mathbb{P}(X \geq n) \geq \mathbb{P}(X > n)$ so that 2. yields the second inequality. $\triangle$

∇ **Exercise 2.7 (König-Huighens Formula)** Let X be any random variable with finite expected value m.

1. Show that

$$\mathbb{E}\left[(X - \lambda)^2\right] = \mathbb{E}\left[(X - m)^2\right] + (m - \lambda)^2, \quad \lambda \in \mathbb{R}. \tag{2.6}$$

Use the above relation to deduce that the best approximation of X in the sense of least-squares is its expected value.
2. Assume that X takes values in $[a, b]$. Show that

$$\mathrm{Var}\, X \leq \frac{(b - a)^2}{4}.$$

Solution
1. We have

$$(X - \lambda)^2 = (X - m)^2 + 2(X - m)(m - \lambda) + (m - \lambda)^2,$$

from which König-Huighens formula (2.6) derives. This formula is the stochastic equivalent of the Pythagorean theorem in geometry. Further,

$$\inf_{\lambda \in \mathbb{R}} \mathbb{E}\left[(X - \lambda)^2\right] = \mathbb{E}\left[(X - m)^2\right] = \mathrm{Var}\, X.$$

2. Question 1 induces that

$$\mathrm{Var}\, X = \mathbb{E}\left[(X - m)^2\right] \leq \mathbb{E}\left[(X - \lambda)^2\right],$$

and we have

$$\mathbb{E}\left[(X - \lambda)^2\right] = \sup_{a \leq x \leq b} (x - \lambda)^2 \int_a^b d\mathbb{P}_X(x).$$

Setting $\lambda = (a + b)/2$, we have

$$\sup_{x \in [a,b]} \left(x - \frac{a + b}{2}\right)^2 = \left(\frac{b - a}{2}\right)^2,$$

which yields the desired inequality.   △

∇ **Exercise 2.8 (Geometric Probabilities)** Draw at random a chord on a circle with center 0 and radius r, as shown in Fig. 2.14. Let X be the random variable equal to the length of this chord.

Fig. 2.14 Geometric
probabilities (Exercise 2.8)

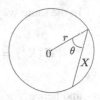

1. Determine the distribution of X. Compute its expected value and its variance.
2. Show that $\mathbb{P}(X > r\sqrt{3}) = 1/3$.

Solution

1. We have $X = 2r \cos \theta$, where $\theta \sim \mathcal{U}(-\pi/2, \pi/2)$ is the angle between the chord and the radius, so

$$F_X(x) = \mathbb{P}(2r \cos \theta \leq x) = \mathbb{P}(\text{Arccos}\frac{x}{2r} \leq \theta \leq \frac{\pi}{2}) + \mathbb{P}(\frac{-\pi}{2} \leq \theta \leq -\text{Arccos}\frac{x}{2r}).$$

Since, by Property 4 of distribution functions,

$$\mathbb{P}(\text{Arccos}\frac{x}{2r} \leq \theta \leq \frac{\pi}{2}) = F_\theta\left(\frac{\pi}{2}\right) - F_\theta\left(\text{Arccos}\frac{x}{2r}\right)$$

and

$$\mathbb{P}(\frac{-\pi}{2} \leq \theta \leq -\text{Arccos}\frac{x}{2r}) = F_\theta\left(-\text{Arccos}\frac{x}{2r}\right) - F_\theta\left(-\frac{\pi}{2}\right),$$

we get

$$F_X(x) = 1 - \frac{1}{\pi}\text{Arccos}\frac{x}{2r} + \frac{-1}{\pi}\text{Arccos}\frac{x}{2r} - 0 = 1 - \frac{2}{\pi}\text{Arccos}\frac{x}{2r}.$$

Differentiating with respect to x yields

$$f_X(x) = \frac{2}{\pi\sqrt{4r^2 - x^2}},$$

because $(\text{Arccos})'(x) = 1/\sqrt{1 - x^2}$. The expected value of X is

$$\mathbb{E}\,X = \mathbb{E}\,(2r \cos \theta) = \frac{2r}{\pi} \int_{-\pi/2}^{\pi/2} \cos \theta\, d\theta = \frac{4r}{\pi}.$$

In the same way,

$$\mathbb{E}\left(X^2\right) = \frac{4r^2}{\pi} \int_{-\pi/2}^{\pi/2} \cos^2(\theta)d\theta = \frac{4r^2}{\pi} \int_{-\pi/2}^{\pi/2} \frac{1 + \cos(2\theta)}{2} d\theta = 2r^2,$$

and hence $\mathbb{V}\mathrm{ar}\, X = \mathbb{E}\left(X^2\right) - (\mathbb{E}\, X)^2 = 2r^2\left(1 - 8/\pi^2\right)$.

2. We have

$$\mathbb{P}(X > r\sqrt{3}) = \mathbb{P}\left(\cos\theta > \frac{\sqrt{3}}{2}\right) = \frac{1}{\pi} \int_{-\pi/6}^{\pi/6} d\theta,$$

so $\mathbb{P}(X > r\sqrt{3}) = 1/3$.   △

▽ Exercise 2.9 (A Mixture of Symmetrical Distributions)

1. Let Y be a random variable with density f_Y on $\mathbb{R}$. Let $m \in \mathbb{R}$. Show that if f_Y is symmetric with respect to m, that is $f_Y(m - y) = f_Y(m + y)$, then $\mathbb{E}\, Y = m$.
2. Let f_1 and f_2 be positive real functions such that $\int_{\mathbb{R}}[f_1(x) + f_2(x)]\, dx = 1$, with respective supports $[\alpha, \beta]$ and $[\gamma, \delta]$, where $\alpha < \beta < \gamma < \delta$. Suppose that f_1 is symmetric with respect to $m_1 = (\alpha + \beta)/2$ and f_2 with respect to $m_2 = (\gamma + \delta)/2$. Let X be the random variable with density plotted in Fig. 2.15, that is

$$f(x) = \begin{cases} f_1(x) & \text{if } \alpha \le x \le \beta, \\ f_2(x) & \text{if } \gamma \le x \le \delta, \\ 0 & \text{otherwise.} \end{cases}$$

Show that $m_1 < \mathbb{E}\, X < m_2$.

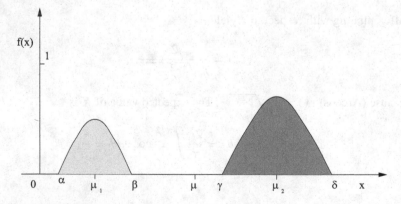

Fig. 2.15 An example of a mixture of symmetrical densities—Exercise 2.9

3. Compute $(m - m_1)/(m_2 - m)$ as a function of f.

Solution

1. We compute by the linear changes of variables $u = m - y$ and $v = y - m$

$$\int_{-\infty}^{+\infty} (y - m) f_Y(y) dy = \int_{-\infty}^{m} (y - m) f_Y(y) dy + \int_{m}^{+\infty} (y - m) f_Y(y) dy$$

$$= \int_{+\infty}^{0} u f_Y(m - u) du + \int_{0}^{+\infty} v f_Y(v + m) dv = 0.$$

Therefore,

$$\mathbb{E} Y = \int_{-\infty}^{+\infty} (y - m) f_Y(y) dy + \int_{-\infty}^{+\infty} m f_Y(y) dy = m.$$

2. In the same way, $\int_{\alpha}^{\beta} (x - m_1) f(x) dx = 0$ and $\int_{\gamma}^{\delta} (x - m_2) f(x) dx = 0$, so

$$\mathbb{E} X = m_1 \int_{\alpha}^{\beta} f(x) dx + m_2 \int_{\gamma}^{\delta} f(x) dx = m_1 c + m_2 (1 - c),$$

where $c \in]0, 1[$, because the support of f is $[\alpha, \beta] \cup [\gamma, \delta]$. Thus, $\mathbb{E} X \in]m_1, m_2[$.
3. We compute

$$m - m_1 = m_2 \int_{\gamma}^{\delta} f(x) dx - m_1 \left(1 - \int_{\alpha}^{\beta} f(x) dx\right)$$

$$= m_2 \int_{\gamma}^{\delta} f(x) dx - m_1 \int_{\gamma}^{\delta} f(x) dx,$$

and similarly $m - m_2$ by symmetry, so that

$$\frac{m - m_1}{m_2 - m} = \frac{m_2 \int_{\gamma}^{\delta} f(x) dx - m_1 \int_{\gamma}^{\delta} f(x) dx}{m_2 \int_{\alpha}^{\beta} f(x) dx - m_1 \int_{\alpha}^{\beta} f(x) dx} = \frac{\int_{\gamma}^{\delta} f(x) dx}{\int_{\alpha}^{\beta} f(x) dx},$$

as required. $\triangle$

∇ **Exercise 2.10 (Transformations of the Log-Normal Distribution)** Let X be a random variable with log-normal distribution with parameters 0 and 1.

1. Set $Z = aX + b$, where $a \in \mathbb{R}_+^*$ and $b \in \mathbb{R}$. Identify the pairs (a, b) for which the random variable Z has a log-normal distribution too.
2. Set $T = X^r$, for a given $r \in \mathbb{N}^*$. Determine the distribution of T.

Solution

1. The density of X is $f_X(x) = \mathbb{1}_{\mathbb{R}^*_+}(x)e^{-(\log x)^2/2}/x\sqrt{2\pi}$. Applying Example 2.59, the density of Z is

$$f_Z(z) = \frac{1}{\sqrt{2\pi}(z-b)}e^{-(\log[(z-b)/a])^2/2}\mathbb{1}_{[b,+\infty[}(z),$$

that is the density of a log-normal distribution for any a if $b = 0$. Then, its parameters are $\log a$ and 1.

2. The random variable X has a log-normal distribution with parameters 0 and 1, so by definition $Y = \log X$ has a standard Gaussian distribution. Since $T = e^{rY}$ and $rY \sim \mathcal{N}(0, r^2)$, we conclude that T has a log-normal distribution with parameters 0 and r^2. $\qquad\qquad\qquad\qquad\qquad\qquad\qquad\qquad\qquad\qquad\qquad\triangle$

$\triangledown$ Exercise 2.11 (Lack of Memory of the Exponential Distributions)

1. Let $T \sim \mathcal{E}(\lambda)$. Show that

$$\mathbb{P}(T > t + s) = \mathbb{P}(T > t)\mathbb{P}(T > s), \quad s \in \mathbb{R}^*_+, \ t \in \mathbb{R}^*_+.$$

2. Let T be a nonnegative random variable satisfying the above equation, with a continuous distribution function. Set $\mathbb{P}(T > 1) = \alpha$, and suppose that $0 < \alpha < 1$.
   (a) Write $\mathbb{P}(T > nt)$ as a function of $\mathbb{P}(T > t)$ and n, for $t \in \mathbb{R}^*_+$ and $n \in \mathbb{N}^*$.
   (b) Show that if $\mathbb{P}(T > 1/p)$, then $\mathbb{P}(T > q/p)$ for all positive integers p and q, and finally that $\mathbb{P}(T > t)$ for all $t \in \mathbb{R}^*_+$.
   (c) What is the distribution of T?
3. Show that the hazard rate of a nonnegative random variable T with density f is constant if and only if the distribution of T is a translated exponential distribution.

Solution

1. As in Example 2.89, we compute

$$\mathbb{P}(T > t + s \mid T > t) = e^{-\alpha s} = \mathbb{P}(T > s).$$

Note that the reliability R of the exponential distribution thus appears as a solution of Cauchy's functional equation

$$R(s + t) = R(s)R(t), \quad s > 0, \ t > 0.$$

2. (a) By Theorem 1.32,

$$\mathbb{P}(T > nt) = \mathbb{P}(T > t)\mathbb{P}[T > (n-1)t + t \mid T > t]$$

$$= \mathbb{P}(T > t)\mathbb{P}[T > (n-1)t] = \cdots = \mathbb{P}(T > t)^n.$$

(b) Taking $n = p$ and $t = 1/p$ in a. yields $\mathbb{P}(T > 1) = \mathbb{P}(T > 1/p)^p$, so
that $\mathbb{P}(T > 1/p) = \alpha^{1/p}$. Taking $n = q$ and $t = 1/p$ in (a) again yields
$\mathbb{P}(T > q/p) = \mathbb{P}(T > 1/p)^q$, so that $\mathbb{P}(T > p/q) = \alpha^{q/p}$.

Since the reliability $t \to \mathbb{P}(T > t)$ and the power function $t \to t^\alpha$ are
continuous from the right,

$$\mathbb{P}(T > t) = \mathbb{P}(T > 1)^t = \alpha^t, \quad t \in \mathbb{R}_+^*.$$

(c) Since $\mathbb{P}(T > t) = \alpha^t$, finally $\mathbb{P}(T \leq t) = 1 - \alpha^t$, that is the distribution
function of the exponential distribution with parameter $-\log\alpha$.
3. By definition, $h(x) = f(x)/[1 - F(x)]$, for all x, where F is the distribution
function of T. If $h(x) = c$ for all $x \in \mathbb{R}_+$, then some $x_0 \in \mathbb{R}_+$ exists such that
$-\log(1 - F(x)) = c(x - x_0)$. Hence $F(x) = 1 - e^{-c(x-x_0)}$, meaning that $T - x_0$
has an exponential distribution with parameter c. The converse is clear. $\triangle$

∇ **Exercise 2.12 (Paley-Zygmund Inequality)** Let X be any nonnegative random
variable with finite variance. Show that

$$\mathbb{P}(X \geq \lambda \mathbb{E}\, X) \geq (1-\lambda)^2 \frac{(\mathbb{E}\, X)^2}{\mathbb{E}\,(X^2)}, \quad \lambda \in [0, 1].$$

Solution
We get by using Hölder's inequality

$$\left(\mathbb{E}\,[X\mathbb{1}_{(X \geq \lambda \mathbb{E}\, X)}]\right)^2 \leq \mathbb{E}\,(X^2)\mathbb{E}\,[\mathbb{1}_{(X \geq \lambda \mathbb{E}\, X)}].$$

First, $\mathbb{E}\,[X\mathbb{1}_{(X \geq \lambda \mathbb{E}\, X)}] = \mathbb{E}\, X - \mathbb{E}\,[X\mathbb{1}_{(X < \lambda \mathbb{E}\, X)}]$ and $\mathbb{E}\,[X\mathbb{1}_{(X < \lambda \mathbb{E}\, X)}] \leq \lambda \mathbb{E}\, X$, so
$\mathbb{E}\,[X\mathbb{1}_{(X \geq \lambda \mathbb{E}\, X)}] \geq (1-\lambda)\mathbb{E}\, X$. Since $\mathbb{E}\,[\mathbb{1}_{(X \geq \lambda \mathbb{E}\, X)}] = \mathbb{P}(X \geq \lambda \mathbb{E}\, X)$, the desired
inequality follows.

Such inequalities are known as inequalities of the Chebyshev type. $\triangle$

∇ **Exercise 2.13 (Characteristic Function of the Cauchy Distribution)** Let Y be
a random variable with symmetric exponential distribution with parameter $\alpha > 0$
(also called Laplace), with density defined by $f(y) = \alpha e^{-\alpha|y|}/2$ on $\mathbb{R}$; see
Exercise 3.6 for details.

Determine the characteristic function of Y and then the characteristic function of
a random variable with Cauchy distribution $\mathcal{C}(\alpha)$.

Solution
We have

$$\varphi_Y(t) = \frac{\alpha}{2} \int_{\mathbb{R}_-} e^{ity} \frac{\alpha}{2} e^{\alpha y} dy + \frac{\alpha}{2} \int_{\mathbb{R}_+} e^{ity} \frac{\alpha}{2} e^{-\alpha y} dy$$

$$= \frac{\alpha}{2(it+\alpha)} - \frac{\alpha}{2(it-\alpha)} = \frac{\alpha^2}{\alpha^2+t^2}.$$

Since the function $t \rightarrow \alpha^2/(\alpha^2+t^2)$ is integrable, Fourier's inversion formula yields

$$\frac{\alpha}{2} e^{-\alpha|y|} = \frac{1}{2\pi} \int_{\mathbb{R}} e^{-ity} \frac{\alpha^2}{\alpha^2+t^2} dt,$$

from which it derives that the characteristic function of a Cauchy distribution $\mathcal{C}(\alpha)$ is given by $\varphi(t) = \alpha e^{-\alpha|t|}$ for $t \in \mathbb{R}$.   △

Random Vectors

3

While Chap. 2 investigates one real random variable at a time, the present chapter is dedicated to studying simultaneously several random variables, which are multivariate random variables, also called random vectors.

Many properties of the random vectors are similar to the properties of the real random variables. Thus we will extend notions seen in Chap. 2 and add properties linked to the relations between variables such as covariance, independence, entropy, and order statistics. We will insist on the effective determination of the distribution of functions of several random variables, among which sums play an important part. A full section is devoted to Gaussian vectors.

3.1 Relations Between Random Variables

Studying relations between random variables is both interesting by itself for applications and a necessary first step to the investigation of random vectors.

Unless otherwise stated, all considered random variables are supposed to be defined on the same probability space $(\Omega, \mathcal{F}, \mathbb{P})$.

3.1.1 Covariance

The definition of variance for vectors is based on the notion of covariance of random variables.

Definition 3.1 Let X and Y be two square integrable real random variables. Their covariance is defined as

$$\mathbb{C}\text{ov}(X, Y) = \mathbb{E}[(X - \mathbb{E}X)(Y - \mathbb{E}Y)] = \mathbb{E}(XY) - (\mathbb{E}X)(\mathbb{E}Y).$$

Their (linear) correlation coefficient is

$$\rho_{X,Y} = \frac{\mathbb{C}\mathrm{ov}\,(X,Y)}{\sqrt{\mathbb{V}\mathrm{ar}\,X}\sqrt{\mathbb{V}\mathrm{ar}\,Y}}.$$

Two variables with a null covariance are said to be uncorrelated. Clearly, the expectation of their product is the product of their respective expectations.

Properties of Covariance and Correlation Coefficients
1. $\mathbb{C}\mathrm{ov}\,(X,X) = \mathbb{V}\mathrm{ar}\,X$ and $\mathbb{C}\mathrm{ov}\,(X,Y) = \mathbb{C}\mathrm{ov}\,(Y,X)$.
2. $\mathbb{C}\mathrm{ov}\,(aX+bY,Z) = a\,\mathbb{C}\mathrm{ov}\,(X,Z)+b\,\mathbb{C}\mathrm{ov}\,(Y,Z)$, for all $(a,b) \in \mathbb{R}^2$. Therefore, $|\rho_{X,Y}| = |\rho_{aX+b,cY+d}|$, for all $a \in \mathbb{R}^*$, $c \in \mathbb{R}^*$, $b \in \mathbb{R}$, $d \in \mathbb{R}$; the correlation coefficient depends neither on the unity nor on the origin.
3. $\mathbb{C}\mathrm{ov}\,(X,Y)^2 \leq \mathbb{V}\mathrm{ar}\,X\mathbb{V}\mathrm{ar}\,Y$, by Schwarz's inequality, and hence, $\rho_{X,Y} \in [-1,1]$.

The correlation coefficient is a measure of the degree of linearity between two variables.

Proposition 3.2 *If X and Y are two square integrable real random variables, then*

$$|\rho_{X,Y}| = 1 \text{ if and only if } Y = aX + b, \text{ with } (a,b) \in \mathbb{R}^* \times \mathbb{R}.$$

Proof We can suppose that $\mathbb{E}\,X = 0$. If $Y = aX + b$, then $\mathbb{E}\,Y = b$ and $\mathbb{C}\mathrm{ov}\,(X,Y) = a\mathbb{E}\,(X^2) = a\mathbb{V}\mathrm{ar}\,X$, and hence, $|\rho_{X,Y}| = 1$.

Conversely, we can also suppose that $\mathbb{E}\,Y = 0$. Thus, if $|\rho_{X,Y}| = 1$, then $[\mathbb{E}\,(XY)]^2 = \mathbb{E}\,(X^2)\mathbb{E}\,(Y^2)$. Therefore,

$$\mathbb{V}\mathrm{ar}\left[Y - \frac{\mathbb{E}\,(XY)}{\mathbb{E}\,(X^2)}X\right] = \mathbb{V}\mathrm{ar}\,(Y) - a^2\mathbb{V}\mathrm{ar}\,(X) = 0,$$

so this variable is a.s. constant, and hence, $Y = aX$. In the general case where the expectation of X and Y is not null, $a = \mathbb{C}\mathrm{ov}\,(X,Y)/\mathbb{V}\mathrm{ar}\,X$. $\square$

3.1.2 Independence of Random Variables

The notion of independence of random variables is based on the notion of independence of their generated σ-algebras and hence of collections of events. Independence of σ-algebras and of events is investigated in Chap. 1.

Definition 3.3 Let $X_i \colon (\Omega, \mathcal{F}, \mathbb{P}) \rightarrow (\Omega_i, \mathcal{F}_i)$, for $i \in [\![1,d]\!]$, be d random variables. They are said to be independent if the d generated σ-algebras $\sigma(X_1), \ldots,$

$\sigma(X_d)$ are independent, that is

$$\mathbb{P}(X_1 \in B_1, \ldots, X_d \in B_d) = \prod_{i=1}^{d} \mathbb{P}(X_i \in B_i), \quad B_1 \in \mathcal{F}_1, \ldots, B_d \in \mathcal{F}_d.$$

If $X_1, \ldots, X_d$ are independent, then $g_1(X_1), \ldots, g_d(X_d)$ are also independent, for all Borel functions $g_1, \ldots, g_d$.

If the generated σ-algebra $\sigma(X_1, \ldots, X_d)$ and a σ-algebra $\mathcal{G}$ included in $\mathcal{F}$ are independent, we will also say that $X_1, \ldots, X_d$ and $\mathcal{G}$ are independent.

If $X_1, \ldots, X_d$ are independent, all with the same distribution P, they are typically referred to as i.i.d. (independent identically distributed) with (common) distribution P.

▷ *Example 3.4 (Indicator Functions)* Events of a σ-algebra $\mathcal{F}$ are independent if and only if their indicator functions are independent random variables. Also, d σ-algebras $\mathcal{F}_1, \ldots, \mathcal{F}_d$ included in $\mathcal{F}$ are independent if and only if the d coordinate random variables $c_i : (\Omega, \mathcal{F}, \mathbb{P}) \to (\Omega_i, \mathcal{F}_i)$ are independent. ◁

Random variables are said to be independent conditionally to a non- null event B if they are independent for the probability $\mathbb{P}(\cdot \mid B)$.

For discrete random variables, X and Y with respective values $\{x_i : i \in I\}$ and $\{y_j : j \in J\}$, independence amounts to

$$\mathbb{P}(X = x_i, Y = y_j) = \mathbb{P}(X = x_i)\mathbb{P}(Y = y_j), \quad i \in I, j \in J.$$

▷ *Example 3.5 (Independent Uniform Variables)* Two individuals have booked an appointment. Each of them arrives at random, say between noon and 1 p.m. Let us compute the probability that the first arrived waits for the other at least a quarter of an hour.

Let X and Y be the times of the arrivals of the two individuals, in minutes. They are i.i.d. with distribution $\mathcal{U}(0, 60)$, so

$$\mathbb{P}(|X - Y| > 15) = \mathbb{P}(X + 15 < Y) + \mathbb{P}(Y + 15 < X) =$$

$$= 2\mathbb{P}(X < Y - 15) = 2 \int_{15}^{60} \int_{0}^{y-15} \frac{1}{(60)^2} \, dx dy$$

$$= \frac{2}{(60)^2} \int_{15}^{60} (y - 15) dy = \frac{(45)^2}{(60)^2} = 9/16.$$

The first arrived will wait at least 15 minutes with probability 9/16. ◁

Below we present various criteria of independence. First, non-correlation is a consequence of independence.

Proposition 3.6 *Independent square integrable random variables are uncorrelated.*

Proof We prove the proposition for discrete random variables.

Let us set $X = \sum_{i \in I} x_i \mathbb{1}_{A_i}$, where $A_i = (X = x_i)$, and $Y = \sum_{j \in J} y_j \mathbb{1}_{B_j}$, where $B_j = (Y = y_j)$. We can write

$$XY = \sum_{(i,j) \in I \times J} x_i y_j \mathbb{1}_{A_i} \mathbb{1}_{B_j} = \sum_{(i,j) \in I \times J} x_i y_j \mathbb{1}_{A_i \cap B_j}.$$

Since X and Y have finite expectations, we get

$$\mathbb{E}(XY) = \sum_{(i,j) \in I \times J} x_i y_j \mathbb{P}(A_i \cap B_j) \overset{(1)}{=} \sum_{i \in I} \sum_{j \in J} x_i y_j \mathbb{P}(A_i) \mathbb{P}(B_j)$$

$$= \left[\sum_{i \in I} x_i \mathbb{P}(A_i) \right] \left[\sum_{j \in J} y_j \mathbb{P}(B_j) \right] = (\mathbb{E}X)(\mathbb{E}Y),$$

(1) by independence of X and Y. □

On the contrary, two uncorrelated variables are not independent in general, as shown in the next example.

▷ *Example 3.7 (Uncorrelated Dependent Variables)* Let $Y \sim \mathcal{U}(0,1)$. The moments of Y have been computed in Example 2.58. If $X = Y(Y + a)$ where $a \in \mathbb{R}$, we have

$$\mathrm{Cov}(X, Y) = \mathbb{E}(Y^3) + a\mathbb{E}(Y^2) - (\mathbb{E}Y)\mathbb{E}(Y^2) - a(\mathbb{E}Y)^2$$

$$= \frac{1}{4} + \frac{a}{3} - \frac{1}{6} - \frac{a}{4} = \frac{1+a}{12}.$$

For $a = -1$, these variables are uncorrelated, but they are dependent since $X = Y(Y - 1)$ is a function of Y, clearly measurable for the σ-algebra generated by Y.

◁

Conditional distributions for discrete variables are defined in terms of probabilities of events.

Definition 3.8 Let X and T be two discrete random variables with respective values $\{x_i : i \in I\}$ and $\{t_j : j \in J\}$. The distribution of X conditional on $(T = t_j)$ is given by $\{\mathbb{P}(X = x_i \mid T = t_j) : i \in I\}$.

▷ *Example 3.9 (Continuation of Example 1.30)* Let X and T be the respective indicator functions of the events A and B. Their common distribution is a Bernoulli distribution with parameter $1/2$. The distribution of X conditional on T is given by

$$\mathbb{P}(X = i \mid T = j) = \frac{\mathbb{P}(X = i, T = j)}{\mathbb{P}(T = j)}, \quad i, j = 0, 1.$$

For instance, $(X = 0, T = 0) =$"the result is odd and larger than 3"="the result is 5," so $\mathbb{P}(X = 0 \mid T = 0) = \mathbb{P}(\overline{B} \mid \overline{A}) = 1/3$. We compute similarly $\mathbb{P}(X = 1 \mid T = 1) = 1/3$ and $\mathbb{P}(X = 1 \mid T = 0) = \mathbb{P}(X = 0 \mid T = 1) = 2/3$. ◁

For discrete variables, a simple connection can be made between independence and conditional distributions, thanks to the definitions.

Proposition 3.10 *Let X and T be two discrete random variables with respective values $\{x_i : i \in I\}$ and $\{t_j : j \in J\}$. They are independent if and only if the conditional distribution of X given $(T = t_j)$ is equal to the distribution of X for all j, and symmetrically for T.*

Independence does not imply conditional independence, as shown in the next example.

▷ *Example 3.11 (Independent But Conditionally Dependent Variables)* Let X_1 and X_2 be two i.i.d. variables with Bernoulli distribution $\mathcal{B}(1/2)$ taking values in $\{-1, 1\}$. We have

$$\mathbb{P}(X_1 = 1 \mid X_1 + X_2 = 0) = \frac{\mathbb{P}(X_1 = 1, X_1 + X_2 = 0)}{\mathbb{P}(X_1 + X_2 = 0)}$$

$$= \frac{\mathbb{P}(X_1 = 1)\mathbb{P}(X_2 = -1)}{\mathbb{P}(X_1 + X_2 = 0)} = \frac{(1/2)^2}{1/2} = \frac{1}{2}.$$

We compute similarly $\mathbb{P}(X_2 = 1 \mid X_1 + X_2 = 0) = 1/2$ and $\mathbb{P}(X_1 = 1, X_2 = 1 \mid X_1 + X_2 = 0) = 0$, so X_1 and X_2 are dependent conditionals on $X_1 + X_2$.

Note that $\mathbb{P}(X_1 = -1 \mid X_1 + X_2 = -2) = 1$. ◁

3.1.3 Stochastic Order Relation

The stochastic order relations are useful tools in particular for investigating the reliability or the survival function of systems.

Definition 3.12 Let X and Y be two real random variables. Then X is said to be stochastically larger than Y, and we will write $X \overset{st}{\geq} Y$, if

$$\mathbb{P}(X > x) \geq \mathbb{P}(Y > x), \quad x \in \mathbb{R}. \tag{3.1}$$

Fig. 3.1 Systems with two
parallel and series
components—Example 3.14

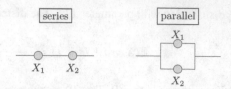

In other words, $X \overset{st}{\geq} Y$ if the reliability of X is higher than the reliability of Y, or $R_X(t) \geq R_Y(t)$ for all $t \in \mathbb{R}$; see Definition 2.87. Note that $X \overset{st}{\geq} Y$ does not imply $X(\omega) \geq Y(\omega)$ for all $\omega \in \Omega$, and only distributions are involved.

▷ *Example 3.13* Let $X \sim \mathcal{E}(\lambda)$ and $Y \sim \mathcal{E}(\mu)$ be two exponential variables. If $\lambda \leq \mu$, then $X \overset{st}{\geq} Y$.

On the contrary, if $X \sim \mathcal{N}(0, \sigma^2)$ and $Y \sim \mathcal{N}(0, \tau^2)$ are two normal variables with $\sigma \geq \tau$, then $X \overset{st}{\geq} Y$.   ◁

▷ *Example 3.14 (Series and Parallel Systems)* Consider two systems with two components as shown in Fig. 3.1. Let T_1 and T_2 be the lifetimes of the components, assumed to be independent, with respective distribution functions F_1 and F_2.

If the components are connected in parallel, the lifetime is $T = \max(T_1, T_2)$. Its distribution function is $F_T(t) = F_1(t)F_2(t)$, because

$$\mathbb{P}(\max(T_1, T_2) \leq t) = \mathbb{P}(T_1 \leq t, T_2 \leq t) = \mathbb{P}(T_1 \leq t)\mathbb{P}(T_2 \leq t).$$

If the components are connected in series, the lifetime of the system is $T' = \min(T_1, T_2)$. Its distribution function is $F_{T'}(t) = 1 - [1 - F_1(t)][1 - F_2(t)]$, because

$$1 - F_{T'}(t) = \mathbb{P}(T_1 > t, T_2 > t) = \mathbb{P}(T_1 > t)\mathbb{P}(T_2 > t) = [1 - F_1(t)][1 - F_2(t)].$$

Moreover, $F_1(t) + F_2(t) - 2F_1(t)F_2(t) = F_1(t)[1 - F_2(t)] + F_2(t)[1 - F_1(t)] \geq 0$, so $T' \overset{st}{\leq} T$. In other words, $\max(T_1, T_2) \overset{st}{\geq} \min(T_1, T_2)$, and a parallel system is more reliable than a series system.   ◁

Proposition 3.15 *Let X and Y be two real random variables. If $X \overset{st}{\geq} Y$, then $\mathbb{E} X \geq \mathbb{E} Y$.*

Proof First, suppose that X and Y are nonnegative. Integrating both sides of (3.1) and using (2.5) p. 86, for $p = 1$, yield $\mathbb{E} X \geq \mathbb{E} Y$.

For general random variables, if $X \overset{st}{\geq} Y$, then $X^+ \overset{st}{\geq} Y^+$ and $Y^- \overset{st}{\geq} X^-$, so that $\mathbb{E} X^+ \geq \mathbb{E} Y^+$ and $\mathbb{E} Y^- \geq \mathbb{E} X^-$ from which $\mathbb{E} X \geq \mathbb{E} Y$ follows.   □

This necessary condition can be replaced by a sufficient condition by considering expectation of general functions.

Theorem 3.16 *Let X and Y be two real random variables. Then $X \overset{st}{\geq} Y$ if and only if $\mathbb{E}[h(X)] \geq \mathbb{E}[h(Y)]$ for all non-decreasing Borel functions h.*

Proof Let h^{-1} denote the generalized inverse of h, defined by $h^{-1}(y) = \inf\{t \in \mathbb{R} : h(t) > y\}$. We have

$$\mathbb{P}[h(X) > x] = \mathbb{P}[X > h^{-1}(x)] \geq \mathbb{P}[Y > h^{-1}(x)] = \mathbb{P}[h(Y) > x], \quad x \in \mathbb{R},$$

so that $h(X) \overset{st}{\geq} h(Y)$, and hence, $\mathbb{E}[h(X)] \geq \mathbb{E}[h(Y)]$ by applying Theorem 3.16.

The proof of the converse is straightforward by considering the non-decreasing function $\mathbb{1}_{[x,+\infty[}$.   □

Note that many other stochastic order relations exist.

3.1.4 Entropy

The different entropic quantities measure the degree of uncertainty of the studied phenomenon, either intrinsic or relative to another. They can be expressed in terms of random variables, even if they depend only on their distributions. It is simpler to present first the discrete case.

Definition 3.17 Let X and Y be two discrete random variables with respective values $\{x_i : i \in I\}$ and $\{y_j : j \in J\}$.

The Shannon entropy of X is that of its distribution, that is

$$\mathcal{H}(X) = - \sum_{i \in I} \mathbb{P}(X = x_i) \log \mathbb{P}(X = x_i).$$

The (joint) entropy of (X, Y) is that of its distribution, that is

$$\mathbb{S}(X, Y) = - \sum_{i \in I} \sum_{j \in J} \mathbb{P}(X = x_i, Y = y_j) \log \mathbb{P}(X = x_i, Y = y_j).$$

The entropy of X conditional on Y is

$$\mathbb{S}(X \mid Y) = \mathbb{S}(X, Y) - \mathbb{S}(Y).$$

Suppose that $\{x_i : i \in I\} \subset \{y_j : j \in J\}$. The entropy of X relative to Y (or Kullback–Leibler information) is that of their distributions, that is

$$\mathbb{K}(X \mid Y) = \sum_{i \in I} \mathbb{P}(X = x_i) \log \frac{\mathbb{P}(X = x_i)}{\mathbb{P}(Y = x_i)}.$$

We can also write

$$\mathbb{S}(X \mid Y) = -\sum_{i \in I} \sum_{j \in J} \mathbb{P}(X = x_i, Y = y_j) \log \frac{\mathbb{P}(X = x_i, Y = y_j)}{\mathbb{P}(Y = y_j)}. \tag{3.2}$$

▷ *Example 3.18 (Entropy of a Geometric Distribution)* The entropy of a random variable with geometric distribution $\mathcal{G}(p)$ is

$$-\sum_{k \geq 1} pq^{k-1} \log(pq^{k-1}) = -p \log p \sum_{k \geq 1} q^{k-1} - p \log q \sum_{k \geq 1} (k-1) q^{k-1}$$

$$= -\log p - \frac{q}{p} \log q,$$

where $q = 1 - p$.   ◁

Properties of the entropy of discrete random variables derive easily from the properties of entropy of probabilities shown in Chap. 1.

Properties of Entropy
1. The entropy of a random variable X taking N values is maximum (and equals $\log N$) if X has a uniform distribution.
2. $\mathbb{S}(X \mid Y) = 0$ if and only if $X = g(Y)$. Indeed, let us consider conditional entropy under the form of (3.2). If it is null, then necessarily, $\mathbb{P}(X = x_i, Y = y_j) = \mathbb{P}(Y = y_j)$ for all (i, j), meaning that X is a deterministic function of Y. The converse is clear.
3. $\mathbb{S}(X \mid Y) \leq \mathbb{S}(X)$, with equality if and only if the variables are independent. Indeed,

$$\mathbb{S}(X) = -\sum_{i \in I} \sum_{j \in J} \mathbb{P}(X = x_i, Y = y_j) \log \mathbb{P}(X = x_i)$$

$$= \sum_{i \in I} \sum_{j \in J} \mathbb{P}(X = x_i, Y = y_j) \log \frac{\mathbb{P}(X = x_i)\mathbb{P}(Y = y_j)}{\mathbb{P}(X = x_i, Y = y_j)},$$

which, since $\log x \leq x - 1$ for all positive real numbers x, yields

$$\mathbb{S}(X) \leq \sum_{i \in I} \sum_{j \in J} \mathbb{P}(X = x_i, Y = y_j) \left(\frac{\mathbb{P}(X = x_i)\mathbb{P}(Y = y_j)}{\mathbb{P}(X = x_i, Y = y_j)} - 1 \right) = 0.$$

This is equivalent to $\mathbb{P}(X = x_i, Y = y_j) = \mathbb{P}(X = x_i)\mathbb{P}(Y = y_j)$ for all $(i, j) \in I \times J$, that is to the independence of the two variables.

4. $\mathbb{S}(X, Y) \leq \mathbb{S}(X) + \mathbb{S}(Y)$, with equality if and only if the variables are independent, thanks to both the definition and Property 3.

Properties 2 and 3 can be interpreted as follows: the quantity of information brought on X by the knowledge of Y decreases the uncertainty on X. At the limit (that is if X is a function of Y), this uncertainty is null.

▷ *Example 3.19 (Information Theory)*   A system of information is composed of a source, a canal of transmission, and a receptor. A part of the message (or signal) emitted by the source is modified during the transmission to the receptor by the presence of noise in the canal of transmission.

The entropy is a means for evaluating the rate of transmission through the canal. Let the variable X represent the source, and Y the receptor. Then $\mathbb{S}(X)$ is the entropy of the source, $\mathbb{S}(Y)$ that of the receptor, and $\mathbb{S}(X, Y)$ that of the global system. The conditional entropy $\mathbb{S}(Y \mid X)$ is the entropy of the receptor when the emitted message is known, and thus measures the noise. Finally, $\mathbb{S}(X \mid Y)$ is the entropy of the source when the received message is known, which measures what can be learned on the emitted message from the received message.

A measure of the capacity of transmission of the canal is given by

$$C = \max[\mathbb{S}(X) - \mathbb{S}(X \mid Y)],$$

where the maximum is taken over all the possible distributions of X and Y.

Without noise, $\mathbb{P}(X = x_i, Y = y_j) = 0$ if $i \neq j$, so $\mathbb{S}(X \mid Y) = \mathbb{S}(Y \mid X) = 0$ and $\mathbb{S}(X, Y) = \mathbb{S}(X) = \mathbb{S}(Y)$. The uncertainty of the receptor is equal to the uncertainty of the source. The loss of information through the canal of transmission is minimum and $C = \max \mathbb{S}(X) = \log N$.

On the contrary, if no correlation exists between the emitted message and the received message—that is if every x_i can be transformed into any y_j with the same probability, then $\mathbb{P}(X = x_i, Y = y_j) = p_i$ for all j, with $\sum_{i \in I} p_i = 1/N$. This induces that $\mathbb{S}(X \mid Y) = \mathbb{S}(X)$ and $\mathbb{S}(Y \mid X) = \mathbb{S}(Y)$. The system transmits no information, the loss of information through the canal of transmission is maximum, and $C = 0$.   ◁

Shannon entropy extends to random variables with densities. Its interpretation in terms of information is less easy.

Definition 3.20 Let X and Y be two random variables with respective positive densities f_X and f_Y.

• The entropy of X is the entropy of its distribution, that is

$$\mathbb{S}(X) = \mathbb{E}[-\log f_X(X)] = -\int_{\mathbb{R}} f_X(u) \log f_X(u) du.$$

- The entropy of X relative to Y (or Kullback–Leibler information, or divergence) is that of their distributions, that is

$$\mathbb{K}(X \mid Y) = \int_{\mathbb{R}} f_X(u) \log \frac{f_X(u)}{f_Y(u)} du.$$

▷ *Example 3.21 (Entropy of an Exponential Distribution)* The entropy of a random variable $X \sim \mathcal{E}(\lambda)$ is

$$\int_{\mathbb{R}_+} \lambda e^{-\lambda x} \log(\lambda e^{-\lambda x}) \, dx = \log \lambda \int_{\mathbb{R}_+} \lambda e^{-\lambda x} \, dx - \lambda \int_{\mathbb{R}_+} x \lambda e^{-\lambda x} \, dx = \log \lambda + 1,$$

because $\int_{\mathbb{R}_+} f_X(x) \, dx = 1$ and $\mathbb{E} X = 1/\lambda$.   ◁

Proposition 3.22 *The Kullback–Leibler information is nonnegative and is null if and only if the variables have the same distribution.*

Proof Since $\int_{\mathbb{R}} f_Y(u) du = \int_{\mathbb{R}} f_X(u) du$, we can write

$$\mathbb{K}(X \mid Y) = \int_{\mathbb{R}} \frac{f_X(u)}{f_Y(u)} \log \left[\frac{f_X(u)}{f_Y(u)} \right] f_Y(u) du + \int_{\mathbb{R}} \left[1 - \frac{f_X(u)}{f_Y(u)} \right] f_Y(u) du.$$

Since $x \log x \geq 1 - x$ with equality for $x = 1$, the conclusion follows.   □

For continuous random variables, the entropy $\mathbb{S}(X)$ can be negative or infinite. Thus, the notion of maximum can only be considered under constraints, leading to maximum entropy methods.

▷ *Example 3.23 (Maximum of Entropy Under Constraints)* The maximum of $\mathbb{S}(X)$ for $a \leq X \leq b$, where $a < b$ are fixed real numbers, is $\log(b - a)$, obtained for a variable with uniform distribution $\mathcal{U}(a, b)$.

The maximum of $\mathbb{S}(X)$ for $X \geq 0$ and $\mathbb{E} X = a$, where a is a fixed positive real number, is $\log(ea)$, obtained for a variable with exponential distribution $\mathcal{E}(1/a)$.

The maximum of $\mathbb{S}(X)$ for $\mathbb{E} X = 0$ and $\mathbb{E}(X^2) = \sigma^2$, where σ^2 is a fixed positive real number, is $\log(\sqrt{2\pi e}\sigma)$, obtained for a variable with normal distribution $\mathcal{N}(0, \sigma^2)$.

The proof of all these results is based on the Lagrange multipliers method. Extension to random vectors is presented in Exercise 3.12.   ◁

3.2   Characteristics of Random Vectors

Let $(\Omega, \mathcal{F}, \mathbb{P})$ be a probability space. In probability theory, Borel functions

$$X = (X_1, \ldots, X_d)' : (\Omega, \mathcal{F}, \mathbb{P}) \longrightarrow (\mathbb{R}^d, \mathcal{B}(\mathbb{R}^d))$$
$$\omega \longrightarrow (X_1(\omega), \ldots, X_d(\omega))'$$

are referred to as d-variate real random variables or as d-dimensional real random vectors, if $d > 1$. We will simply call them random vectors. Random variables so appear as one-dimensional random vectors, say $d = 1$.

All coordinate functions X_j, for $j \in [\![1, d]\!]$, of a random vector X function are real random variables. Investigating X amounts to investigate simultaneously the d phenomena or (one-dimensional) variables $X_1, \ldots, X_d$ linked to a unique random experiment. Among many examples are the simultaneous measures of physical characteristics of a population, such as weight, height, age, etc.

In general, the distribution of the sum—or product or some other function—of n dependent variables $X_1, \ldots, X_n$ cannot be identified from the distributions of the variables only. In other words, the distribution of the vector $(X_1, \ldots, X_n)$ as a whole is necessary to the determination of the distribution of its functions.

3.2.1   Product of Probability Spaces

In order to investigate the occurrence of events linked to several simultaneous random experiments, a probability space modeling all these experiments must be constructed. Let us first detail this construction in the countable case.

Suppose two random experiments modeled by two countable probability spaces, say $(\Omega_1, \mathcal{P}(\Omega_1), \mathbb{P}_1)$ and $(\Omega_2, \mathcal{P}(\Omega_2), \mathbb{P}_2)$, are observed simultaneously. Outcomes in such an experiment are a pair of points (ω_1, ω_2) with $\omega_1 \in \Omega_1$ and $\omega_2 \in \Omega_2$. In other words, the sample space of this experiment is the Cartesian product $\Omega = \Omega_1 \times \Omega_2$, and the set of events is $\mathcal{P}(\Omega) = \mathcal{P}(\Omega_1) \times \mathcal{P}(\Omega_2)$. On this space, a probability $\mathbb{P}$—referred to as the tensorial product of $\mathbb{P}_1$ and $\mathbb{P}_2$ and denoted by $\mathbb{P}_1 \otimes \mathbb{P}_2$,—can be defined by setting $\mathbb{P}(A) = \sum_{\omega \in A} \mathbb{P}(\{\omega\})$ for all $A \subset \Omega$, where

$$\mathbb{P}(\{\omega\}) = \mathbb{P}_1 \otimes \mathbb{P}_2(\{\omega\}) = \mathbb{P}_1(\{\omega_1\})\mathbb{P}_2(\{\omega_2\}), \quad \omega = (\omega_1, \omega_2) \in \Omega.$$

Every event linked to the first experiment can be represented either as a subset A_1 of Ω_1 or as the subset $A_1 \times \Omega_2$ of Ω, and we have

$$\mathbb{P}(A_1 \times \Omega_2) = \sum_{\omega_1 \in A_1} \sum_{\omega_2 \in \Omega_2} \mathbb{P}_1(\{\omega_1\})\mathbb{P}_2(\{\omega_2\}) = \mathbb{P}_1(A_1).$$

Moreover,

$$\mathbb{P}(A_1 \times A_2) = \mathbb{P}[(A_1 \times \Omega_2) \cap (\Omega_1 \times A_2)] = \sum_{\omega_1 \in A_1} \sum_{\omega_2 \in A_2} \mathbb{P}_1(\{\omega_1\})\mathbb{P}_2(\{\omega_2\})$$

$$= \mathbb{P}_1(A_1)\mathbb{P}_2(A_2) = \mathbb{P}(A_1 \times \Omega_2)\mathbb{P}(\Omega_1 \times A_2).$$

The product of n general probability spaces is similarly defined.

Definition 3.24 Let $(\Omega_1, \mathcal{F}_1), \ldots, (\Omega_n, \mathcal{F}_n)$ be measurable spaces. The product $\prod_{i=1}^{n}(\Omega_i, \mathcal{F}_i)$ is defined as $(\prod_{i=1}^{n} \Omega_i, \otimes_{i=1}^{n} \mathcal{F}_i)$, for the Cartesian product $\prod_{i=1}^{n} \Omega_i = \{(\omega_1, \ldots, \omega_n) : \omega_i \in \Omega_i\}$ and the σ-algebra generated by the rectangles $\otimes_{i=1}^{n} \mathcal{F}_i = \sigma(\{A_1 \times \cdots \times A_n : A_i \in \mathcal{F}_i\})$.

This σ-algebra is the smallest such that the projections on the coordinates are measurable. We will write $\prod_{i=1}^{n}(\Omega, \mathcal{F}) = (\Omega^n, \mathcal{F}^{\otimes n}) = (\Omega, \mathcal{F})^n$, with the particular cases $\mathcal{P}(\Omega) \times \mathcal{P}(\Omega) = \mathcal{P}(\Omega \times \Omega)$ for any finite Ω, and $(\mathbb{R}, \mathcal{B}(\mathbb{R}))^n = (\mathbb{R}^n, \mathcal{B}(\mathbb{R}^n))$, that is $\mathcal{B}(\mathbb{R})^{\otimes n} = \mathcal{B}(\mathbb{R}^n)$.

The next result on the product for probability spaces is stated without proof.

Theorem-Definition 3.25 *If* $(\Omega_1, \mathcal{F}_1, \mathbb{P}_1), \ldots, (\Omega_n, \mathcal{F}_n, \mathbb{P}_n)$ *are probability spaces, a unique probability* $\mathbb{P}$ *defined on* $\prod_{i=1}^{n}(\Omega_i, \mathcal{F}_i)$ *exists, called product probability, such that*

$$\mathbb{P}(A_1 \times \cdots \times A_n) = \prod_{i=1}^{n} \mathbb{P}_i(A_i), \quad A_1 \in \mathcal{F}_1, \ldots, A_n \in \mathcal{F}_n.$$

This probability is denoted by $\otimes_{i=1}^{n} \mathbb{P}_i$. The triple $(\prod_{i=1}^{n} \Omega_i, \otimes_{i=1}^{n} \mathcal{F}_i, \otimes_{i=1}^{n} \mathbb{P}_i)$ is referred to as the product probability space of $(\Omega_1, \mathcal{F}_1, \mathbb{P}_1), \ldots, (\Omega_n, \mathcal{F}_n, \mathbb{P}_n)$.

The product of a finite number of measure spaces can also be defined similarly. In particular, the Lebesgue measure λ_d on $(\mathbb{R}^d, \mathcal{B}(\mathbb{R}^d))$ is the d-times product of the Lebesgue measure λ on $(\mathbb{R}, \mathcal{B}(\mathbb{R}))$.

3.2.2 Distribution of Random Vectors

The image probabilities and distribution functions seen in Chap. 2 for random variables extend straightforwardly to random vectors.

Vectors are usually understood as column vectors. Still, for simplifying notation where no ambiguity may arise, we will write $X = (X_1, \ldots, X_d)$ instead of $X = (X_1, \ldots, X_d)'$ when no matrix calculus is concerned.

▷ *Example 3.26 (Multinomial Distribution)* An urn contains N bowls among which N_i bear number i, for $i \in [\![1, d]\!]$; n bowls are drawn at random and with replacement. Let X_i be the variable equal to the number of bowls bearing number i, and let X be the vector of the numbers of the drawn bowls, that is $X = (X_1, \ldots, X_d)$.

If $\sum_{i=1}^{r} k_i = n$, and $p_i = N_i/N$, then

$$\mathbb{P}[X = (k_1, \ldots, k_d)] = \mathbb{P}(X_1 = k_1, \ldots, X_d = k_d) = \frac{n!}{k_1! \ldots k_d!} p_1^{k_1} \cdots p_d^{k_d}.$$

This defines the multinomial distribution $\mathcal{M}(n; p_1, \ldots, p_d)$.

Note that for $d = 1$, it amounts to a binomial distribution. $\triangleleft$

Definition 3.27 Let $X = (X_1, \ldots, X_d)$ be a d-dimensional random vector. The distribution of X (or joint distribution of $X_1, \ldots, X_d$) is the image probability of $\mathbb{P}$ by X, denoted by $\mathbb{P}_X$ or $\mathbb{P}_{(X_1, \ldots, X_d)}$.

The marginal distributions of X are the distributions of the r-tuples $(X_{i_1}, \ldots, X_{i_r})$ for $1 \leq i_1 < \cdots < i_r \leq d$ and $1 \leq r < d$.

For instance, the marginal distributions of a 3-dimensional vector (X_1, X_2, X_3) are the distributions of $X_1, X_2, X_3, (X_1, X_2), (X_1, X_3)$, and (X_2, X_3).

$\triangleright$ *Example 3.28 (Marginal Distributions of a Multinomial Distribution)* Let $(X_1, \ldots, X_4)$ be a 4-dimensional random vector with multinomial distribution $\mathcal{M}(n; p_1, \ldots, p_4)$. The distribution of (X_1, X_2) is given by

$$\mathbb{P}(X_1 - k_1, X_2 - k_2) -$$

$$= \sum_{k_3=0}^{n-k_1-k_2} \mathbb{P}(X_1 = k_1, X_2 = k_2, X_3 = k_3, X_4 = n - k_1 - k_2 - k_3)$$

$$= \frac{n!}{(n - k_1 - k_2)! k_1 k_2!} p_1^{k_1} p_2^{k_2} \sum_{k_3-0}^{n-k_1-k_2} \binom{n - k_1 - k_2}{k_3} p_3^{k_3} \left(1 - \sum_{i=1}^{3} p_i\right)^{n - \sum_{i-1}^{3} k_i}$$

$$= \frac{n!}{(n - k_1 - k_2)! k_1! k_2!} p_1^{k_1} p_2^{k_2} (1 - p_1 - p_2)^{n - k_1 - k_2},$$

that is a $\mathcal{M}(n; p_1, p_2)$ distribution. In the same way, it can be shown that $X_i \sim \mathcal{B}(n, p_i)$, for $i \in [\![1, 4]\!]$. $\triangleleft$

The distribution function of a random vector is a multi-dimensional real function.

Definition 3.29 Let X be a d-dimensional random vector. The function F_X defined on $\mathbb{R}^d$ by

$$F_X(x_1, \ldots, x_d) = \mathbb{P}_X(]-\infty, x_1] \times \cdots \times]-\infty, x_d]) = \mathbb{P}(X_1 \leq x_1, \ldots, X_d \leq x_d)$$

is called the distribution function of X.

Properties of Multi-dimensional Distribution Functions

1. F_X is increasing in each of the variables.
2. F_X is continuous from the right in each of the variables.
3. $F_X(x_1, \ldots, x_d)$ converges to 0 when x_j tends to $-\infty$ for at least one j, since $]-\infty, x_1] \times \cdots \times]-\infty, x_d]$ converges then to the empty set. In the same way, $F_X(x_1, \ldots, x_d)$ converges to 1 when x_j tends to $+\infty$ for all $j \in [\![1, d]\!]$, since $]-\infty, x_1] \times \cdots \times]-\infty, x_d]$ converges then to $\mathbb{R}^d$.
4. Let us set

$$\Delta_{ab}^i F_X(x_1, \ldots, x_d) = F_X(x_1, \ldots, x_{i-1}, b, x_{i+1}, \ldots, x_d)$$
$$-F_X(x_1, \ldots, x_{i-1}, a, x_{i+1}, \ldots, x_d),$$

with $a_i \leq b_i$, for $i \in [\![1, d]\!]$. Then

$$\Delta_{a_1 b_1}^1 \Delta_{a_2 b_2}^2 \cdots \Delta_{a_d b_d}^d F_X(x_1, \ldots, x_d) = \mathbb{P}_X(]a_1, b_1] \times \cdots \times]a_d, b_d]) \geq 0.$$

5. The distribution function of a random vector characterizes its distribution, because $\mathcal{B}(\mathbb{R}^r)$ is generated by the open rectangles.
6. If F_X is continuous, then $\mathbb{P}(X = x) = 0$, for all $x \in \mathbb{R}^d$, but the converse does not hold in general for $d > 1$.
7. If F is the distribution function of a pair (X, Y), then the distribution function of X is given by $F_X(x) = \lim_{y \to +\infty} F(x, y)$. This result carries over to random vectors of any dimension and to any marginal distributions.

A multi-dimensional distribution function F is the distribution function of a random vector if and only if it satisfies Properties 1–4 The first three properties are sufficient for random variables, but not for random vectors, as shown in the next example.

▷ *Example 3.30* Let $F : \mathbb{R}^2 \to \mathbb{R}$ be defined by

$$F(x, y) = \begin{cases} 1 & \text{if } x \geq 0, \ y \geq 0 \text{ and } x + y \geq 1, \\ 0 & \text{otherwise.} \end{cases}$$

This function satisfies Properties 1–3, but not 4. It is the distribution function of no random vector X. Indeed, we would then have

$$\mathbb{P}(X \in]0, 1] \times]0, 1]) = \Delta_{01}^1 \Delta_{01}^2 F(x, y) = \Delta_{01}^1 F(x, 1) - \Delta_{01}^1 F(x, 0)$$
$$= F(1, 1) - F(0, 1) - F(1, 0) + F(0, 0) = -1,$$

which is absurd.   ◁

A random vector X is said to be integrable if the expectation of one of its norms is finite, for example, its Euclidean norm $\mathbb{E}\left[(\sum_{i=1}^{d} X_i^2)^{1/2}\right]$. Since

$$\left(\sum_{i=1}^{d} X_i^2\right)^{1/2} \leq \sum_{i=1}^{d} |X_i| \leq d\left(\sum_{i=1}^{d} X_i^2\right)^{1/2},$$

X is integrable if and only if all $X_1, \ldots, X_d$ are integrable.

Definition 3.31 Let $X = (X_1, \ldots, X_d)'$ be a random vector. If $\mathbb{E}\,X_i$ is finite for all $i \in [\![1, d]\!]$, then the vector $\mathbb{E}\,X = (\mathbb{E}\,X_1, \ldots, \mathbb{E}\,X_d)'$ is called the expectation (or mean) of X.

The expectation of a function of a random vector is defined similarly. The quantities $\mathbb{E}\,(X_1^{n_1} \ldots X_d^{n_d})$ for $n_i \in \mathbb{N}$—when finite—are referred to as the moments of X. If the expectation of X is the null vector, X is said to be centered.

Definition 3.32 A random vector is said to be discrete if the set of all its values is a.s. countable. It is said to be continuous if its distribution function is continuous. Finally, a real random vector is said to be absolutely continuous if its distribution is absolutely continuous with respect to the Lebesgue measure on $\mathbb{R}^d$.

Thus, a real random vector is absolutely continuous if $d\,\mathbb{P}_X(x) = f(x)\,dx$, where $f : \mathbb{R}^d \to \mathbb{R}_+$. The function f is called the density of X, or joint density of $X_1, \ldots, X_d$, and is typically denoted by f_X, or by f where no ambiguity may arise.

Properties of Densities of Random Vectors
1. $\int_{\mathbb{R}^d} f(x)\,dx = 1$.
2. A function f is the density of X if and only if $\mathbb{E}\,[h(X)] = \int_{\mathbb{R}^d} h(x) f(x)\,dx$ for all bounded Borel functions $h : \mathbb{R}^d \to \mathbb{R}$.
3. $\mathbb{P}(X \in B) = \int_B f(x)\,dx$ for all $B \in \mathcal{B}(\mathbb{R}^d)$; in particular,

$$F_X(x_1, \ldots, x_d) = \int_{-\infty}^{x_1} \ldots \int_{-\infty}^{x_d} f(t_1, \ldots, t_d)\,dt_1 \ldots dt_d.$$

Conversely, when this quantity is well-defined,

$$f_X(x_1, \ldots, x_d) = \frac{\partial^d}{\partial x_1 \ldots \partial x_d} F_X(x_1, \ldots, x_d), \quad \lambda - \text{a.e.},$$

for all continuity points $(x_1, \ldots, x_d)$ of f_X.
4. If $X = (X_1, X_2)$ is a 2-dimensional random vector with density f_X, then both X_1 and X_2 have (marginal) densities given by $f_{X_1}(x_1) = \int_{\mathbb{R}} f_X(x_1, x_2)\,dx_2$

and $f_{X_2}(x_2) = \int_{\mathbb{R}} f_X(x_1, x_2)\, dx_1$. This carries over to any random vector $X = (X_1, \ldots, X_d)$ and r-tuple $(X_{i_1}, \ldots, X_{i_r})$.

▷ *Example 3.33 (Uniform 2-Dimensional Distribution)* Let (X_1, X_2) be a random vector with density f. Its distribution is uniform on $[a_1, b_1] \times [a_2, b_2]$, where $a_1 < b_1$ and $a_2 < b_2$, if

$$f(x_1, x_2) = \frac{1}{(b_1 - a_1)(b_2 - a_2)} \mathbb{1}_{[a_1,b_1] \times [a_2,b_2]}(x_1, x_2).$$

Its marginal distributions are $X_1 \sim \mathcal{U}([a_1, b_1])$ and $X_2 \sim \mathcal{U}([a_2, b_2])$.
For $[a_1, b_1] \times [a_2, b_2] = [0, 1]^2$, its distribution function is

$$
\begin{aligned}
F_{(X_1,X_2)}(x_1, x_2) &= \int_0^{x_1} \int_0^{x_2} \mathbb{1}_{[0,1]}(u) \mathbb{1}_{[0,1]}(v)\, du\, dv \\
&= \int_{\mathbb{R}_+^2} \mathbb{1}_{[0,1] \cap [0,x_1]}(u) \mathbb{1}_{[0,1] \cap [0,x_2]}(v)\, du\, dv \\
&= \min(x_1, 1) \min(x_2, 1) \mathbb{1}_{\mathbb{R}_+^2}(x_1, x_2).
\end{aligned}
$$

Conversely, the density f is obtained by differentiating $F_{(X_1,X_2)}(x_1, x_2)$ with respect to each variable. ◁

The analytical tools presented in Chap. 2 for random variables extend to random vectors as follows.

Definition 3.34 Let $X = (X_1, \ldots, X_d)$ be a d-dimensional random vector.

1. If X takes values in $\mathbb{N}^d$, the generating function of X is defined by

$$g_X(t_1, \ldots, t_d) = \mathbb{E}\,(t_1^{X_1} \ldots t_d^{X_d}), \quad t \in [-1, 1]^d.$$

2. The characteristic function of X is defined by

$$\varphi_X(t) = \mathbb{E}\,(\exp i < t, X >) = \mathbb{E}\left[\exp i \Big(\sum_{j=1}^d t_j X_j \Big) \right] \quad t \in \mathbb{R}^d.$$

3. If X takes values in $\mathbb{R}_+^d$, the Laplace transform of X is defined by

$$\psi_X(t) = \mathbb{E}\,(\exp < t, X >), \quad t \in \mathbb{R}_+^d.$$

Most properties of these functionals derive immediately from the properties presented in Chap. 2 for $d = 1$. Moreover,

$$\varphi_{(X_1,\ldots,X_r)}(t_1, \ldots, t_r) = \varphi_X(t_1, \ldots, t_r, 0, \ldots, 0), \quad 1 < r < d,$$

and, for $k_1 \in \mathbb{N}, \dots, k_d \in \mathbb{N}$,

$$\frac{\partial}{\partial x_1^{k_1} \dots x_d^{k_d}} \varphi_X(0, \dots, 0) = i^{k_1} \dots i^{k_d} \mathbb{E}\,(X_1^{k_1} \dots X_d^{k_d}), \qquad (3.3)$$

when these quantities are finite.

The expectation of a random vector is a vector; its variance is a matrix.

Definition 3.35 Let $X = (X_1, \dots, X_d)$ be a d-dimensional random vector. If all the X_i are square integrables, then the covariance (or dispersion) matrix of X is the matrix of the covariances of its coordinates, that is

$$\Gamma_X = \mathbb{E}\,[(X - \mathbb{E}\,X)(X - \mathbb{E}\,X)'] = \Big(\mathbb{C}\text{ov}\,(X_i, X_j)\Big)_{1 \le i, j \le k}.$$

When all the X_i are not a.s. constant, the correlation matrix of X is defined as the matrix of the correlation coefficients of its coordinates, that is

$$\big(\rho_{X_i, X_j}\big)_{1 \le i, j \le d} = \left(\frac{\mathbb{C}\text{ov}\,(X_i, X_j)}{\sqrt{\mathbb{V}\text{ar}\,X_i\,\mathbb{V}\text{ar}\,X_j}}\right)_{1 \le i, j \le d}.$$

A centered vector whose variance–covariance matrix is the identity matrix is referred to as standard.

The expectation and the variance–covariance matrix of a d-dimensional random vector can be computed using its characteristic function. Indeed, (3.3) induces

$$i(\mathbb{E}\,X)' = \text{grad}\,\varphi_X(0), \quad \text{and} \quad \mathbb{E}\,(X_i X_j) = \frac{\partial^2}{\partial t_i \partial t_j} \varphi_X(0),$$

when these moments are finite.

Properties of Covariance Matrices
1. $\Gamma_X = \mathbb{E}\,[(X - \mathbb{E}\,X)(X - \mathbb{E}\,X)'] = \mathbb{E}\,(XX') - (\mathbb{E}\,X)(\mathbb{E}\,X)'$.
2. $\Gamma_{X+a} = \Gamma_X$ for all $a \in \mathbb{R}^d$.
3. Γ_X is a nonnegative definite symmetric matrix. Indeed, since $\lambda'(X - \mathbb{E}\,X)$ is a random variable, the quantity

$$\lambda \Gamma_X \lambda' = \mathbb{E}\,[\lambda'(X - \mathbb{E}\,X)(X - \mathbb{E}\,X)'\lambda] = \mathbb{E}\,[\lambda'(X - \mathbb{E}\,X)]^2, \quad \lambda \in \mathbb{R}^d$$

is a variance, so is nonnegative.

By definition, the image of any d-dimensional random vector by any Borel function taking values in $\mathbb{R}^{d'}$ is a d'-dimensional random vector. The linear transforms are especially useful; their expectation and covariance matrices derive straightforwardly from the definitions.

Proposition 3.36 *Let X be a d-dimensional random vector. Let A denote the matrix of a linear function from $\mathbb{R}^d$ to $\mathbb{R}^{d'}$, and let V be a vector of $\mathbb{R}^{d'}$. Then the vector $Y = AX+V$ is a d'-dimensional random vector. Its expectation is $\mathbb{E}\, Y = A\mathbb{E}\, X+V$, and its covariance matrix is $\Gamma_Y = A\Gamma_X A'$.*

In particular, if $d' = 1$ and if $Y = a'X + v$, where $a' = (a_1, \dots, a_d) \in \mathbb{R}^d$ and $v \in \mathbb{R}$, then Y is a random variable, with $\mathbb{E}\, Y = a'\mathbb{E}\, X + v$ and $\mathrm{Var}\, Y = a'\Gamma_X a$.

3.2.3   Independence of Random Vectors

Several random vectors may be investigated simultaneously.

Definition 3.37 Let X be a square integrable d-dimensional random vector, and let Y be a square integrable d'-dimensional random vector. The variance–covariance matrix of X and Y is defined as

$$Cov(X, Y) = \mathbb{E}\,[(X - \mathbb{E}\, X)(Y - \mathbb{E}\, Y)'] = \Big(\mathbb{C}\mathrm{ov}\,(X_i, Y_j)\Big)_{1 \le i \le d, 1 \le j \le d'}.$$

Two random vectors X and Y are said to be uncorrelated if $Cov(X, Y)$ is the null matrix.

Properties of Covariance Matrices of Two Vectors
1. $Cov(X, X)$ is the variance–covariance matrix Γ_X of X.
2. $Cov(X, Y) = Cov(Y, X)'$, the transpose matrix.
3. Set $Z' = (X, Y)' = (X_1, \dots, X_d, Y_1, \dots, Y_{d'})$. Then

$$\Gamma_Z = \begin{pmatrix} \Gamma_X & Cov(X, Y) \\ Cov(Y, X) & \Gamma_Y \end{pmatrix} \begin{matrix} \updownarrow d \\ \updownarrow d' \end{matrix}$$
$$\underset{\xleftrightarrow{\quad d \quad}}{} \underset{\xleftrightarrow{\quad d' \quad}}{}$$

4. The variance–covariance matrix of a random vector whose coordinates are independent is a diagonal matrix. The converse does not hold true in general but does for Gaussian vectors, as will be shown in Sect. 3.4 below.

Recall that, according to Definition 3.3, n random vectors $X_1, \dots, X_n$, with respective dimensions $d_1, \dots, d_n$, are independent if

$$\mathbb{P}(X_1 \in B_1, \dots, X_n \in B_n) = \prod_{i=1}^n \mathbb{P}(X_i \in B_i), \quad B_1 \in \mathcal{B}(\mathbb{R}^{d_1}), \dots, B_n \in \mathcal{B}(\mathbb{R}^{d_n}).$$

Other criteria of independence, often more easy to deal with, can be stated.

Theorem 3.38 *Let $X_1, \ldots, X_n$ be n random vectors with dimensions $d_1, \ldots, d_n$. The four following statements are equivalent:*

1. *$X_1, \ldots, X_n$ are independent.*
2. *$\mathbb{P}_{(X_1, \ldots, X_n)} = \overset{n}{\underset{i=1}{\bigotimes}} \mathbb{P}_{X_i}$, meaning that the (joint) distribution of the vector is the product of its marginal distributions.*
3. *$\mathbb{E}\left[\prod_{i=1}^n h_i(X_i)\right] = \prod_{i=1}^n \mathbb{E}[h_i(X_i)]$, for all bounded real-valued Borel functions h_i defined on $\mathbb{R}^{d_i}$, with $i \in [\![1, n]\!]$.*
4. *$F_{(X_1, \ldots, X_n)}(x_1, \ldots, x_n) = \prod_{i=1}^n F_{X_i}(x_i)$ for all $x_1 \in \mathbb{R}^{d_1}, \ldots, x_n \in \mathbb{R}^{d_n}$.*
5. *For n random vectors with densities, $f_{(X_1, \ldots, X_n)}(x_1, \ldots, x_n) = \prod_{i=1}^n f_{X_i}(x_i)$, for all $x_1 \in \mathbb{R}^{d_1}, \ldots, x_n \in \mathbb{R}^{d_n}$.*

Thus, two random vectors $(X_1, \ldots, X_d)$ and $(Y_1, \ldots, Y_{d'})$ are independent if the $d + d'$ variables $X_1, \ldots, X_d, Y_1, \ldots, Y_{d'}$ are independent.

Proof We prove the theorem for two vectors.
$1 \Rightarrow 2$ since

$$\iint_{B_1 \times B_2} d\mathbb{P}_{(X_1, X_2)} = \mathbb{P}(X_1 \in B_1, X_2 \in B_2) = \mathbb{P}(X_1 \in B_1)\mathbb{P}(X_2 \in B_2)$$

$$= \int_{B_1} d\mathbb{P}_{X_1} \int_{B_2} d\mathbb{P}_{X_2} = \iint_{B_1 \times B_2} d\mathbb{P}_{X_1} d\mathbb{P}_{X_2}.$$

$2 \Rightarrow 3$ since

$$\mathbb{E}[h_1(X_1)h_2(X_2)] = \iint_{\mathbb{R}^{d_1+d_2}} h_1(x_1)h_2(x_2) d\mathbb{P}_{(X_1, X_2)}(x_1, x_2)$$

$$= \iint_{\mathbb{R}^{d_1+d_2}} h_1(x_1)h_2(x_2) d\mathbb{P}_{X_1}(x_1) d\mathbb{P}_{X_2}(x_2)$$

$$\overset{(1)}{=} \int_{\mathbb{R}^{d_1}} h_1(x_1) d\mathbb{P}_{X_1}(x_1) \int_{\mathbb{R}^{d_2}} h_2(x_2) d\mathbb{P}_{X_2}(x_2)$$

$$= \mathbb{E}[h_1(X_1)]\mathbb{E}[h_2(X_2)],$$

(1) by Fubini's theorem.

$3 \Rightarrow 4$ since $F_{(X_1, X_2)}(x_1, x_2) = \mathbb{E}[\mathbb{1}_{(X_1 \in]-\infty, x_1])}\mathbb{1}_{(X_2 \in]-\infty, x_2])}]$.

$4 \Rightarrow 1$ since the rectangles generate $\mathcal{B}(\mathbb{R}^{d_1+d_2})$; see Theorem 1.13.

$5 \Leftrightarrow 2$ by definition, since $d\mathbb{P}_{(X_1, X_2)}(x_1, x_2) = f_{(X_1, X_2)}(x_1, x_2) dx_1 dx_2$ and, for $i = 1, 2$, $d\mathbb{P}_{X_i}(x_i) = f_{X_i}(x_i) dx_i$. $\qquad \square$

Another criterion of independence is given by characteristic functions.

Proposition 3.39 *Let $X_1, \ldots, X_n$ be n random vectors. They are independent if and only if the characteristic function of their joint distribution is the product of their marginal characteristic functions, that is*

$$\varphi_{(X_1, \ldots, X_n)}(t_1, \ldots, t_n) = \prod_{j=1}^{n} \varphi_{X_j}(t_j), \quad (t_1, \ldots, t_n) \in \mathbb{R}^{d_1} \times \cdots \times \mathbb{R}^{d_n}.$$

Proof For simplifying notation, we prove the proposition for n random variables.
  The direct implication derives directly from criterion 3 of Theorem 3.38.
  Conversely, Proposition 2.68 and Fubini's theorem jointly imply

$$\mathbb{P}(x_1 < X_1 \le y_1, \ldots, x_n < X_n \le y_n) =$$

$$= \lim_{t \to +\infty} \int_{-t}^{t} \cdots \int_{-t}^{t} \prod_{j=1}^{n} \frac{e^{-iu_j x_j} - e^{-iu_j y_j}}{iu_j} \varphi_{(X_1, \ldots, X_n)}(u_1, \ldots, u_n) \, du_1 \ldots du_n$$

$$= \prod_{j=1}^{n} \left[\lim_{t \to +\infty} \int_{-t}^{t} \frac{e^{-iu_j x_j} - e^{-iu_j y_j}}{iu_j} \varphi_{X_j}(u_j) \, du_j \right] = \prod_{j=1}^{n} \mathbb{P}(x_j < X_j \le y_j).$$

The conclusion follows from applying criterion 4 of Theorem 3.38 when x_j tends to $-\infty$ for all j. $\qquad \square$

The same property is also satisfied by the other transforms.

3.3   Functions of Random Vectors

The distribution of functions of random vectors can be deduced from the distribution of these vectors, when certain conditions are satisfied. We will focus on order statistics of independent variables and sums of random variables or vectors, before considering general functions.

3.3.1   Order Statistics

The distribution of a ranked random vector can easily be deduced from the distribution of its components, when they are i.i.d. This is a particularly useful tool in statistics; when a sample of variable values is arrayed in ascending order of magnitude, these ordered values are known as ordered statistics.

  Let $X_1, \ldots, X_n$ be i.i.d. random variables defined on $(\Omega, \mathcal{F}, \mathbb{P})$ with common density f and distribution function F. Let $\omega \in \Omega$ be such that $X_i(\omega) \ne X_j(\omega)$, for $1 \le i < j \le n$. Let $\mathcal{S}_n$ denote the set of all permutations of $[\![1, n]\!]$, and σ_ω the permutation of $\mathcal{S}_n$ that ranks the values $X_1(\omega), \ldots, X_n(\omega)$, that is such that $X_{\sigma_\omega(1)}(\omega) < \cdots < X_{\sigma_\omega(n)}(\omega)$,. Set $\sigma(i) : \omega \to \sigma_\omega(i)$.

Theorem-Definition 3.40 *The function* $\Sigma \colon (\Omega, \mathcal{F}) \to \mathcal{S}_n$ *such that* $\Sigma(\omega) = \sigma_\omega$ *is defined* $\mathbb{P}$-*a.s. and is measurable. Moreover,* $X_{\sigma(1)}, \ldots, X_{\sigma(n)}$ *are real random variables, and the random vector* $(X_{\sigma(1)}, \ldots, X_{\sigma(n)})$ *is called the order (or ordinal) statistics of* $X_1, \ldots, X_n$.

Note that $X_{\sigma(1)}, \ldots, X_{\sigma(n)}$ are randomly indexed random variables.

Proof First, let us show that Σ is $\mathbb{P}$-a.s. defined. The set on which it is not defined is $N = \{\omega \in \Omega : \exists i \neq j, \ 1 \leq i, \ j \leq n \text{ such that } X_i(\omega) = X_j(\omega)\}$.

We have $N = \cup_{1 \leq i \neq j \leq n}(X_i = X_j)$, so $N \in \mathcal{F}$. Since the random variables $X_1, \ldots, X_n$ are continuous and i.i.d., we have $\mathbb{P}(X_i = X_j) = 0$. Therefore, N is null and the conclusion follows.

Second, let us show that $(\Sigma = s) \in \mathcal{F}$ for all $s \in \mathcal{S}_n$. We have

$$(\Sigma = s) = \{\omega \in \Omega \setminus N : X_{s(1)}(\omega) < \cdots < X_{s(n)}(\omega)\} = (X_{s(1)} < \cdots < X_{s(n)}),$$

so $(\Sigma = s) \in \mathcal{F}$, and hence, Σ is measurable.

Finally, all the $X_{\sigma(i)}$ are random variables as composed of random variables X_i and measurable functions $\sigma(i)$. $\qquad\qquad\qquad\qquad\qquad\qquad\qquad\qquad\qquad \square$

For simplification, we will write $X_{(i)}$ instead of $X_{\sigma(i)}$.

Theorem 3.41 *The density of the order statistics* $(X_{(1)}, \ldots, X_{(n)})$ *is*

$$n! \prod_{i=1}^{n} f(x_i) \mathbb{1}_C(x_1, \ldots, x_n),$$

where $C = \{(x_1, \ldots, x_n) \in \mathbb{R}^n : x_1 < \cdots < x_n\}$.

Proof For all $B \in \mathcal{B}(\mathbb{R}^n)$,

$$\big((X_{\sigma(1)}, \ldots, X_{\sigma(n)}) \in B\big) = \{\omega \in \Omega \setminus N : X_{\sigma(1)}(\omega), \ldots, X_{\sigma(n)}(\omega)) \in B\}$$

$$= \bigcup_{s \in \mathcal{S}_n} \big[\big((X_{s(1)}, \ldots, X_{s(n)}) \in B\big) \cap (\sigma = s) \big],$$

so

$$\mathbb{P}\big((X_{(1)}, \ldots, X_{(n)}) \in B\big) = \sum_{s \in \mathcal{S}_n} \mathbb{P}\big(((X_{s(1)}, \ldots, X_{s(n)}) \in B) \cap (\sigma = s)\big).$$

Since $(\sigma = s) = (X_{s(1)} < \cdots < X_{s(n)}) = ((X_{s(1)}, \ldots, X_{s(n)}) \in C)$, we get

$$\mathbb{P}\big((X_{(1)}, \ldots, X_{(n)}) \in B\big) = \sum_{s \in \mathcal{S}_n} \mathbb{P}\big((X_{s(1)}, \ldots, X_{s(n)}) \in B \cap C\big)$$

$$\overset{(1)}{=} |\mathcal{S}_n| \mathbb{P}_{X_1}^{\otimes n}(B \cap C) = n! \mathbb{P}_{X_1}^{\otimes n}(B \cap C)$$

$$= n! \int_B \mathbb{1}_C(x_1, \ldots, x_n) f(x_1) \ldots f(x_n)\, dx_1 \ldots dx_n,$$

(1) because the distribution of an n-tuple does not depend on the order of its components and the n random variables are i.i.d.   □

The distributions of $X_{(k)}$ for $k \in [\![1, n]\!]$ follow by taking marginal distributions of the random vector $(X_{(1)}, \ldots, X_{(n)})$. Their densities are

$$f_{(k)}(x) = \frac{n!(n-k)!}{(k-1)!} F^{k-1}(x)(1 - F(x))^{n-k} f(x), \quad k \in [\![1, n]\!]. \tag{3.4}$$

In particular, the density of $X_{(1)} = \min_{1 \le i \le n} X_k$ is $f_{(1)} = nf(1 - F)^{n-1}$, and the density of $X_{(n)} = \max_{1 \le i \le n} X_i$ is $f_{(n)} = nf F^{n-1}$.

The distribution function of $X_{(k)}$ can be inferred as the primitive of $f_{(k)}$, but it is simpler to determine it directly. The event $(X_{(k)} \le x)$ is the set of all ω such that at least k indices i exist for which $X_i(\omega) \le x$. The random variable that counts the number of these indices is $\sum_{i=1}^{n} \mathbb{1}_{(X_i \le x)}$, with binomial distribution $\mathcal{B}(n, F(x))$; therefore,

$$F_{(k)}(x) = \mathbb{P}(X_{(k)} \le x) = \sum_{i=k}^{n} \binom{n}{i} F(x)^i [1 - F(x)]^{n-i}. \tag{3.5}$$

In particular, $F_{(1)}(x) = 1 - [1 - F(x)]^n$ and $F_{(n)}(x) = F(x)^n$.

Investigation of the asymptotic behavior—as n tends to infinity—of $X_{(1)}$ and $X_{(n)}$ is referred to as the theory of extreme values.

3.3.2 Sums of Independent Variables or Vectors

The determination of the distributions of sums of random variables or vectors is based on the notion of convolution.

Definition 3.42 The convolution of two elements is defined as follows:

• Let f and g be two nonnegative or integrable Borel functions defined on $\mathbb{R}^d$, with $d \geq 1$. The convolution (product) of f and g is the function on $\mathbb{R}^d$ defined by

$$f * g(x) = \int_{\mathbb{R}^d} f(x - y)g(y)dy = \int_{\mathbb{R}^d} f(y)g(x - y)dy, \quad x \in \mathbb{R}^d.$$

• Let μ and ν be two measures on $(\mathbb{R}^d, \mathcal{B}(\mathbb{R}^d))$, with $d \geq 1$. The convolution of μ and ν is the measure $\mu * \nu$ on $\mathbb{R}^d$ defined by

$$\mu * \nu(B) = \mu \otimes \nu(\{(x, y) \in \mathbb{R}^d \times \mathbb{R}^d : x + y \in B\}), \quad B \in \mathcal{B}(\mathbb{R}^d).$$

The convolution of measures applies to Borel sets $B \in \mathcal{B}(\mathbb{R}^d)$ under the form

$$\mu * \nu(B) = \int_{\mathbb{R}^d} \mu(\{t - x : t \in B\})d\nu(x) = \int_{\mathbb{R}^d \times \mathbb{R}^d} \mathbb{1}(x + y)d\mu(x)d\mu(y),$$

or to bounded Borel functions $h \colon \mathbb{R}^d \to \mathbb{R}$ under the form

$$\int_{\mathbb{R}^d} h(z)d\mu * \nu(z) = \iint_{\mathbb{R}^d \times \mathbb{R}^d} h(x + y)d\mu(x)d\nu(y). \quad (3.6)$$

Proposition 3.43 *If μ and ν are absolutely continuous with respect to the Lebesgue measure λ on $\mathbb{R}^d$, with respective densities f and g, then $\mu * \nu$ has the density $f * g$ with respect to λ.*

Proof Since by the change of variables $z = x + y$, $y = y$,

$$\iint_{\mathbb{R}^d \times \mathbb{R}^d} h(x + y)f(x)\,dxg(y)dy = \int_{\mathbb{R}^d} h(z)\left[\int_{\mathbb{R}^d} f(z - y)g(y)dy\right]dz.$$

The result follows according to (3.6).   □

Properties of Convolution
1. The convolution operator is commutative, associative, and distributive with respect to addition.
2. If $f \in L^1$ and $g \in L^p$, with $p \geq 1$, then $f * g$ is λ-a.e. finite, belongs to L^p, and $\|f * g\| \leq \|f\|_1 \|g\|_p$.
3. The convolution of a measure μ, on $(\mathbb{R}^d, \mathcal{B}(\mathbb{R}^d))$, by the Dirac measure δ_a, $a \in \mathbb{R}^d$, gives $\delta_a * \mu(B) = \tau_a \mu(B) = \mu(B - a)$.

For discrete measures, the convolution takes the form of a sum. Let μ and ν be two measures defined on a countable space $\Omega = \{\omega_n : n \in \mathbb{Z}\}$ with respective values $\mu(\omega_n) = \mu_n$ and $\nu(\omega_n) = \nu_n$. Their convolution is given by $(\mu * \nu)_n = $

$\sum_{k\in\mathbb{Z}} \mu_k \nu_{n-k}$ for $n \in \mathbb{Z}$. In particular, if $\Omega = \{\omega_n : n \in \mathbb{N}\}$, then

$$(\mu * \nu)_n = \sum_{k=0}^{n} \mu_k \nu_{n-k}, \quad n \in \mathbb{N}. \tag{3.7}$$

The convolution of a function $f \in L^1$ and of a mass function F—or of the associated Lebesgue–Stieltjes measure λ_F—is referred to as their Lebesgue–Stieltjes convolution and is given by

$$f * F(x) = \int_{\mathbb{R}^d} f(x - t)dF(t) = \int_{\mathbb{R}^d} f(x - t)d\lambda_F(t).$$

If f is a mass function too, the above convolution is a commutative operation.

Further, if F and G are two distribution functions defined on $\mathbb{R}$, their convolution is defined by

$$F * G(x) = \int_{\mathbb{R}} G(x - t)dF(t) = \int_{\mathbb{R}} F(x - t)dG(t), \quad x \in \mathbb{R}.$$

In particular, if F and G are distribution functions of positive random variables— and hence null on $\mathbb{R}_-$, then

$$F * G(x) = \int_0^x G(x - t)dF(t) = \int_0^x F(x - t)dG(t), \quad x \in \mathbb{R}_+.$$

The distribution of a sum of independent random vectors derives easily from the following result on Fourier or Laplace transforms of convolution of distributions.

Theorem 3.44 *The Fourier (Laplace) transform of the convolution of two elements is the product of their Fourier (Laplace) transforms.*

Proof We prove the theorem for two integrable functions f and g and the Fourier transform

$$\widehat{f * g}(t) = \int_{\mathbb{R}^d} e^{itx} f * g(x)\, dx = \int_{\mathbb{R}^d} e^{itx} \int_{\mathbb{R}^d} f(x - y)g(y)dy\, dx$$

$$\overset{(1)}{=} \int_{\mathbb{R}^d} g(y)e^{ity}dy \int_{\mathbb{R}^d} f(x - y)e^{it(x-y)}\, dx$$

$$= \widehat{g}(t) \int_{\mathbb{R}^d} f(u)e^{itu}du = \widehat{g}(t)\widehat{f}(t),$$

(1) by Fubini's theorem.   □

Theorem 3.45 *Let $X_1, \ldots, X_n$ be n independent d-dimensional random vectors. Then:*

1. $\mathbb{P}_{X_1+\cdots+X_n} = \mathbb{P}_{X_1} * \cdots * \mathbb{P}_{X_n}$.
2. $F_{X_1+\cdots+X_n} = F_{X_1} * \cdots * F_{X_n}$.
3. *If, moreover, $X_1, \ldots, X_n$ have densities $f_1, \ldots, f_n$, then their sum $X_1 + \cdots + X_n$ has a density too, say f, with $f = f_1 * \cdots * f_n$.*

Proof For two variables:

1. It is an immediate consequence of the definition of convolution.
2. We compute

$$\mathbb{P}(X_1 + X_2 \leq x) = \int_{\mathbb{R}} \mathbb{P}(X_2 \leq x - u)\mathbb{P}(X_1 \in du) = \int_{\mathbb{R}} F_{X_2}(x - u)dF_{X_1}(u).$$

3. Let $h \colon \mathbb{R} \to \mathbb{R}$ be any bounded Borel function. We have

$$\mathbb{E}\left[h(X_1 + X_2)\right] = \iint_{\mathbb{R}^2} h(x_1 + x_2)d\mathbb{P}_{(X_1, X_2)}(x_1, x_2)$$

$$= \iint_{\mathbb{R}^2} h(x_1 + x_2)d\mathbb{P}_{X_1}(x_1)d\mathbb{P}_{X_2}(x_2)$$

$$\overset{(1)}{=} \iint_{\mathbb{R}^2} h(x)d(\mathbb{P}_{X_1} * \mathbb{P}_{X_2})(x),$$

(1) by (3.6). Applying Theorem 3.43 yields the conclusion. □

▷ *Example 3.46 (Sum of i.i.d. Uniform Variables)* Let X and Y be two i.i.d. variables with uniform distribution on $[0, 1]$. The density of $S = X + Y$, which takes values in $[0, 2]$, is

$$f_S(s) = f_X * f_Y(s) = \int_{\mathbb{R}} \mathbb{1}_{[0,1]}(s')\mathbb{1}_{[0,1]}(s - s')ds' = \int_{\mathbb{R}} \mathbb{1}_{[s-1,s]\cap[0,1]}(s')ds'.$$

If $0 \leq s \leq 1$, then $s - 1 \leq 0$, and hence, $[s - 1, s] \cap [0, 1] = [0, s]$, so that

$$\int_{\mathbb{R}} \mathbb{1}_{[s-1,s]\cap[0,1]}(s')ds' = \int_0^s ds' = s.$$

If $1 \leq s \leq 2$, then $[s - 1, s] \cap [0, 1] = [s - 1, s]$ and

$$\int_{\mathbb{R}} \mathbb{1}_{[s-1,s]\cap[0,1]}(s')ds' = \int_{s-1}^s ds' = 2 - s.$$

Therefore, $f_S(s) = \mathbb{1}_{[0,1]}(s) + (2 - s)\mathbb{1}_{[1,2]}(s)$. This is the tent function.

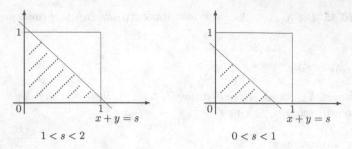

Fig. 3.2 Sum of i.i.d. variables with distribution $\mathcal{U}(0, 1)$—Example 3.46

Note that this can be obtained geometrically, by considering the intersection of the unit square of $\mathbb{R}^2$ and of the semi-plane under the line $x + y = s$; the quantity $\mathbb{P}(S \leq s)$ is the surface of this area, as shown in Fig. 3.2.   ◁

▷ *Example 3.47 (Sum of Independent Exponential Variables)* Let $X \sim \mathcal{E}(\lambda)$ and $Y \sim \mathcal{E}(\mu)$ be two independent random variables, with $\lambda \neq \mu$. The density of their sum is

$$f_{X+Y}(s) = \lambda\mu \int_{\mathbb{R}_+} \mathbb{1}_{\mathbb{R}_+}(s - t)\mathbb{1}_{\mathbb{R}_+}(t)e^{-\lambda(s-t)}e^{-\mu t}\, dt$$

$$= \lambda\mu e^{-\lambda s}\mathbb{1}_{\mathbb{R}_+}(s) \int_0^s e^{(\lambda - \mu)t}\, dt = \frac{\lambda\mu}{\lambda - \mu}(e^{-\mu s} - e^{-\lambda s})\mathbb{1}_{\mathbb{R}_+}(s).$$

This distribution is referred to as the hypo-exponential distribution with parameters λ and μ. Therefore, the sum of two exponential variables with different parameters is not exponentially distributed. In the case of equal parameters, the sum is Erlang distributed; see Example 3.53 below.   ◁

The distribution of the sum of discrete independent variables is also given by the convolution of their distributions. For instance for two variables with positive integer values, (3.7) becomes

$$\mathbb{P}(X + Y = n) = \sum_{k=0}^n \mathbb{P}(X = k, Y = n - k) = \sum_{k=0}^n \mathbb{P}(X = k)\mathbb{P}(Y = n - k).$$

See Example 3.50 below for an application to Poisson variables.

The variance of the sum of any variables can be written in terms of the variances and covariances of the variables.

Proposition 3.48 *Let $X_1, \ldots, X_d$ be random variables with finite variances. Then*

$$\mathbb{V}\text{ar} \sum_{i=1}^{d} X_i = \sum_{i=1}^{d} \mathbb{V}\text{ar}\, X_i + \sum_{i \neq j} \mathbb{C}\text{ov}\,(X_i, X_j).$$

If moreover $X_1, \ldots, X_d$ are uncorrelated, then

$$\mathbb{V}\text{ar}\,(X_1 + \cdots + X_d) = \mathbb{V}\text{ar}\, X_1 + \cdots + \mathbb{V}\text{ar}\, X_d.$$

In particular, for two variables, we get

$$\mathbb{V}\text{ar}\,(X + Y) = \mathbb{V}\text{ar}\, X + \mathbb{V}\text{ar}\, Y + 2\mathbb{C}\text{ov}\,(X, Y).$$

Similarly, for vectors,

$$\Gamma_{\sum_{i=1}^{n} X_i} = \sum_{i=1}^{n} \Gamma_{X_i} + 2 \sum_{1 \le i < j \le n} Cov(X_i, X_j).$$

In both above relations, the covariance terms are null when the variables or vectors are uncorrelated.

Proof For two variables, we have

$$\mathbb{V}\text{ar}\,(X_1 + X_2) = \mathbb{E}\,[(X_1 + X_2 - \mathbb{E}\, X_1 - \mathbb{E}\, X_2)^2] =$$
$$= \mathbb{E}\,[(X_1 - \mathbb{E}\, X_1)^2 + (X_2 - \mathbb{E}\, X_2)^2 + 2(X_1 - \mathbb{E}\, X_1)(X_2 - \mathbb{E}\, X_2)].$$

The case of uncorrelated variables follows by applying Proposition 3.6. □

The generating function, the Laplace transform, and the characteristic function transform sums into products. Thus, since each of them characterize distributions, the distribution of a sum of variables or of vectors can be determined by using ordinary product of functions instead of convolution.

Theorem 3.49 *Let $X_1, \ldots, X_n$ be independent d-dimensional random vectors.*

1. *The characteristic function of their sum is equal to the product of their characteristic functions.*
2. *If $X_1, \ldots, X_n$ are positive, the Laplace transform of their sum is equal to the product of their Laplace transforms.*
3. *If $X_1, \ldots, X_n$ take integer values, the generating function of their sum is equal to the product of their generating functions.*

In particular, for i.i.d. vectors,

$$g_{\sum_{i=1}^n X_i}(t) = [g_{X_1}(t)]^n \quad \text{and} \quad \varphi_{\sum_{i=1}^n X_i}(t) = [\varphi_{X_1}(t)]^n, \quad t \in \mathbb{R}.$$

Proof Points 1 and 2 derive directly from Theorem 3.44.
3 For two variables, we compute

$$g_{X_1+X_2}(t) = \mathbb{E}\left(t^{X_1+X_2}\right) = \mathbb{E}\left(t^{X_1}\right)\mathbb{E}\left(t^{X_2}\right) = g_{X_1}(t)g_{X_2}(t).$$

The case of vectors is similar.   □

As shown in the next example, Point 3 of Theorem 3.49 for the generating functions is a powerful tool for computing the distribution of sums of classical discrete variables.

▷ *Example 3.50 (Sums of Discrete Variables)* Let $X \sim \mathcal{P}(\lambda)$ and $Y \sim \mathcal{P}(\mu)$ be two independent variables. Let us show that $X + Y \sim \mathcal{P}(\lambda + \mu)$ by using two different methods:

1. By convolution,

$$\mathbb{P}(X + Y = n) = \sum_{k=0}^n \mathbb{P}(X = k)\mathbb{P}(Y = n - k) = \sum_{k=0}^n e^{-\lambda}\frac{\lambda^k}{k!}e^{-\mu}\frac{\mu^{n-k}}{(n-k)!}$$

$$= \frac{e^{-(\lambda+\mu)}}{n!}\sum_{k=0}^n \binom{n}{k}\lambda^k\mu^{n-k} = e^{-(\lambda+\mu)}\frac{(\lambda+\mu)^n}{n!}.$$

2. The generating function of $X + Y$ is

$$g_{X+Y}(t) = \mathbb{E}\left(t^X t^Y\right) = \mathbb{E}\left(t^X\right)\mathbb{E}\left(t^Y\right) \overset{(1)}{=} e^{\lambda(t-1)}e^{\mu(t-1)} = e^{(\lambda+\mu)(t-1)},$$

(1) using Example 2.65. We recognize the generating function of the Poisson distribution with parameter $\lambda + \mu$.

Let now $X \sim \mathcal{B}(n, p)$ and $Y \sim \mathcal{B}(m, p)$ be two independent variables. Then $X + Y \sim \mathcal{B}(m + n, p)$. Indeed, using Example 2.70, we get

$$g_{X+Y}(t) = [pt + (1 - p)]^n[pt + (1 - p)]^m = [pt + (1 - p)]^{n+m}, \quad t \in \mathbb{R}.$$

Similarly, for negative binomial variables, if $X \sim \mathcal{B}_-(r_1, p)$ and $Y \sim \mathcal{B}_-(r_2, p)$ are independent, then $X + Y \sim \mathcal{B}_-(r_1 + r_2, p)$.   ◁

In particular, the sum of n i.i.d. random variables with common distribution $\mathcal{B}(p)$ has a binomial distribution $\mathcal{B}(n, p)$. Still, all binomial variables are not sums of Bernoulli variables.

▷ *Example 3.51 (A Binomial Variable that Is Not a Sum of Bernoulli Variables)* Let $\Omega = \{a, b, c\}$. Setting $\mathbb{P}(\{a\}) = 1/9$ and $\mathbb{P}(\{b\}) = \mathbb{P}(\{c\}) = 4/9$ defines a probability $\mathbb{P}$ on $(\Omega, \mathcal{P}(\Omega))$.

Let X be the random variable defined on Ω by $X(a) = 0$, $X(b) = 1$, and $X(c) = 2$. Then $\mathbb{P}(X = 0) = P(\{a\}) = 1/9 = (1/3)^2$, $\mathbb{P}(X = 1) = \mathbb{P}(\{b\}) = 4/9 = 2 \cdot (2/3) \cdot (1/3)$ and $\mathbb{P}(X = 2) = P(\{c\}) = 4/9 = 2 \cdot (2/3)^2$, so X has a binomial distribution $\mathcal{B}(2, 2/3)$.

Clearly, X cannot be the sum of two random variables with distribution $\mathcal{B}(2/3)$ because no subset of Ω has probability 2/3. Of course, we can enlarge the probability space, in order to obtain that.   ◁

As shown in the next examples, Point 1 of Theorem 3.49 for the characteristic functions is a powerful tool for determining the distribution of sums of random variables with densities.

▷ *Example 3.52 (Sums of Independent Normal Variables)* Let $X \sim \mathcal{N}(m_1, \sigma_1^2)$ and $Y \sim \mathcal{N}(m_2, \sigma_2^2)$ be two independent variables. Then $X + Y \sim \mathcal{N}(m_1 + m_2, \sigma_1^2 + \sigma_2^2)$. Indeed, using Example 2.71, the characteristic function of $X + Y$ is

$$\varphi_{X+Y}(t) = e^{itm_1} e^{-\sigma_1^2 t^2/2} e^{itm_2} e^{-\sigma_2^2 t^2/2} = e^{it(m_1+m_2)} e^{-(\sigma_1^2+\sigma_2^2)t^2/2}, \quad t \in \mathbb{R}$$

and characterizes the distribution. This result does not carry over to dependent variables, as shown below in Example 3.68.   ◁

Note that a normal random variable can be proven to decompose into a sum of i.i.d. variables only if these variables also have normal distributions.

▷ *Example 3.53 (Sums of i.i.d. Exponential Variables)* Let $X_1, \ldots, X_n$ be i.i.d. random variables with distribution $\mathcal{E}(\lambda)$. Set $S_k = X_1 + \cdots + X_k$, for $1 \leq k \leq n$.

By the linear change of variables $s_1 = t_1$, $s_2 = t_1 + t_2, \ldots$, $s_n = t_1 + \cdots + t_n$, we get the density of the vector $(S_1, S_2, \ldots, S_n)$ from the density $f_{(X_1, \ldots, X_n)}(t_1, \cdots, t_n) = \lambda^n e^{-\lambda(t_1 + \cdots + t_n)} \mathbb{1}_{\mathbb{R}_+^n}(t_1, \cdots, t_n)$, that is

$$f_{(S_1, \ldots, S_n)}(s_1, \ldots, s_n) = \lambda^n e^{-\lambda s_n} \mathbb{1}_{(0 < s_1 < \cdots < s_n)}.$$

The distribution of S_n follows by taking the n-th marginal distribution, that is

$$f_{S_n}(t) = \frac{(\lambda t)^{n-1}}{(n-1)!} \lambda e^{-\lambda t} \mathbb{1}_{\mathbb{R}_+}(t).$$

Thus, S_n has an Erlang distribution $\mathcal{E}(n, \lambda)$.   ◁

Similarly, if $X \sim \gamma(a_1, b)$ and $Y \sim \gamma(a_2, b)$ are independent, then $X + Y \sim \gamma(a_1+a_2, b)$; in particular, if $X \sim \chi^2(m)$ and $Y \sim \chi^2(n)$, then $X+Y \sim \chi^2(m+n)$. If X_1 and X_2 are two i.i.d. standard normal variables, then X_1^2 and X_2^2 are i.i.d. with distribution $\chi^2(1)$, so $X_1^2 + X_2^2 \sim \chi^2(2)$. Further, since the $\chi^2(2)$ distribution is an exponential distribution $\mathcal{E}(1/2)$, we obtain that $\sqrt{X_1^2 + X_2^2}$ has a Weibull distribution $\mathcal{W}(2, 1/\sqrt{2})$.

The distribution of weighted sums or of products of variables can also be determined using the above results.

▷ *Example 3.54 (Product of Log-Normal Distributions)* Let X_1 and X_2 be two independent variables with log-normal distributions with respective parameters $(0, 1)$ and (m, σ^2). Let us determine the distribution of the product $T = X_1 X_2$.

We know that $X_i = \exp Y_i$, where $Y_1 \sim \mathcal{N}(0, 1)$ and $Y_2 \sim \mathcal{N}(m, \sigma^2)$, so $T = \exp(Y_1 + Y_2)$. Since $Y_1 + Y_2 \sim \mathcal{N}(m, 1 + \sigma^2)$, we obtain that T has a log-normal distribution with parameters $(m, 1 + \sigma^2)$.   ◁

Using characteristic functions, one can easily show that if X and Y are two i.i.d. variables with Cauchy distribution $\mathcal{C}(\alpha)$, then $aX + bY$ also has a Cauchy distribution $\mathcal{C}((a + b)\alpha)$; this is an example of a stable distribution.

Variables may be dependent even if the distribution of their sum is the convolution of their distributions. Nevertheless, the following result is a criterion of independence expressed in terms of sums.

Corollary 3.55 *Random vectors $X_1, \dots, X_n$ with dimension d are independent if and only if the distribution of any linear combination $t_1 X_1 + \cdots + t_n X_n$ is the convolution of the distributions of $t_1 X_1, \dots, t_n X_n$ for all $t_1, \dots, t_n$ in $\mathbb{R}^d$.*

Proof We prove the corollary for two variables. The direct implication is clear. Conversely,

$$\varphi_{(X,Y)}(t_1, t_2) = \mathbb{E}\left[e^{i(t_1 X + t_2 Y)}\right] = \varphi_{t_1 X + t_2 Y}(1) \stackrel{(1)}{=} \varphi_X(t_1)\varphi_Y(t_2), \quad (t_1, t_2) \in \mathbb{R}^2,$$

(1) by Theorem 3.49. Then Proposition 3.39 yields the conclusion.   □

Finally, let us prove a classical inequality involving sums.

Theorem 3.56 (Kolmogorov) *Let $X_1, \dots, X_n$ be independent centered random variables with finite variances $\sigma_1^2, \dots, \sigma_n^2$. Then, for all $\varepsilon > 0$,*

$$\mathbb{P}\left[\max(|X_1|, \dots, |X_1 + \cdots + X_n|) > \varepsilon\right] \leq \sum_{k=1}^{n} \frac{\sigma_k^2}{\varepsilon^2}.$$

Proof Let us set $S_n = X_1 + \cdots + X_n$, and consider for $2 \leq k < n$ the events $E = (\max(|S_1|, \ldots, |S_n|) > \varepsilon)$ and $E_k = (|S_k| > \varepsilon, |S_i| < \varepsilon, i = 1, \ldots, k-1)$. We have

$$\mathbb{E}\,(\mathbb{1}_{E_k} S_n^2) = \mathbb{E}\,(\mathbb{1}_{E_k} S_k^2) + \mathbb{E}\,[\mathbb{1}_{E_k}(S_n - S_k)^2] + 2\mathbb{E}\,[\mathbb{1}_{E_k} S_k(S_n - S_k)]$$

$$\overset{(1)}{=} \mathbb{E}\,(\mathbb{1}_{E_k} S_k^2) + \mathbb{P}(E_k)\mathbb{E}\,[(S_n - S_k)^2],$$

(1) since $\mathbb{1}_{E_k}$ and $\mathbb{1}_{E_k} S_k$ do not depend on $S_n - S_k = X_{k+1} + \cdots + X_n$.

Therefore, $\mathbb{E}\,(\mathbb{1}_{E_k} S_k^2) \leq \mathbb{E}\,(\mathbb{1}_{E_k} S_n^2)$, for all k. Moreover, $|S_k| > \varepsilon$ on E_k, so

$$\mathbb{P}(E_k) \leq \frac{1}{\varepsilon^2}\mathbb{E}\,(\mathbb{1}_{E_k} S_k^2) \leq \frac{1}{\varepsilon^2}\mathbb{E}\,(\mathbb{1}_{E_k} S_n^2).$$

Finally, since E is the disjoint union of the E_k, we compute

$$\mathbb{P}(E) = \sum_{k=2}^{n-1} \mathbb{P}(E_k) \leq \frac{1}{\varepsilon^2}\sum_{k=2}^{n-1} \mathbb{E}\,(\mathbb{1}_{E_k} S_n^2) = \frac{1}{\varepsilon^2}\mathbb{E}\,(\mathbb{1}_E S_n^2) \leq \frac{1}{\varepsilon^2}\mathbb{E}\,(S_n^2),$$

from which the conclusion follows because $\mathbb{E}\,(S_n^2) = \mathbb{V}\mathrm{ar}\,S_n = \sum_{k=1}^{n} \sigma_k^2$. □

3.3.3 Determination of Distributions

Let $X = (X_1, \ldots, X_d)$ be random vectors with densities, and let $\psi : \mathbb{R}^d \to \mathbb{R}^d$ be any Borel function. The methods described in Chap. 2 for random variables carry over to random vectors for identifying the density of $Z = \psi(X_1, \ldots, X_d)$ from the density of X. We will give many examples after stating some necessary theoretical results.

If f is the density of X on $D_1 \subset \mathbb{R}^d$ and if $\psi : D_1 \to D_2$ is a diffeomorphism, then Theorem 2.22 yields directly

$$f_Z(z) = |J_{\psi^{-1}}(z)| f_X(\psi^{-1}(z))\mathbb{1}_{D_2}(z), \tag{3.8}$$

since $\mathbb{P}_Z$ is then the image measure of $\mathbb{P}_X$ by ψ.

If ψ is not a diffeomorphism from D_1 into D_2, Theorem 2.1 can still be used by considering a partition of D_1 into K disjoint subsets, say $D_1(1), \ldots, D_1(K)$, such that $\lambda(D_1) = \lambda[\cup_{k=1}^{K} D_1(k)]$ and that the restrictions $\psi_k : D_1(k) \to \psi(D_1(k))$ of ψ on $D_1(k)$ are all diffeomorphisms for all k. Thus, we get

$$f_Z(z) = \sum_{k=1}^{K} |J_{\psi_k^{-1}}(z)| f_X(\psi_k^{-1}(z))\mathbb{1}_{\psi(D_1(k))}(z). \tag{3.9}$$

Just as for random variables, computing either $\mathbb{E}[h(Z)] = \int_{\mathbb{R}^d} h(z) f_Z(z) dz$ for any bounded Borel function h or the distribution function F_Z and then its partial derivatives constitutes alternative methods.

▷ *Example 3.57 (Distribution of a Function of i.i.d. Exponential Variables)* Let X, Y, Z be three i.i.d. variables with distribution $\mathcal{E}(\lambda)$. Let us determine the distribution of $(X + Y + Z, X + Z, -Z)$.

Let $\psi : D_1 = \mathbb{R}_+^* \times \mathbb{R}_+^* \times \mathbb{R}_+^* \to D_2 = \{(u, v, w) \in \mathbb{R}^3 : u > v > -w > 0\}$ be defined by $\psi(x, y, z) = (x + y + z, x + z, -z)$. This is a diffeomorphism with inverse function $\psi^{-1}(u, v, w) = (v + w, u - v, -w)$ and Jacobian 1. The joint density of (X, Y, Z) is

$$f_{(X,Y,Z)}(x, y, z) = \lambda^3 e^{-\lambda(x+y+z)} \mathbb{1}_{D_1}(x, y, z),$$

and hence, (3.8) yields $f(u, v, w) = \lambda^3 e^{-\lambda u} \mathbb{1}_{D_2}(u, v, w)$.

Considering for any bounded Borel function $h : \mathbb{R}^3 \to \mathbb{R}$,

$$\mathbb{E}[h(X, Y, Z)] = \lambda^3 \iiint_{D_1} h(x + y + z, x + z, -z) e^{-\lambda(x+y+z)} \, dx dy dz$$

$$= \lambda^3 \iiint_{D_2} h(u, v, w) e^{-\lambda u} \, du dv dw,$$

would lead to the same result.   ◁

Similarly, if $X \sim \gamma(a_1, b)$ and $Y \sim \gamma(a_2, b)$ are independent, one can show for instance that $X/(X + Y) \sim \beta_1(a_1, a_2)$ and $X + Y \sim \gamma(a_1 + a_2, b)$ are independent.

Let now X and Y be two vectors with respective dimensions d and d' and joint distribution with density $f_{(X,Y)}$. Let $u : \mathbb{R}^{d+d'} \to \mathbb{R}^{d''}$ be a Borel function. For determining the distribution of $U = u(X, Y)$, one can determine first the distribution of (U, V), where $V = Y$, by using a change of variables. The distribution of U is then obtained by taking the marginal distribution. The variable V is referred to as an auxiliary variable.

For example, the density of the sum $X + Y$ of two random variables can be found as the second marginal distribution of $(X, X + Y)$; see Example 3.53.

The density of U can also be determined by using the following relation, valid for any bounded Borel function $h : \mathbb{R}^{d''} \to \mathbb{R}$,

$$\mathbb{E}[h(U)] = \int_{\mathbb{R}^{d+d'}} h(u(x, y)) f_{(X,Y)}(x, y) \, dx dy = \int_{\mathbb{R}^d} h(u) f_U(u) du.$$

▷ *Example 3.58 (Distribution of the Quotient of Two Variables)* Let X and Y be two variables whose joint distribution has a density $f_{(X,Y)}$. Set $U = X/Y$, well-defined because Y has a continuous distribution function.

Let us consider the auxiliary variable $V = Y$ and the change of variable $(u, v) = \psi(x, y) = (x/y, y)$. The function ψ is a diffeomorphism from $\mathbb{R} \times \mathbb{R}_+^*$ into $\mathbb{R} \times \mathbb{R}_+^*$ with Jacobian y and from $\mathbb{R} \times \mathbb{R}_-^*$ into $\mathbb{R} \times \mathbb{R}_-^*$ with Jacobian $-y$. Moreover, $\psi^{-1}(u, v) = (uv, v)$, with Jacobian v. Since the Lebesgue measure on $\mathbb{R}^2$ of $\mathbb{R} \times \{0\}$ is zero, (3.9) yields $f_{(U,V)}(u, v) = f_{(X,Y)}(uv, v)|v|$, and hence,

$$f_U(u) = \int_{\mathbb{R}} f_{(X,Y)}(uv, v)|v|dv$$

is the density of $U = X/Y$. ◁

The above example yields alternative definitions of Cauchy, Fisher, and Student distributions.

Proposition 3.59

1. *If $X \sim \mathcal{N}(0, 1)$ and $Y \sim \mathcal{N}(0, 1)$ are independent, then $U = X/Y \sim \mathcal{C}(1)$.*
2. *If $X, Y_1, \ldots, Y_n$ are i.i.d. with distribution $\mathcal{N}(0, \sigma^2)$, then*

$$\frac{X}{\sqrt{\sum_{i=1}^n Y_i^2/n}} \sim \tau(n).$$

3. *If $X_1 \sim \chi^2(n)$ and $X_2 \sim \chi^2(m)$ are independent, then*

$$\frac{X_1/n}{X_2/m} \sim F(n, m).$$

Proof Indeed, $f(x, y) = e^{-(x^2+y^2)/2}/2\pi$, so

$$f_U(u) = \frac{2}{2\pi} \int_{\mathbb{R}_+} e^{-(u^2+1)y^2/2} y dy = \frac{1}{\pi} \frac{1}{1+u^2}.$$

The other cases are similar. □

▷ *Example 3.60 (Distribution of the Product of Two Variables)* Let X and Y be two variables whose joint distribution has a density $f_{(X,Y)}$. Set $U = XY$. Let us consider again the auxiliary variable $V = Y$ and the change of variable $(u, v) = \psi(x, y) = (xy, y)$, with $\psi^{-1}(u, v) = (u/v, v)$ whose Jacobian is $1/v$, for $v \neq 0$. Hence, by the same argument as in Example 3.58, (3.9) yields $f_{(U,V)}(u, v) = f_{(X,Y)}(u/v, v)/|v|$ for $v \neq 0$, and

$$f_U(u, v) = \int_{\mathbb{R}} f_{(X,Y)}(u/v, v) \frac{1}{|v|} dv$$

is the density of $U = XY$. ◁

The knowledge of the type of distribution allows for more intricate computations.

▷ *Example 3.61 (Distribution of the Logarithm of a Product)* Let X and Y be two
i.i.d. variables with uniform distribution on $[-1, 1]$. Let us determine the density of
$Z = \log|XY|$. Applying Example 3.60 yields the density of $U = XY$, that is

$$f_U(u) = \int_{\mathbb{R}} \frac{1}{4|y|} \mathbb{1}_{[-1,1]}\left(\frac{u}{y}\right) \mathbb{1}_{[-1,1]}(y) dy$$

$$= \mathbb{1}_{[-1,1]}(u)\left(\int_{|u|}^1 \frac{1}{4|y|}dy + \int_{-1}^{-|u|} \frac{1}{-4|y|}dy\right) = \mathbb{1}_{[-1,1]}(u)\int_{|u|}^1 \frac{1}{2|y|}dy,$$

or $f_U(u) = -\mathbb{1}_{[-1,1]}(u)\log|u|/2$. So, the distribution function of Z is

$$F_Z(z) = \mathbb{P}(\log|U| \leq z) = \mathbb{P}(-e^z \leq U < 0) + \mathbb{P}(0 < U \leq e^z)$$

$$= \int_{-e^z}^0 \frac{-1}{2}\log|v|\mathbb{1}_{[-1,1]}(v)dv + \int_0^{e^z} \frac{-1}{2}\log|v|\mathbb{1}_{[-1,1]}(v)dv$$

$$= \int_{\max(-1,-e^z)}^0 \frac{-1}{2}\log(-v)dv + \int_0^{\min(1,e^z)} \frac{-1}{2}\log v\, dv$$

$$= -\int_0^{\min(1,e^z)} \log v\, dv.$$

Integrating by parts yields $F_Z(z) = e^z(1-z)$ if $z < 0$, and $F_Z(z) = 1$ if $z \geq 0$.
Thus, the density of Z is $f_Z(z) = F_Z'(z) = -ze^z\mathbb{1}_{\mathbb{R}_-}(z)$. ◁

3.4 Gaussian Vectors

Gaussian vectors play a part as important in the field of random vectors as the
Gaussian variables play in the field of random variables.

Definition 3.62 A d-dimensional random vector $X = (X_1, \ldots, X_d)'$ is called
a Gaussian vector if all the linear combinations of its coordinates are Gaussian
variables, in other words if

$$a'X = \sum_{i=1}^d a_i X_i \sim \mathcal{N}(m_a, \sigma_a^2), \quad a = (a_1, \ldots, a_d)' \in \mathbb{R}^d,$$

where (m_a, σ_a^2) depends on a.

If $X = (X_1, \ldots, X_d)'$ is a Gaussian vector, then all $X_1, \ldots, X_d$ are Gaussian variables and, for any linear or affine function $f : \mathbb{R}^d \to \mathbb{R}^{d'}$, the image $f(X)$ is a d'-dimensional Gaussian vector.

A Gaussian vector X is referred to as standard if $\mathbb{E}\, X = 0$ and $\Gamma_X = I_d$, the identity matrix of $\mathbb{R}^d$.

▷ *Example 3.63 (Some Typical Gaussian Vectors)* Any random vector a.s. constant, with Dirac distribution on $\mathbb{R}^d$, is a degenerated Gaussian vector.

Any vector $(X_1, \ldots, X_d)'$, where $X_1, \ldots, X_d$ are i.i.d. with distribution $\mathcal{N}(0, 1)$, is a standard Gaussian vector.

Many other examples are available; for instance $(X_1, X_2)'$ such that $X_1 \sim \mathcal{N}(0, 1)$ and $X_2 = 0$ a.s. is a Gaussian vector.   ◁

The expression of the characteristic function of a Gaussian vector derives immediately from that of a Gaussian variable. It will be used below for obtaining explicit expressions of densities of Gaussian vectors.

Theorem 3.64 *Let X be a Gaussian vector with expectation M and variance–covariance matrix Γ_X. The characteristic function of X is*

$$\varphi_X(t) = \exp\left(it'M - \frac{1}{2}t'\Gamma_X t\right) \quad t \in \mathbb{R}^d.$$

Thus, the characteristic function of a Gaussian vector X—and hence its distribution—is characterized by its expectation M and its variance–covariance matrix Γ_X. We will write $X \sim \mathcal{N}_d(M, \Gamma_X)$. Since this is also the joint distribution of $X_1, \ldots, X_d$, we will write either $X = (X_1, \ldots, X_d)'$ or $X = (X_1, \ldots, X_d)$ when only distributions are involved.

Proof If $t \in \mathbb{R}^d$, then $t'X$ is a Gaussian variable, with expectation $t'M$ and variance $t'\Gamma_X t$. Therefore,

$$\varphi_{t'X}(u) = \mathbb{E}\left[\exp(iut'X)\right] = \exp\left(iut'M - \frac{u^2}{2}t'\Gamma_X t\right).$$

Setting $u = 1$ yields the desired formula.   □

▷ *Example 3.65 (A Bidimensional Gaussian Vector)* Let $X \sim \mathcal{N}_2(M, \Gamma_X)$ with $M = (1, 2)'$ and

$$\Gamma_X = \begin{pmatrix} 1 & -2 \\ -2 & 5 \end{pmatrix}.$$

If $Y_1 = X_1 + 3X_2 + 1$ and $Y_2 = 2X_1 - 5X_2 - 1$, then, thanks to Proposition 3.36, $Y = (Y_1, Y_2)' \sim \mathcal{N}(M_Y, \Gamma_Y)$, where $M_Y = (8, -9)'$ and

$$\Gamma_Y = \begin{pmatrix} 34 & -75 \\ -75 & 169 \end{pmatrix}.$$

Note that $\varphi_X(t_1, t_2) = \exp[i(t_1 + 2t_2) - (t_1^2 - 4t_1t_2 + 5t_2^2)/2]$ for $(t_1, t_2) \in \mathbb{R}^2$.

The independence of the coordinates of a Gaussian vector can be read on its variance–covariance matrix.

Theorem 3.66 *Let $X = (X_1, \ldots, X_d) \sim \mathcal{N}_d(M, \Gamma_X)$ be a d-dimensional Gaussian vector. Then, the variables $X_1, \ldots, X_d$ are independent if and only if Γ_X is a diagonal matrix.*

Proof If X_i and X_j are independent, then $\mathrm{Cov}(X_i, X_j) = 0$.
Conversely, let us set $\Gamma_X = \mathrm{diag}(\sigma_i^2)$ and $M = (m_1, \ldots, m_d)$. We have

$$\varphi_X(t) = \varphi_{(X_1, \ldots, X_d)}(t_1, \ldots, t_d)$$

$$= \exp\left(it_1m_1 + \cdots + it_dm_d - \frac{1}{2}\sigma_1^2 t_1^2 - \cdots - \frac{1}{2}\sigma_d^2 t_d^2\right)$$

$$= \prod_{j=1}^{d} \exp\left(it_jm_j - \frac{1}{2}\sigma_j^2 t_j^2\right) = \varphi_{X_1}(t_1) \ldots \varphi_{X_d}(t_d),$$

and hence, according to Proposition 3.39, $X_1, \ldots, X_d$ are independent. □

The density of a standard Gaussian vector follows immediately; see also Fig. 3.3.

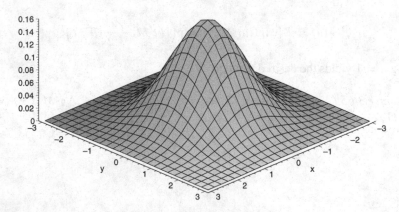

Fig. 3.3 Density of the distribution $\mathcal{N}_2(0, I)$

Corollary 3.67 *If $X \sim \mathcal{N}_d(0, I_d)$, then its density is*

$$f_X(x) = (2\pi)^{-d/2} \exp\left(-\sum_{j=1}^{d} \frac{x_j^2}{2}\right), \quad x \in \mathbb{R}^d.$$

Proof The variance–covariance matrix of X is $\Gamma = I_d$, which is a diagonal matrix, and hence, $X_1, \ldots, X_d$ are independent. Therefore, $f_X = f_{(X_1, \ldots, X_d)} = f_{X_1} \cdots f_{X_d}$, and the result follows.   □

In general, uncorrelated Gaussian variables are not independent and do not constitute a Gaussian vector, as shown in the next example.

▷ *Example 3.68 (A Pair of Gaussian Variables that Is Not a Gaussian Vector)* Let $X_1 \sim \mathcal{N}(0, 1)$. Let $\varepsilon \sim \mathcal{B}(1/2)$ take values in $\{-1, 1\}$ and be independent of X_1. Set $X_2 = \varepsilon X_1$. We have

$$\mathbb{P}(X_2 \leq x) = \mathbb{P}(X_1 \leq x, \varepsilon = 1) + \mathbb{P}(-X_1 \leq x, \varepsilon = -1)$$

$$= \frac{1}{2}\mathbb{P}(X_1 \leq x) + \frac{1}{2}\mathbb{P}(-X_1 \leq x).$$

Since the standard Gaussian distribution is symmetric, $-X_1 \sim \mathcal{N}(0, 1)$, and hence, $X_2 \sim \mathcal{N}(0, 1)$ too.

Since ε and X_1 are independent, $\text{Cov}(X_1, X_2) = \mathbb{E}(\varepsilon X_1^2) = \mathbb{E}(\varepsilon)\mathbb{E}(X_1^2) = 0$, so X_1 and X_2 are indeed uncorrelated and Gaussian. Since, for instance, $\mathbb{P}(X_1 + X_2 = 0) = \mathbb{P}(\varepsilon = -1) = 1/2$, the variable $X_1 + X_2$ is not a Gaussian variable, and hence, (X_1, X_2) is not a Gaussian vector.   ◁

The density of any Gaussian vector derives from the density of a standard Gaussian vector as stated in the following theorem. The proof is based on a linear algebra result which we state without proof.

Lemma 3.69 *Let Γ be a $d \times d$ symmetric nonnegative definite matrix with rank r. Then:*

1. *A $d \times r$ matrix with rank r exists such that $\Gamma = BB'$.*
2. *An orthonormal basis of eigen-vectors of Γ exists in which Γ is written as a diagonal matrix D such that $D_{ii} = 0$ for $i > r$ and $D_{ii} > 0$ for $i \leq r$.*

In particular, it is always possible to transform a Gaussian vector into a Gaussian vector whose coordinates are uncorrelated. Also, all given real vectors and nonnegative symmetric matrix are the expectation and variance–covariance matrix of a Gaussian vector.

Theorem 3.70 *Let $M \in \mathbb{R}^d$, and let Γ be a $d \times d$ symmetric nonnegative definite matrix with rank r. Let B be such that $\Gamma = BB'$.*

1. If $Y \sim \mathcal{N}_d(0, I_d)$, then $M + BY \sim \mathcal{N}_d(M, \Gamma)$.
2. If $X \sim \mathcal{N}_d(M, \Gamma)$, then an r-dimensional standard Gaussian vector Y exists such that $X = M + BY$.

Proof

1. Let us set $U = M + BY$. Since U is a linear transform of a Gaussian vector, it is a Gaussian vector too, with $\mathbb{E}\,U = M$ and $\Gamma_U = B(\Gamma_Y)B' = BB' = \Gamma$.
2. We can suppose without loss of generality that $M = 0$.

Let $(v_1, \dots, v_d)$ be an orthonormal basis of eigen-vectors of Γ associated with the eigen-values $(\lambda_1^2, \dots, \lambda_r^2, 0, \dots, 0)$.
Set $Z = (Z_1, \dots, Z_d)$, where $Z_i = <X, v_i>$. Let V be the $d \times d$ matrix whose columns are the vectors v_i, and set $\Gamma_Z = V'\Gamma V$; this matrix is a diagonal matrix, with $\mathbb{V}\text{ar}\,Z_i = \lambda_i^2$ for $i \leq r$ and $Z_i = 0$ a.s. for $r < i \leq d$.
Set $Y = (Z_1, \dots, Z_r)$. We have $Y \sim \mathcal{N}_r(0, I_r)$, and $X = BY$, with necessarily, according to 1, $\Gamma = BB'$. $\square$

▷ *Example 3.71 (Continuation of Example 3.65)* We have $\Gamma = BB'$, where

$$B = \begin{pmatrix} 0 & 1 \\ 1 & -2 \end{pmatrix} \quad \text{and} \quad B^{-1} = \begin{pmatrix} 2 & 1 \\ 1 & 0 \end{pmatrix}.$$

Therefore, setting $Y_1 = 2X_1 + X_2 - 4$ and $Y_2 = X_1 - 1$, we get $Y = (Y_1, Y_2)' \sim \mathcal{N}(0, I_2)$ because $Y = B^{-1}(X - M)$.   ◁

The determinant of a matrix Γ is not zero if and only if Γ is invertible or equivalently if none of its eigen-values are null. The vector is then referred to as non-degenerated.

Theorem 3.72 *Let $X \sim \mathcal{N}_d(M, \Gamma)$. The Gaussian vector X has a density with respect to the Lebesgue measure if and only if $\det \Gamma \neq 0$, and then*

$$f_X(x) = (2\pi)^{-d/2}(\det \Gamma)^{-1/2} \exp[-\frac{1}{2}(x - M)'\Gamma^{-1}(x - M)]$$

$$= (2\pi)^{-d/2}(\det \Gamma)^{-1/2} \exp[-\frac{1}{2} \sum_{i,j=1}^{d} \gamma_{ij}(x_i - m_i)(x_j - m_j)],$$

for $x \in \mathbb{R}^d$, where $M = (m_i)$ and $\Gamma^{-1} = (\gamma_{ij})$.

Proof Let $Y \sim \mathcal{N}_d(0, I_d)$. According to Corollary 3.67, the density of Y is

$$f_Y(y) = (2\pi)^{-d/2} \exp\left(-\frac{1}{2} \sum_{i=1}^{d} y_j^2\right).$$

Set $Z = M + BY$, where $\Gamma = BB'$. Then $Z \sim \mathcal{N}_d(M, \Gamma)$, and the linear change of variables $z = M + By$ yields

$$f_Z(z) = (2\pi)^{-d/2} \det(B^{-1}) \exp\left[-\frac{1}{2}(z-M)'(B^{-1})'B^{-1}(z-M)\right].$$

Moreover, $\Gamma = BB'$, so $\Gamma^{-1} = (BB')^{-1} = (B^{-1})'B^{-1}$, and hence, $\det(B^{-1}) = (\det \Gamma)^{-1/2}$. $\square$

Finally, let us state a stochastic Pythagorean theorem that proves to be particularly useful in statistics.

Theorem 3.73 (Cochran) *Let $X \sim \mathcal{N}_d(M, \sigma^2 I_d)$.*

1. *The coordinates of X in any orthonormal basis of $\mathbb{R}^d$ constitute a Gaussian vector with the same distribution.*
2. *Any transform of X by an isometry of $\mathbb{R}^d$ is a Gaussian vector with the same distribution.*
3. *Suppose that $M = 0$. Let $E_1,\ldots,E_p$ be pairwise orthogonal subspaces of $\mathbb{R}^d$, with respective dimensions $r_1,\ldots,r_p$, and such that $\mathbb{R}^d = E_1 \oplus \cdots \oplus E_p$. The orthogonal projections $\Pi_{E_1}(X), \ldots, \Pi_{E_p}(X)$ of X on $E_1,\ldots,E_p$ are independent, and the random variables*

$$\|\Pi_{E_1}(X)\|^2, \ldots, \|\Pi_{E_p}(X)\|^2$$

are independent with respective distributions $\sigma^2\chi^2(r_1),\ldots, \sigma^2\chi^2(r_p)$.

Proof We can suppose without loss of generality that $M = 0$.

1. Let $(e_1, \ldots, e_d)$ denote an orthonormal basis of $\mathbb{R}^d$. The coordinates of X in this basis are $Y = (<X, e_1>, \ldots, <X, e_d>)'$, and we have

$$\varphi_Y(t) = \mathbb{E}\left(\exp i \sum_{j=1}^{d} t_j <X, e_j>\right) = \mathbb{E}\left(\exp i <X, \sum_{j=1}^{d} t_j e_j>\right).$$

The characteristic function of X is

$$\varphi_X(t) = \mathbb{E}(\exp i <t, X>) = \exp\left(-\frac{1}{2}\sigma^2 \sum_{j=1}^{d} t_j^2\right),$$

so $\varphi_Y(t) = \exp(-\sigma^2 \sum_{j=1}^d t_j^2/2)$, meaning that $Y \sim X$.

2. The lines (and columns) of the matrix A of any isometry of $\mathbb{R}^d$ are orthonormal. Moreover, $AA' = I_d$, from which the conclusion follows by using Point 1.

3. An orthonormal basis of $\mathbb{R}^d$, say $(e_1, \ldots, e_d)$, exists such that E_j is spanned by $(e_{j_1}, \ldots, e_{j_{r_1}})$.

Then $\Pi_{E_j}(X) = \sum_{k=r_{j-1}+1}^{r_j} < X, e_{j_k} > e_{j_k}$, with Euclidean norm $\|\Pi_{E_j}(X)\|^2 = \sum_{k=r_{j-1}+1}^{r_j} < X, e_{j_d} >^2$. The result follows by considering the above vector Y. $\square$

3.5   Exercises and Complements

∇ **Exercise 3.1 (Matching Problem—Continuation of Exercises 1.3 and 2.3)** Let X_i be the indicator function of the event "The sportsman i has taken back his own bag." Write S_n in terms of $X_1, \ldots, X_n$ to compute the expectation and the variance of S_n.

Solution
We have $S_n = X_1 + \cdots + X_n$. On the one hand, $\mathbb{P}(X_i = 1) = 1/n = 1 - \mathbb{P}(X_i = 0)$, so $\mathbb{E}\, X_i = 1/n$ and $\mathbb{E}\, S_n = 1$. On the other hand, $\mathbb{V}\text{ar}\, X_i = (n-1)/n^2$ and $\mathbb{P}(X_i X_j = 1) = 1/n(n-1) = 1 - \mathbb{P}(X_i X_j = 0)$, so

$$\mathbb{C}\text{ov}\,(X_i, X_j) = \frac{1}{n(n-1)} - \frac{1}{n} \times \frac{1}{n} = \frac{1}{n^2(n-1)}$$

and $\mathbb{V}\text{ar}\, S_n = \sum_{i=1}^n \mathbb{V}\text{ar}\, X_i - \sum_{i \neq j} \mathbb{C}\text{ov}\,(X_i, X_j) = 1$. $\triangle$

∇ **Exercise 3.2 (Mutual Entropy)**   Let X and Y be two random variables defined on the same probability space.

(A) Suppose X and Y are discrete, with respective values $\{x_i : i \in I\}$ and $\{y_j : j \in J\}$. Their mutual entropy (or information) is

$$\mathcal{S}(X, Y) = \sum_{i \in I} \sum_{j \in J} \mathbb{P}(X = x_i, Y = y_j) \log \frac{\mathbb{P}(X = x_i, Y = y_j)}{\mathbb{P}(X = x_i)\mathbb{P}(Y = y_j)}.$$

1. Write $\mathbb{S}(X, Y)$ in terms of the joint, marginal, and conditional entropies of X and Y.
2. Show that $\mathcal{S}(X, Y)$ is a nonnegative quantity. Under which condition on X and Y is it null?

(B) Suppose X and Y have joint density f and respective marginal densities f_X and f_Y.
1. Define a mutual entropy for these continuous variables.
2. (a) Use the convexity of the logarithm for showing that $\mathcal{S}(X, Y)$ is a nonnegative quantity.
   (b) Use a passage from discrete to continuous frameworks for showing that $\mathcal{S}(X, Y)$ is finite.

Solution

(A) 1. We have $\mathcal{S}(X, Y) = \mathbb{S}(X) - \mathbb{S}(X \mid Y)$. By definition, $\mathbb{S}(X \mid Y) = \mathbb{S}(X, Y) - \mathbb{S}(Y)$, so, using Property 2 of entropy,

$$\mathbb{S}(X \mid Y) = \mathbb{S}(X) + \mathbb{S}(Y) - \mathbb{S}(X, Y).$$

2. We know by Property 3 of entropy that $\mathbb{S}(X \mid Y) \leq \mathbb{S}(X)$, so $\mathcal{S}(X, Y)$ is indeed nonnegative and is null if and only if X and Y are independent.

(B) 1. Analogously, if f_X and f_Y are defined and not null on the same support $E \in \mathcal{B}(\mathbb{R})$, we can define

$$\mathcal{S}(X, Y) = \iint_{E \times E} f(x, y) \log \frac{f(x, y)}{f_X(x) f_Y(y)} \, dx dy.$$

2. (a) We can write

$$\mathcal{S}(X, Y) = -\iint_{E \times E} f(x, y) \log \frac{f_X(x) f_Y(y)}{f(x, y)} \, dx dy.$$

Since $\log x \leq x - 1$ for x positive, it follows that

$$\mathcal{S}(X, Y) \geq -\iint_{E \times E} f(x, y) \left[\frac{f_X(x) f_Y(y)}{f(x, y)} - 1 \right] dx dy$$

$$= -\iint_{E \times E} f_X(x) f_Y(y) \, dx dy + \iint_{E \times E} f(x, y) \, dx dy$$

$$= -1 + 1 = 0.$$

(b) When Δx and Δy jointly tend to 0, we have

$$\log \frac{\mathbb{P}(X \in \Delta x, Y \in \Delta y)}{\mathbb{P}(X \in \Delta x) \mathbb{P}(Y \in \Delta y)} \longrightarrow \log \frac{f(x, y)}{f_X(x) f_Y(y)},$$

and this quantity remains finite, contrary to each of its terms.

Thus, the mutual information $\mathcal{S}(X, Y)$ is finite, whereas the different entropies $\mathbb{S}(X)$, $\mathbb{S}(Y)$, and $\mathbb{S}(X, Y)$ may not be when the variables are continuous. $\triangle$

▽ **Exercise 3.3 (The Ballot Theorem)**

1. An urn contains a bowls bearing number 0 and b bowls bearing number 2, where
   $a > b$. All the bowls are drawn at random and without replacement. Let X_i be
   the random variable equal to the number of the bowls drawn at the i-th drawing,
   for $i \in [\![1, n]\!]$, where $n = a + b$.
   (a) Compute the expectation of $S_m = \sum_{i=1}^{m} X_i$ for $m \in [\![1, n]\!]$.
   (b) Show that for all $b' \in [\![0, b]\!]$,

   $$\mathbb{P}(S_m < m, \ m \in [\![1, n]\!] | S_{2b} = 2b') = \mathbb{P}(S_m < m, \ m \in [\![1, 2b]\!] | S_{2b} = 2b').$$

   (c) Show by induction and using (b) that

   $$\mathbb{P}(S_m < m, \ m \in [\![1, n]\!] \mid S_n = 2b) = 1 - 2b/n. \tag{3.10}$$

2. Application: in a ballot, two candidates A and B are confronted. Candidate A
   wins. Compute the probability that A has led the ballot all along the counting of
   the votes.

Solution
1.(a) Since $\mathbb{P}(S_n = 2b) = 1$, we have $\mathbb{E} S_n = 2b$. Moreover, the variables
   X_i are exchangeable—that is $(X_1, \ldots, X_n) \sim (X_{(1)}, \ldots, X_{(n)})$, so $\mathbb{E} X_i = \mathbb{E} S_n/n = 2b/n$. Therefore, $\mathbb{E} S_m = 2bm/n$.
   (b) We know that $S_n = 2b$. Hence, if $S_{2b} = 2b'$, then necessarily, $S_m < m$ for all
   m such that $n \geq m \geq 2b + 1$, from which the desired equality follows.
   (c) For (3.10) to hold for $n = 1$, it is necessary that $a = 1$ and $b = 0$ and so
   $\mathbb{P}(X_1 < 1) = \mathbb{P}(X_1 = 0) = 1$.
   Suppose that (3.10) holds for all $m < n$. On the first hand,

   $$\mathbb{P}(S_m < m, \ m \in [\![1, n]\!]) =$$

   $$= \sum_{b'=0}^{b} \mathbb{P}(S_m < m, \ m \in [\![1, n]\!] \mid S_{2b} = 2b')\mathbb{P}(S_{2b} = 2b'),$$

   and on the other hand, since $\mathbb{P}(S_n = 2b) = 1$, we have

   $$\mathbb{P}(S_n < m, \in [\![1, n]\!]) = \mathbb{P}(S_n < m, m \in [\![1, n]\!] \mid S_n = 2b).$$

   Using the induction hypothesis for $n = 2b$, we get

   $$\mathbb{P}(S_m < m, \ m \in [\![1, 2b]\!] \mid S_{2b} = 2b') = 1 - b'/b.$$

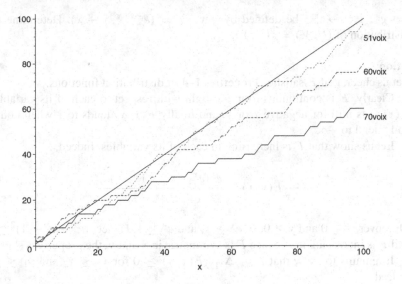

Fig. 3.4 Illustration of the ballot theorem

Thus,

$$\mathbb{P}(S_m < m, m \in [\![1, n]\!]) \stackrel{(1)}{=} \sum_{b'=0}^{b}(1 - b'/b)\mathbb{P}(S_{2b} = 2b') = 1 - \mathbb{E}\, S_{2b}/2b$$

$$\stackrel{(2)}{=} 1 - \frac{2b}{n},$$

(1) by 1(b) and (2) by 1(a).

2. Considering the votes as bowls in the urn, and the counting as drawing these bowls, the searched probability is $(a - b)/(a + b)$, where a is the number of votes for A and b that for B; see Fig. 3.4 for an illustration.

Note that this constitutes an example of a random walk, whose definition will be given in Chap. 4. △

▽ **Exercise 3.4 (Distribution of a Pair)** Let $F : \mathbb{R} \times \mathbb{R} \to \mathbb{R}$ be the function defined by

$$F(x, y) = 1 - e^{-x} - e^{-y} + (e^x + e^y - 1)^{-1}\mathbb{1}_{\mathbb{R}_+^2}(x, y).$$

1. Show that F is the distribution function of a pair of random variables, say (X, Y).
2. Give the density $f(x, y)$ of (X, Y). Are these variables independent?
3. Determine the two marginal distribution functions. Give the distributions of X and Y.

4. Let $g : \mathbb{R}^2 \to \mathbb{R}^2$ be defined by $g(x, y) = (e^{-x+y}, y + x)$. Determine the distribution of $(U, V) = g(X, Y)$.

Solution

1. Let us check that F satisfies Properties 1–4 of distribution functions.

   Clearly, F is continuous from the right with respect to each of its variables, $F(x, y)$ is null for negative x or y, and finally $F(x, y)$ tends to 1 when both x and y tend to $+\infty$.

   Let us show that F is increasing in each of its variables. Indeed,

$$\frac{\partial}{\partial x} F(x, y) = \frac{e^x(e^x + e^y - 1)^2 - e^x}{(e^x + e^y - 1)^2}.$$

Moreover, $x \geq 0$ and $y \geq 0$, so $e^x \geq 1$, and $e^y \geq 1$. Hence, $(e^x + e^y - 1)^2 \geq 1$ and F is increasing in x. Since F is symmetric in x and y, the property is shown.

   It remains to show that $\Delta_{x_1 x_2} \Delta_{y_1 y_2} F(x, y) \geq 0$ for $x_1 \leq x_2$ and $y_1 \leq y_2$. Indeed,

$$\Delta_{x_1 x_2} \Delta_{y_1 y_2} F(x, y) =$$

$$= F(x_2, y_2) - F(x_1, y_2) - F(x_2, y_1) + F(x_1, y_1)$$

$$= -\frac{e^{x_2} - e^{x_1}}{(e^{x_2} + e^{y_2} - 1)(e^{x_1} + e^{y_2} - 1)} + \frac{e^{x_2} - e^{x_1}}{(e^{x_2} + e^{y_1} - 1)(e^{x_1} + e^{y_1} - 1)}.$$

It is well-known that $e^{x_2} - e^{x_1} \geq 0$ and

$$(e^{x_2} + e^{y_1} - 1)(e^{x_1} + e^{y_1} - 1) \leq (e^{x_2} + e^{y_2} - 1)(e^{x_1} + e^{y_2} - 1).$$

Therefore, F is a 2-dimensional distribution function.

2. The searched density is

$$f(x, y) = \frac{\partial^2}{\partial x \partial y} F(x, y) = \frac{2e^x e^y}{(e^x + e^y - 1)^3} \neq f_1(x) f_2(y),$$

so the joint density is not the product of the marginal densities, and hence, the variables are dependent.

3. We have $F_X(x) = \lim_{y \to +\infty} F(x, y) = 1 - e^{-x}$, for all $x \geq 0$. Symmetrically, $F_Y(y) = 1 - e^{-y}$ for $y \geq 0$. Hence, X and Y have the same exponential distribution with parameter 1.

4. Let us set $u = e^{-x+y}$ and $v = x + y$, that is $x = (v - \log u)/2$ and $y = (v + \log u)/2$, with $u > 0$. The Jacobian of this change of variables is

$$\det \begin{pmatrix} -e^{-x+y} & 1 \\ e^{-x+y} & 1 \end{pmatrix} = -2e^{-x+y} = -2u.$$

Hence, the density of the pair (U, V) is

$$f_{(U,V)}(u, v) = \frac{1}{2e^{-x+y}} f(x, y) = [u(e^{(v-\log u)/2} + e^{(v+\log u)/2} - 1)^3]^{-1} e^{2v} \mathbb{1}_{D_2}(u, v),$$

where $D_2 = \{(u, v) \in \mathbb{R}_+^2 : e^{-v} < u < e^v\}$. $\triangle$

∇ **Exercise 3.5 (Reliability of a Two-Component Device)**   A device has two components with respective lifetimes T_1 and T_2 with joint density given by

$$f_{(T_1,T_2)}(x, y) = ce^{-(x+2y)} \mathbb{1}_{\mathbb{R}_+^2}(x, y).$$

1. Compute the normalization constant c.
2. Determine the joint distribution function. Are the two variables independent? Give their marginal distribution functions and densities.
3. Suppose first that the components are in series and then that they are in parallel; see Fig. 3.1 p. 100. Determine the distribution function and density of the lifetime of the system. Compute the mean time to failure.
4. Check that the system with components in parallel is the most reliable one.

Solution
1. We have

$$\iint_{\mathbb{R}_+^2} ce^{-(x+2y)} dx dy = \int_0^{+\infty} ce^{-2y} dy = c/2.$$

For f to be a density, this quantity has to be equal to 1, so $c = 2$.
2. We have for $x \geq 0$ and $y \geq 0$,

$$F_{(T_1,T_2)}(x, y) = \mathbb{P}(T_1 \leq x, T_2 \leq y) = \int_0^x \int_0^y 2e^{-(w+2v)} dw dv$$

$$= (1 - e^{-2y}) \int_0^x e^{-w} dw = (1 - e^{-2y})(1 - e^{-x}).$$

Hence, using Theorem 3.38, T_1 are T_2 independent. Therefore, $F_{T_1}(x) = \mathbb{P}(T_1 \leq x, T_2 \in \mathbb{R}_+) = 1 - e^{-x}$ for $x \geq 0$, and $F_{T_2}(y) = 1 - e^{-2y}$ for $y \geq 0$, from which it follows that $f_{T_1}(x) = e^{-x}$ and $f_{T_2}(y) = 2e^{-2y}$ for $x \geq 0$ and $y \geq 0$. We can check that for $x \geq 0$ and $y \geq 0$,

$$f_{(T_1,T_2)}(x, y) = 2e^{-(x+2y)} = e^{-x} 2e^{-2y} = f_{T_1}(x) f_{T_2}(y).$$

3. The lifetime of the system is given in Example 3.14. The distribution function of T is

$$F_T(t) = 1 - \mathbb{P}(T_1 > t) \mathbb{P}(T_2 > t) = 1 - e^{-3t},$$

so $f_T(t) = 3e^{-3t}$ for $t \geq 0$. Hence, $T \sim \mathcal{E}(3)$ and $\mathbb{E}\,T = 1/3$.

For components in parallel, we have $F_{T'}(t) = (1-e^{-t})(1-e^{-2t})$, so $f_{T'}(t) = e^{-t} + 2e^{-2t} - 3e^{-3t}$ for $t \geq 0$. Hence, $\mathbb{E}\,T' = 1 + 1/2 - 1/3 = 7/6$.

4. We compute $R_T(t) = e^{-3t}$ and $R_{T'}(t) = e^{-t} + e^{-2t} - e^{-3t}$.

The parallel system is more reliable than the series system, as already shown in Example 3.14.   △

▽ **Exercise 3.6 (The Laplace Distribution)** Let X_1 and X_2 be two i.i.d. variables with exponential density with parameter λ.

1. Determine the density of $Y = X_1 - X_2$.
2. Determine the density of $Z = Y + a$, where $a \in \mathbb{R}$.

Solution

1. The function $\psi : \mathbb{R}_+ \to \mathbb{R}_-$ defined by $\psi(x) = -x$ is a diffeomorphism, and hence, using (2.4) p. 68, the density of $-X_2$ is $f(x) = \lambda e^{\lambda x} \mathbb{1}_{\mathbb{R}_-}(x)$.

   According to Theorem 3.45, the density of Y is the convolution of the densities of X_1 and $-X_2$, that is

$$f_Y(y) = \lambda^2 \int_{\mathbb{R}} e^{\lambda(y-x)} \mathbb{1}_{[y,+\infty[}(x) e^{-\lambda x} \mathbb{1}_{\mathbb{R}_+}(x)\, dx,$$

   and hence, $f_Y(y) = \lambda e^{-\lambda |y|}/2$, for all $y \in \mathbb{R}$, because if $y > 0$,

$$e^{\lambda y} \int_y^{+\infty} \lambda^2 e^{-2\lambda x}\, dx = \frac{\lambda e^{-\lambda y}}{2},$$

   and if $y < 0$,

$$e^{\lambda y} \int_{\mathbb{R}_+} \lambda^2 e^{-2\lambda x}\, dx = \frac{\lambda e^{\lambda y}}{2}.$$

2. A translation yields the density of Z, that is $f_Z(z) = \lambda e^{-\lambda |z-a|}/2$, for all $z \in \mathbb{R}$.

The distribution is referred to as the symmetric exponential or Laplace distribution with parameters a and λ; see Fig. 3.5 for an example. It can be regarded as an ordinary exponential distribution together with its reflection about a.   △

▽ **Exercise 3.7 (Polar Coordinates and Rayleigh Distribution)** Let X and Y be any two random variables with joint density f. Let f_X and f_Y denote their respective marginal densities. Let (R, Θ) denote the polar coordinates of the point

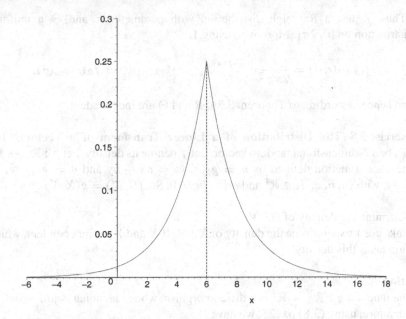

Fig. 3.5 Density of the Laplace distribution with $a = 6$ and $\lambda = 0.25$

with Cartesian coordinates (X, Y).

1. Determine the joint and marginal densities of R and Θ.
2. Suppose $(X, Y) \sim \mathcal{N}_2(0, \sigma^2 I_2)$. Show that R and Θ are independent.

Solution

1. The function $\phi : \mathbb{R}_+ \times [0, 2\pi[\to \mathbb{R}^* \times \mathbb{R}$ defined by $\phi(r, \theta) = (r\cos\theta, r\sin\theta)$ is a diffeomorphism whose Jacobian is r. So, using (3.8) p. 125, with $\psi = \phi^{-1}$, we get

$$f_{(R,\Theta)}(r, \theta) = rf(r\cos\theta, r\sin\theta)\mathbb{1}_{\mathbb{R}_+}(r)\mathbb{1}_{[0,2\pi[}(\theta).$$

Therefore,

$$f_R(r) = \int_0^{2\pi} rf(r\cos\theta, r\sin\theta)d\theta \quad \text{and} \quad f_\Theta(\theta) = \int_0^{+\infty} rf(r\cos\theta, r\sin\theta)dr.$$

2. For a standard Gaussian vector (X, Y), we get

$$f_R(r) = \frac{r}{\sigma^2}e^{-r^2/2\sigma^2}\mathbb{1}_{\mathbb{R}_+}(r) \quad \text{and} \quad f_\Theta(\theta) = \frac{1}{2\pi}\mathbb{1}_{[0,2\pi[}(\theta).$$

Thus, R has a Rayleigh distribution with parameter σ and Θ a uniform distribution on $[0, 2\pi[$. Moreover, using 1,

$$f_{(R,\Theta)}(r, \theta) = \frac{r}{2\pi\sigma^2} e^{-r^2/2\sigma^2} \mathbb{1}_{\mathbb{R}_+}(r) \mathbb{1}_{[0,2\pi[}(\theta) = f_R(r) f_\Theta(\theta),$$

and hence, according to Theorem 3.39, R and Θ are independent. $\triangle$

∇ **Exercise 3.8 (The Distribution of a Linear Transform of a Vector)** Let (X, Y) be a 2-dimensional random vector. Let f denote its density. Let $g : \mathbb{R}^2 \to \mathbb{R}^2$ be the linear function defined by $u = g_1(x, y) = ax + by$ and $v = g_2(x, y) = cx + dy$, with $(a, b, c, d) \in \mathbb{R}^4$ and $ad - bc \neq 0$. Set $(U, V) = g(X, Y)$.

1. Determine the density of (U, V).
2. Make use 1 to determine the density of $X + Y$. If X and Y are independent, which form takes this density?

Solution
1. The function $g : \mathbb{R}^2 \to \mathbb{R}^2$ is a diffeomorphism whose Jacobian is $|ad - bc|^{-1}$. Therefore, using (3.8) p. 125, we have

$$f_{(U,V)}(u, v) = \frac{1}{|ad - bc|} f\left(\frac{du - bv}{ad - bc}, \frac{-cu + av}{ad - bc}\right).$$

2. Setting $a = b = 1$ and $c = -d = 1$ yields $2f_{(U,V)}(u, v) = f((u + v)/2, (u - v)/2)$. The density of $X + Y$ is the marginal density of U, that is

$$h_U(u) = \int_{\mathbb{R}} f_{(U,V)}(u, v) dv = \int_{\mathbb{R}} f(y, u - y) dy.$$

If, moreover, X and Y are independent, then $f(x, y) = f_X(x) f_Y(y)$, and we check that $h_U(u) = f_X * f_Y(u)$. $\triangle$

∇ **Exercise 3.9 (The Bidimensional Pareto Distribution)** Let $\alpha > 0$ and $r < s$ be real numbers. Let (X, Y) be a 2-dimensional random vector with Pareto distribution, whose density is

$$f(x, y) = \frac{\alpha(\alpha + 1)(s - r)^\alpha}{(y - x)^{\alpha+2}} \mathbb{1}_{[-\infty,r]}(x) \mathbb{1}_{[s,+\infty[}(y).$$

1. Compute the moments of $Y - X$, when they are finite.
2. Determine the distribution of $U = s - X$, and then of $V = Y - r$.
3. Compute the expectation and the variance of X and of Y.
4. Compute the covariance of X and Y.

Solution

1. Let $k \in \mathbb{N}^*$. For $k < \alpha$, we compute

$$\mathbb{E}\left[(Y - X)^k\right] = \int_{-\infty}^{r} \int_{s}^{+\infty} \frac{\alpha(\alpha + 1)(s - r)^\alpha}{(y - x)^{\alpha + 2 - k}} \, dx dy$$

$$= \alpha(\alpha + 1)(s - r)^\alpha \int_{s}^{+\infty} \int_{-\infty}^{r} \frac{1}{(y - x)^{\alpha + 2 - k}} \, dx dy.$$

Since

$$\int_{-\infty}^{r} \frac{1}{(y - x)^{\alpha + 2 - k}} \, dx = \int_{-\infty}^{y - r} \frac{-1}{u^{\alpha + 2 - k}} \, du = \frac{(y - r)^{-\alpha - 1 + k}}{(\alpha + 1 - k)},$$

we get

$$\mathbb{E}\left[(Y - X)^k\right] = \frac{\alpha(\alpha + 1)(s - r)^k}{[\alpha - (k - 1)](\alpha - k)}.$$

2. For any bounded Borel function, we have

$$\mathbb{E}\left[h(s - X)\right] = \int_{s-r}^{+\infty} h(u) \frac{\alpha(s - r)^\alpha}{u^{\alpha + 1}} \, du.$$

According to Exercise 2.4, $U = s - X$ has a Pareto distribution with parameters α and $s - r$. Symmetrically, we get that V has the same distribution.

3. We know by Exercise 2.4, that for $\alpha > 1$,

$$\mathbb{E}\left(s - X\right) = \mathbb{E}\left(Y - r\right) = \frac{\alpha(s - r)}{\alpha - 1},$$

and for $\alpha > 2$,

$$\mathrm{Var}\left(s - X\right) = \mathrm{Var}\left(Y - r\right) = \frac{\alpha(s - r)^2}{(\alpha - 2)(\alpha - 1)}.$$

Moreover, $\mathbb{E}\left(s - X\right) = s - \mathbb{E}X$; hence, $\mathbb{E}X = \alpha(r - s)/(\alpha - 1)$ and $\mathrm{Var}\,X = \mathrm{Var}\left(s - X\right)$. Similarly, $\mathbb{E}Y = \alpha(s - r)/(\alpha - 1)$ and $\mathrm{Var}\,Y = \mathrm{Var}\left(Y - r\right)$.

4. We can write for $\alpha > 2$,

$$\mathbb{E}\left[(Y - X)^2\right] =$$

$$= \mathbb{E}\left(\left[(Y - \mathbb{E}Y) + \mathbb{E}\left(Y - X\right) - (X - \mathbb{E}X)\right]^2\right)$$

$$= \mathbb{E}\left[(Y - \mathbb{E}Y)^2 + \left[\mathbb{E}\left(Y - X\right)\right]^2 + 2\mathbb{E}\left(Y - \mathbb{E}Y\right)\mathbb{E}\left(Y - X\right)\right.$$

$$\left. + \mathbb{E}\left[(X - \mathbb{E}X)^2\right] - 2\mathbb{E}\left[(Y - \mathbb{E}Y)(X - \mathbb{E}X)\right] - 2\mathbb{E}\left(X - \mathbb{E}X\right)\mathbb{E}\left(Y - X\right).\right.$$

Setting $Z = X - Y$, we obtain

$$\mathbb{E}(Z^2) = \mathrm{Var}\, Y + (\mathbb{E}\, Z)^2 + \mathrm{Var}\, X - 2\mathbb{C}\mathrm{ov}\,(X, Y)$$
$$= 2\mathrm{Var}\, X + (\mathbb{E}\, Z)^2 - 2\mathbb{C}\mathrm{ov}\,(X, Y),$$

from which it follows that

$$\mathbb{C}\mathrm{ov}\,(X, Y) = \mathrm{Var}\, X + \frac{1}{2}[(\mathbb{E}\, Z)^2 - \mathbb{E}(Z^2)] = -\frac{(s - r)^2}{(\alpha - 1)^2(\alpha - 2)}.$$

$$\triangle$$

∇ **Exercise 3.10 (Functions of Order Statistics)**   Let $X_1, \ldots, X_n$ be i.i.d. variables with density f.

1. Determine the density of the pair $(X_{(1)}, X_{(n)})$.
2. Suppose the common distribution of $X_1, \ldots, X_n$ is the uniform distribution on $[0, 1]$. Determine the distribution of the range $E_n = X_{(n)} - X_{(1)}$.

Solution
1. The distribution function of $(X_{(1)}, X_{(n)})$ is for $x < y$,

$$F_{(1,n)}(x, y) = \mathbb{P}(X_{(1)} \leq x, X_{(n)} \leq y) = \mathbb{P}(X_{(n)} \leq y) - \mathbb{P}(X_{(1)} \geq x, X_{(n)} \leq y).$$

Since $\mathbb{P}(X_{(n)} \leq y) = F(y)^n$ for all $y \in \mathbb{R}$, and according to (3.4) p. 116,

$$\mathbb{P}(X_{(1)} \geq x, X_{(n)} \leq y) = \mathbb{P}\Big[\bigcap_{i=1}^{n}(x \leq X_i \leq y)\Big] = \prod_{i=1}^{n} \mathbb{P}(x \leq X_i \leq y)$$
$$= [F(y) - F(x)]^n.$$

The density

$$f_{(1,n)}(x, y) = \frac{\partial}{\partial x \partial y} F_{(1,n)}(x, y) = n(n-1) f(x) f(y) [F(y) - F(x)]^{n-2} \mathbb{1}_{\{x<y\}}(x, y)$$

follows, where $\{x < y\} = \{(x, y) \in \mathbb{R}^2 : x < y\}$.
2. Therefore,

$$f_{(1,n)}(x, y) = \frac{\partial^2}{\partial y \partial x} F_{(1,n)}(x, y)$$

$$= n(n-1)(y-x)^{n-2} \mathbb{1}_{\{0<x<y<1\}}(x, y).$$

The density of $(E_n, X_{(n)})$ follows by the linear change of variables ($u = y - x$, $v = y$) and applying Exercise 3.8, that is

$$f_{(E_n, X_{(n)})}(u, v) = n(n-1)u^{n-2}\mathbb{1}_{\{0<u<v<1\}}(u, v).$$

Finally, the marginal density of E_n is

$$f_{E_n}(u) = \int_{\mathbb{R}} n(n-1)u^{n-2}\mathbb{1}_{\{0<u<v<1\}}(u, v)dv = \mathbb{1}_{[-1,1]}(u)n(n-1)u^{n-2}(1-u).$$

Note that more generally $X_{(k)} - X_{(k-1)} \sim \mathcal{U}(0, 1)$, for $1 < k \leq n$. $\triangle$

∇ **Exercise 3.11 (Functions of a Gaussian Vector)** Let $a \in \mathbb{R}^* \setminus \{1\}$. Let $X = (X_0, \ldots, X_d)'$ be a $(d+1)$-dimensional Gaussian vector such that $X_i \sim \mathcal{N}(0, 1)$ for all i and

$$\Gamma_X = \begin{pmatrix} 1 & a & a & \cdots & a \\ a & 1 & a^2 & \cdots & a^2 \\ a & a^2 & 1 & a^2 & a^2 \\ \vdots & \vdots & \ddots & \ddots & \vdots \\ a & a^2 & & a^2 & 1 & a^2 \\ a & a^2 & \cdots & & a^2 & 1 \end{pmatrix}.$$

1. Determine the distribution of $Y = (X_0, Y_1, \ldots, Y_d)'$ defined by $X_i = aX_0 + \sqrt{1-a^2}Y_i$, of $i \in [\![1, d]\!]$.
2. Write X in terms of Y.
3. (a) Determine the distribution of $S_d = (X_1 + \cdots + X_d)/d$.
   (b) Set $Z = S_d/X_0$. Determine the distribution of $\sqrt{d}(Z - a)/\sqrt{1-a^2}$ and finally of Z.

Solution
1. Since $Y_i = (X_i - aX_0)/\sqrt{1-a^2}$, the vector Y is Gaussian as a linear transform of the Gaussian vector X. It is clearly centered, and we compute $\mathbb{E}(Y_i^2) = (1 + a^2 - 2a^2)/(1-a^2) = 1$, so $Y_i \sim \mathcal{N}(0, 1)$. Moreover,

$$\text{Cov}(Y_i, X_0) = \mathbb{E}(Y_i X_0) = [\mathbb{E}(X_i X_0) - 2a\mathbb{E}(X_0^2)]/(1-a^2) = 0, \quad i \in [\![1, d]\!],$$

and for $i \neq j$,

$$\mathbb{Cov}(Y_i, Y_j) = \mathbb{E}(Y_i Y_j) = \frac{1}{1-a^2}\mathbb{E}[(X_i - aX_0)(X_j - aX_0)]$$

$$= \frac{a^2 - 2a^2 + a^2}{\sqrt{1-a^2}} = 0.$$

According to Corollary 3.67, the random vector Y is a standard Gaussian vector.
2. We have $X = MY$, where

$$
M = \begin{pmatrix}
1 & 0 & \cdots & \cdots & 0 \\
a & \sqrt{1-a^2} & \ddots & & \vdots \\
a & 0 & \sqrt{1-a^2} & \ddots & \vdots \\
\vdots & \vdots & \ddots & \ddots & 0 \\
a & 0 & \cdots & 0 & \sqrt{1-a^2}
\end{pmatrix}.
$$

3.(a) We can write

$$
S_d = aX_0 + \frac{\sqrt{1-a^2}}{d}(Y_1 + \cdots + Y_d),
$$

and hence, $S_d \sim \mathcal{N}(0, a^2 + 1 - a^2/d)$.

(b) We have

$$
\sqrt{\frac{d}{1-a^2}}(Z - a) = \frac{Y_1 + \cdots + Y_d}{\sqrt{d}} \times \frac{1}{X_0}.
$$

According to Example 3.58, this quotient of two standard independent Gaussian variables has the Cauchy distribution $\mathcal{C}(1)$. Therefore, the density of Z is

$$
f_Z(z) = \frac{1}{\pi} \frac{\sqrt{d(1-a)^2}}{1 - a^2 + d(z-a)^2},
$$

thanks to the linear change of variable $z = \sqrt{1-a^2}t/\sqrt{d} + a$ and Example 2.59.   △

▽ **Exercise 3.12 (Maximum of Entropy Under Constraints)** The Shannon entropy of any n-dimensional random vector $X = (X_1, \ldots, X_n)$ with density f is the entropy of its distribution, that is $\mathbb{S}(X_1, \ldots, X_n) = \mathbb{E}[-\log f(X_1, \ldots, X_n)]$.

1. Compute the entropy of a non-degenerated Gaussian vector $\widetilde{X} \sim \mathcal{N}_n(0, \Gamma)$.
2. Show that among the centered n-dimensional random vectors whose variance–covariance matrix is some given Γ, the maximum of entropy is obtained for the Gaussian vector $\widetilde{X}$.

Solution

1. According to Proposition 3.72, the density of $\widetilde{X}$ is

$$f_{\widetilde{X}}(x) = (2\pi)^{-n/2}(\det \Gamma)^{-1/2} \exp[-\frac{1}{2}x'\Gamma^{-1}x], \quad x \in \mathbb{R}^n.$$

Hence, its entropy is

$$\mathbb{S}(\widetilde{X}) = \frac{1}{2}\Big[n\log(2\pi) + \log(\det \Gamma) + \int_{\mathbb{R}^n} x'\Gamma^{-1}x f_{\widetilde{X}}(x)\, dx\Big].$$

Let $\text{Tr}M = \sum_{i=1}^{n} M_{ii}$ denote the trace of any n-dimensional matrix M. We have

$$\int_{\mathbb{R}^n} x'\Gamma^{-1}x f_{\widetilde{X}}(x)\, dx = \mathbb{E}\,(\widetilde{X}'\Gamma^{-1}\widetilde{X}) = \mathbb{E}\,[\text{Tr}(\widetilde{X}'\widetilde{X}\Gamma^{-1})]$$

$$= \text{Tr}[\mathbb{E}\,(\widetilde{X}'\widetilde{X})\Gamma^{-1}] = \text{Tr}(\Gamma\Gamma^{-1}) = n.$$

Therefore, $\mathbb{S}(\widetilde{X}) = \log[\sqrt{(2\pi e)^n \det \Gamma}]$.

2. We can write

$$\mathbb{S}(X) = \int_{\mathbb{R}^n} f(x)\log\frac{f_{\widetilde{X}}(x)}{f(x)}\, dx - \int_{\mathbb{R}^n} f(X)\log f_{\widetilde{X}}(x)\, dx = (i) + (ii).$$

On the one hand, since $\log x \le x - 1$ for all positive x, we obtain

$$0 \le (i) \le \int_{\mathbb{R}^n} f(x)\Big[\frac{f_{\widetilde{X}}(x)}{f(x)} - 1\Big]dx = 0.$$

On the other hand,

$$(ii) = \frac{1}{2}\Big[\log(2\pi) + \log(\det \Gamma) + \int_{\mathbb{R}^n} x'\Gamma^{-1}x f(x)\, dx\Big].$$

Since $\mathbb{E}\,(X'X) = \Gamma$, we obtain in the same way as in 1,

$$\int_{\mathbb{R}^n} x'\Gamma^{-1}x f(x)\, dx = \text{Tr}(\Gamma\Gamma^{-1}) = n.$$

Thus, $(ii) = \mathbb{S}(\widetilde{X})$ and the conclusion follows. $\triangle$

Random Sequences

4

This chapter investigates the foundations of stochastic topology. The main types of convergence of random sequences are defined and compared: almost sure, in mean, quadratic mean, probability, distribution. Different large numbers laws and central limit theorems—doubtless the most remarkable results of probability theory—are presented. Some hints on stochastic simulation are also given.

4.1 Enumerable Sequences

The notion of infinite product probability spaces is required for defining random sequences.

Definition 4.1 Let $(\Omega, \mathcal{F}, \mathbb{P})$ be a probability space. Let $\Omega^{\mathbb{N}}$ denote the set of all sequences of elements of Ω and π_i the projection on the i-th coordinate, for $i \in \mathbb{N}$. The infinite product measurable space $(\Omega, \mathcal{F})^{\mathbb{N}}$ is the measurable space $(\Omega^{\mathbb{N}}, \sigma(\pi_i^{-1}(\mathcal{F}), i \in \mathbb{N}))$, where

$$\sigma(\pi_i^{-1}(\mathcal{F}), i \in \mathbb{N}) = \sigma(\{A_0 \times \cdots \times A_n \times \Omega \times \Omega \times \cdots : A_i \in \mathcal{F}, i \in [\![1, n]\!], n \in \mathbb{N}\})$$

is the σ-algebra generated by the infinite cylinders constructed on the rectangles of $\mathcal{F}^{\otimes n}$ for $n \in \mathbb{N}$.

Theorem 4.2 (Kolmogorov) *For any probability space $(\Omega, \mathcal{F}, \mathbb{P})$, a unique probability, denoted by $\mathbb{P}^{\otimes \mathbb{N}}$, is defined on $(\Omega, \mathcal{F})^{\mathbb{N}}$ by setting*

$$\mathbb{P}^{\otimes \mathbb{N}}(A_0 \times \cdots \times A_n \times \Omega \times \Omega \times \cdots) = \prod_{i=0}^{n} \mathbb{P}(A_i), \quad (A_0, \ldots, A_n) \in \mathcal{F}^n.$$

The resulting probability space is referred to as the infinite product space and is denoted by $(\Omega, \mathcal{F}, \mathbb{P})^{\mathbb{N}}$; see Example 2.44 for an empirical application to the negative binomial distribution.

A real random sequence (or sequence of real random variables) is a stochastic element taking values in $\mathbb{R}^{\mathbb{N}} = \{(x_0, x_1, x_2, \dots) : x_n \in \mathbb{R}, n \in \mathbb{N}\}$.

Let (X_n) be a random sequence defined on $(\Omega, \mathcal{F}, \mathbb{P})^{\mathbb{N}}$. The sequence of real numbers $n \to X_n(\omega)$, for a fixed $\omega \in \Omega$, is referred to as a trajectory of the random sequence. Each X_n is a random variable on $(\Omega, \mathcal{F}, \mathbb{P})$. The image probability of $\mathbb{P}^{\otimes \mathbb{N}}$ by (X_n) is, by definition, characterized by its values at infinite cylinders, so the distribution of (X_n) is characterized by $\mathbb{P}_{(X_{i_1}, \dots, X_{i_n})}$ for all $i_1, \dots, i_n$ and $n \in \mathbb{N}$. In other words, the distribution of a random sequence is determined by its (marginal) finite distributions. The next result follows immediately.

Proposition 4.3 *Two random sequences (X_n) and (Y_n) have the same distribution if and only if*

$$(X_0, \dots, X_n) \sim (Y_0, \dots, Y_n), \quad n \in \mathbb{N}.$$

▷ *Example 4.4 (Gaussian Random Sequences)* Let (X_n) be a random sequence such that $(X_{n_1}, \dots, X_{n_i})$ is a Gaussian vector for all $i \in \mathbb{N}^*$ and $n_1, \dots, n_i$. The sequence (X_n) is called a Gaussian sequence.   ◁

The following definitions are stated here for future use.

Definition 4.5 Let (X_n) be any random sequence. The sequence is said to be:

* Integrable (L^p for $p \in \mathbb{N}^*$, nonnegative, discrete, ...) if each of the random variables X_n is integrable (L^p for $p \in \mathbb{N}^*$, nonnegative, discrete, etc.)
* Equi-integrable if $\sup_{n \geq 0} \mathbb{E}\left[\mathbb{1}_{(|X_n| \geq N)} |X_n|\right] \longrightarrow 0, \ N \to +\infty$
* L^p-bounded if $\sup_{n \geq 0} \|X_n\|_p < +\infty$

Several properties of equi-integrable sequences are illustrated in Exercise 4.10.

▷ *Example 4.6 (An Equi-integrable Sequence)* Let (X_n) be a random sequence. Suppose some $X \in L^1$ exists such that $|X_n| \leq X$ a.s. for all $n \geq 0$. Then

$$\mathbb{E}\left[\mathbb{1}_{(|X_n| \geq N)} |X_n|\right] \leq \mathbb{E}\left[\mathbb{1}_{(X \geq N)} X\right], \quad n \geq 0.$$

Since $\mathbb{E}\left[\mathbb{1}_{(X \geq N)} X\right]$ converges to 0 when N tends to infinity, (X_n) is an equi-integrable random sequence.   ◁

The entropy between times 1 and n of a random sequence $X = (X_n)$ is the entropy of its n-dimensional marginal distribution, that is:

$$\mathbb{H}_n(X) = - \sum_{(x_1,\ldots,x_n) \in E^n} \mathbb{P}(X_1 = x_1, \ldots, X_n = x_n) \log \mathbb{P}(X_1 = x_1, \ldots, X_n = x_n),$$

if X_n takes values in a discrete set $E \subset \mathbb{R}$ for all n, and

$$\mathbb{H}_n(X) = - \int_{\mathbb{R}^n} f_n^X(x_1, \ldots, x_n) \log f_n^X(x_1, \ldots, x_n) dx_1 \ldots dx_n,$$

if X_n takes values in $\mathbb{R}$ and if the distribution of the vector $(X_1, \ldots, X_n)$ has a density f_n^X with respect to the Lebesgue measure on $\mathbb{R}^n$ for all n.

Definition 4.7 Let $X = (X_n)$ be a random sequence. If

$$\frac{\mathbb{H}_n(X)}{n} \longrightarrow \mathbb{H}(X), \quad n \to +\infty,$$

the limit $\mathbb{H}(X)$ is called the (Shannon) entropy rate of the sequence.

Note that if $\mathbb{H}(X)$ is finite, then $\mathbb{H}(X) = \inf \mathbb{H}_n(X)/n$.

▷ *Example 4.8 (Maximum Entropy Rate)* Among all random sequences with the same sequence of covariance matrices, the maximum entropy rate is obtained for a Gaussian sequence. This can be deduced from Exercise 3.12 by letting n tend to infinity. ◁

4.1.1 Sequences of Events

The next defined inferior and superior limit events are a key to investigating sequences of events.

Definition 4.9 Let $(\Omega, \mathcal{F}, \mathbb{P})$ be a probability space. Let (A_n) be a sequence of events.

- The set of all the $\omega \in \Omega$ belonging to infinitely many A_n, that is:

$$\overline{\lim} A_n = \bigcap_{n \geq 0} \bigcup_{k \geq n} A_k$$

$$= \left\{ \omega \in \Omega : \sum_{n \geq 0} \mathbb{1}_{A_n}(\omega) = +\infty \right\} = \{ \omega \in \Omega : \forall n, \exists k \geq n, \; \omega \in A_k \},$$

is called the superior limit of (A_n).

• The set of all the $\omega \in \Omega$ belonging to all A_n but perhaps a finite number, that is:

$$\underline{\lim} \, A_n = \bigcup_{n \geq 0} \bigcap_{k \geq n} A_k$$

$$= \left\{ \omega \in \Omega : \sum_{n \geq 0} \mathbb{1}_{\overline{A_n}}(\omega) < +\infty \right\} = \{ \omega \in \Omega : \exists n, \forall k \geq n, \, \omega \in A_k \},$$

is called the inferior limit of (A_n).

Properties of Superior and Inferior Limits of Events

1. $\underline{\lim} \, A_n \subset \overline{\lim} \, A_n$.
2. A sequence (A_n) is said to converge if $\underline{\lim} \, A_n = \overline{\lim} \, A_n$. In particular, if (A_n) is non-decreasing, then $\overline{\lim} \, A_n = \underline{\lim} \, A_n = \cup_{n \geq 0} A_n$ ($\cap_{n \geq 0} A_n$). Indeed, if the sequence is non-decreasing, then ω belongs to $\cup_{n \geq 0} A_n$ if and only if it belongs to—at least—one A_{n_0}; then it belongs to infinitely many A_n, all but perhaps the first $n_0 - 1$ ones. Similarly, if the sequence is non-increasing, then $\overline{\lim} \, A_n = \underline{\lim} \, A_n = \cap_{n \geq 0} A_n$.
3. If the A_n are pairwise disjoint, then $\overline{\lim} \, A_n = \underline{\lim} \, A_n = \emptyset$.
4. Let (A_n) and (B_n) be two sequences of events. We have $\overline{\lim}(A_n \cup B_n) = \overline{\lim} \, A_n \cup \overline{\lim} \, B_n$. On the contrary, for intersection, we only have

$$\overline{\lim} \, A_n \cap \underline{\lim} \, B_n \subset \overline{\lim}(A_n \cap B_n) \subset \overline{\lim} \, A_n \cap \overline{\lim} \, B_n;$$

the proof uses the latter form of the definition.
5. $\underline{\lim} \, A_n$ and $\overline{\lim} \, \overline{A_n}$ are complementary in Ω. Indeed,

$$\overline{\underline{\lim} \, A_n} = \overline{\bigcup_{n \geq 0} \bigcap_{k \geq n} A_k} = \bigcap_{n \geq 0} \overline{\bigcap_{k \geq n} A_k} = \bigcap_{n \geq 0} \bigcup_{k \geq n} \overline{A_k} = \overline{\lim} \, \overline{A_n}.$$

6. $\mathbb{P}(\underline{\lim} \, A_n) \leq \underline{\lim} \, \mathbb{P}(A_n) \leq \overline{\lim} \, \mathbb{P}(A_n) \leq \mathbb{P}(\overline{\lim} \, A_n)$. The first inequality is obtained by considering the non-decreasing sequence $C_n = \cap_{k \geq n} A_k$ and the second by considering the non-increasing sequence $B_n = \cup_{k \geq n} A_k$.

4.1.2   Independence of Sequences

The notion of independence of (enumerable) sequences of events or of random variables is based on the independence of collections of σ-algebras.

Definition 4.10 Let $(\mathcal{G}_n)$ be a sequence of σ-algebras included in $\mathcal{F}$. The sequence is said to be independent if all finite subcollections are constituted of (mutually) independent σ-algebras, that is, if for all $\{\mathcal{G}_{n_k} : k \in [\![1, N]\!]\}$ drawn from $(\mathcal{G}_n)$,

$$\mathbb{P}(B_{n_1} \cap \cdots \cap B_{n_N}) = \mathbb{P}(B_{n_1}) \ldots \mathbb{P}(B_{n_N}), \quad B_{n_1} \in \mathcal{G}_{n_1}, \ldots, B_{n_N} \in \mathcal{G}_{n_N}.$$

The σ-algebra generated by a sequence of σ-algebras $(\mathcal{G}_n)$ is the smallest containing all sets $B_n \in \mathcal{G}_n$ for all n. The next result is a corollary of Proposition 1.47.

Proposition 4.11 (Associativity Principle) *Let $(\mathcal{G}_n)$ be an independent sequence of σ-algebras included in $\mathcal{F}$. Let I_1 and I_2 be two disjoint subsets of $\mathbb{N}$. Then, the two σ-algebras generated by $(\mathcal{G}_n)_{n \in I_1}$ and by $(\mathcal{G}_n)_{n \in I_2}$ are independent.*

Definition 4.12 Let $(\mathcal{G}_n)$ be a sequence of σ-algebras included in $\mathcal{F}$. The σ-algebra

$$\mathcal{L} = \bigcap_{n \geq 0} \sigma\left(\bigcup_{k \geq n} \mathcal{G}_k\right)$$

is called the asymptotic σ-algebra of $(\mathcal{G}_n)$.

The events of $\mathcal{L}$ are referred to as asymptotic events. When the sequence of the indices is regarded as a sequence of times, the asymptotic events depend only on the events posterior to time n, and this for any integer n.

Theorem 4.13 (Kolmogorov's Zero–One Law) *Let $(\mathcal{G}_n)$ be an independent sequence of σ-algebras included in $\mathcal{F}$. All asymptotic events of $(\mathcal{G}_n)$ are either almost sure or null.*

Proof Let us set $\mathcal{F}_n = \sigma(\cup_{k \geq n} \mathcal{G}_k)$. For any integer n, the σ-algebra $\mathcal{G}_{n-1}$ is independent of $(\mathcal{G}_k)_{k \geq n}$. So, by the associativity principle, it is independent of $\mathcal{F}_n$, and hence also of $\mathcal{L} = \cap_{n \geq 0} \mathcal{F}_n$.

Since $\mathcal{L}$ is independent of $\mathcal{G}_k$ for any integer k, it is also independent of $\mathcal{F}_n = \sigma(\cup_{k \geq n} \mathcal{G}_k)$ for all n, again by the associativity principle. Hence it is independent of $\mathcal{L} = \cap_{n \geq 0} \mathcal{F}_n$. Therefore, every event L of $\mathcal{L}$ is independent of itself, $\mathbb{P}(L) = \mathbb{P}(L \cap L) = \mathbb{P}(L)\mathbb{P}(L)$, that is, $\mathbb{P}(L) = 0$ or 1. $\square$

As for finite collections, the independence of enumerable sequences of events is equivalent to the independence of their generated σ-algebras.

Definition 4.14 A sequence of events (A_n) is said to be independent if the corresponding sequence of generated σ-algebras is independent, that is, if every finite subsequence is constituted of independent events.

The following condition can be proven to be sufficient:

$$\mathbb{P}(A_1 \cap \cdots \cap A_n) = \prod_{i=1}^{n} \mathbb{P}(A_i), \quad n \in \mathbb{N}.$$

Then, considering the decreasing sequence $B_n = \cap_{k \leq n} A_n$ for $n \in \mathbb{N}$ allows to considering the limit that yields

$$\mathbb{P}\Big(\bigcap_{n \geq 0} A_n\Big) = \prod_{n \geq 0} \mathbb{P}(A_n).$$

Theorem 4.15 (Borel–Cantelli Lemma) *Let (A_n) be a sequence of events.*

1. If $\sum_{n \geq 0} \mathbb{P}(A_n) < +\infty$, then $\mathbb{P}(\overline{\lim} A_n) = 0$.
2. If $\sum_{n \geq 0} \mathbb{P}(A_n) = +\infty$ and if (A_n) is independent, then $\mathbb{P}(\overline{\lim} A_n) = 1$.

Therefore, $\overline{\lim} A_n$ is either an almost sure or a null event, whatever be (A_n). Actually, this is a particular case of the Kolmogorov zero–one law; due to its paramount importance in applications, we prove it directly.

Proof

1. By definition, $\overline{\lim} A_n = \cap_{n \geq 0} (\cup_{k \geq n} A_k)$, so

$$0 \leq \mathbb{P}(\overline{\lim} A_n) \leq \mathbb{P}\Big(\bigcup_{k \geq n} A_k\Big) \leq \sum_{k \geq n} \mathbb{P}(A_k), \quad n \in \mathbb{N}.$$

The series with general term $\mathbb{P}(A_n)$ is convergent, so $\sum_{k \geq n} \mathbb{P}(A_k)$ tends to zero when n tends to infinity, and the result follows.
2. We know that $\overline{\overline{\lim}} A_n = \underline{\lim} \overline{A_n} = \cup_{n \geq 0} \cap_{k \geq n} \overline{A_k}$.

Set $B_n = \cap_{k \geq n} \overline{A_n}$ for $n \in \mathbb{N}$. Since the sequence (B_n) is non-decreasing, we have $\mathbb{P}(\underline{\lim} \overline{A_n}) = \mathbb{P}(\cup_{n \geq 0} B_n) = \lim_{n \to +\infty} \mathbb{P}(B_n)$. The event B_n can be decomposed into an intersection of independent events, under the form

$$B_n = \Big(\bigcap_{k=n}^{N-1} \overline{A_k}\Big) \bigcap \Big(\bigcap_{k \geq N} \overline{A_k}\Big), \quad N \geq n+1,$$

and hence $\mathbb{P}(B_n) = \mathbb{P}(\cap_{n \leq k < N} \overline{A_k}) \mathbb{P}(\cap_{k \geq N} \overline{A_k})$.

Since the A_k are independent, $\mathbb{P}(\cap_{n \geq k \geq N-1} \overline{A_k}) = \prod_{k=n}^{N-1} \mathbb{P}(\overline{A_k})$ and $\mathbb{P}(\cap_{k \geq N} \overline{A_k}) \leq 1$, so $\mathbb{P}(B_n) \leq \prod_{k=n}^{N-1} [1 - \mathbb{P}(A_k)]$. Since $1 - x \leq e^{-x}$ for all nonnegative x, we have $1 - \mathbb{P}(A_k) \leq \exp[-\mathbb{P}(A_k)]$ for $k \in \mathbb{N}$, from which it follows that

$$0 \leq \mathbb{P}(B_n) \leq \exp\left[-\sum_{k=n}^{N-1} \mathbb{P}(A_k)\right].$$

The series with general term $\mathbb{P}(A_k)$ diverges, and hence $\sum_{k=n}^{N-1} \mathbb{P}(A_k)$ tends to infinity when N tends to infinity. Therefore $\mathbb{P}(B_n) = 0$, for all n.

Finally, $\mathbb{P}(\underline{\lim}\, \overline{A_n}) = 0 = 1 - \mathbb{P}(\overline{\lim}\, A_n)$ and the result follows. $\qquad\square$

Definition 4.16 A random sequence (X_n) is said to be independent if the sequence of generated σ-algebras $(\sigma(X_n))$ is independent, that is if every finite subsequence is constituted of independent variables.

Note that the independence of $X_0, \ldots, X_n$ for all n is sufficient.

An independent sequence of random variables with the same distribution P is referred to as i.i.d. (independent and identically distributed) with (common) distribution P. We will say that two sequences are independent if all their finite subsequences are independent and that a random variable X and an independent random sequence (X_n) are independent if the sequence $(X, X_0, X_1, \ldots)$ is an independent sequence.

▷ *Example 4.17 (Entropy Rate of i.i.d. Sequences)* The entropy rate of an i.i.d. random sequence with distribution P is equal to the entropy of P. Indeed, an easy proof by induction shows that the entropy of the sequence up to any time n is n times its entropy at time 1. ◁

A sequence of σ-algebras $(\mathcal{F}_n)$ is said to be non-decreasing if $\mathcal{F}_n \subset \mathcal{F}_{n+1}$ for all n.

Definition 4.18 Let $(\Omega, \mathcal{F}, \mathbb{P})$ be a probability space. A non-decreasing sequence of σ-algebras $\mathbf{F} = (\mathcal{F}_n)$ included in $\mathcal{F}$ is called a filtration (or history) of $\mathcal{F}$. We will denote by $\mathcal{F}_\infty$ the generated σ-algebra $\sigma(\cup_{n \geq 0}\mathcal{F}_n)$ for any filtration $(\mathcal{F}_n)$.

In particular, if (X_n) is a random sequence, then the sequence of σ-algebras $(\sigma(X_k, 0 \leq k \leq n))$ is called the natural filtration (or internal history) of the sequence, and $\cap_{n \geq 0}\sigma(X_n, X_{n+1}, \ldots)$ is the tail σ-algebra of the sequence.

The next result appears as a corollary of Kolmogorov's zero–one law.

Corollary 4.19 *Every event of the tail σ-algebra of an independent random sequence has probability either zero or one.*

▷ *Example 4.20 (Series of i.i.d. Sequences)* Let (X_n) be an independent random sequence. Then the series with general term X_n either diverges or converges a.s. Indeed, the event "$\sum_{n \geq 0} X_n$ converges" belongs to the tail σ-algebra of (X_n). ◁

Definition 4.21 Let $\mathbf{F} = (\mathcal{F}_n)$ be a filtration of $\mathcal{F}$. A random sequence (X_n) is said to be $\mathbf{F}$-adapted if X_n is $\mathcal{F}_n$-measurable, or $\mathbf{F}$-predictable if X_n is $\mathcal{F}_{n-1}$-measurable, for all n.

Randomly indexed random variables often appear in applications.

Definition 4.22 Let (S_n) be a random sequence defined on a probability space $(\Omega, \mathcal{F}, \mathbb{P})$. Let N be an a.s. finite random variable taking integer values defined on the same space. The so-called randomly indexed variable $\omega \rightarrow S_{N(\omega)}(\omega)$ is denoted by S_N.

Randomly indexed sums are especially considered, that is, $S_N = \sum_{i=1}^{N} X_i$, where $S_n = X_1 + \cdots + X_n$ for a random sequence (X_n).

Proposition 4.23 *Let (X_n) be an i.i.d. random sequence; let P denote its distribution, with distribution function F, expected value M, and variance s^2. Let N be an a.s. finite random variable independent of (X_n) and taking positive integer values; let m denote its expected value and σ^2 its variance. Then:*

1. *S_N has expected value mM and variance $s^2 m + M^2 \sigma^2$.*
2. *The distribution function of S_N is $F_{S_N}(x) = \sum_{n \geq 1} F^{*(n)}(x) \mathbb{P}(N = n)$.*
3. *The Laplace (characteristic function) of S_N is the composition of the generating function of N and of the Laplace (Fourier) transform of P.*
4. *If X_n takes only integer values, the generating function of S_N is the composition of the generating functions of N and P.*

Proof

1. We have $\mathbb{E}\, S_N = \mathbb{E}\left[\sum_{n \geq 1} \mathbb{1}_{(N=n)} S_n\right] = \sum_{n \geq 1} \mathbb{E}\left[\mathbb{1}_{(N=n)} S_n\right]$. Since N and S_n are independent, we get

$$\mathbb{E}\, S_N = \sum_{n \geq 1} \mathbb{P}(N = n) \mathbb{E}\, S_n = \sum_{n \geq 1} \mathbb{P}(N = n) n \mathbb{E}\, X_1 = (\mathbb{E}\, N)(\mathbb{E}\, X_1) = mM.$$

Since $\mathbb{E}\,(S_n^2) = n \mathbb{V}\text{ar}\, X_1 + n(\mathbb{E}\, X_1)^2$, we compute

$$\mathbb{E}\,(S_N^2) = \sum_{n \geq 1} \mathbb{P}(N = n) \mathbb{E}\,(S_n^2)$$

$$= s^2 \left[\sum_{n \geq 1} n \mathbb{P}(N = n)\right] + M^2 \left[\sum_{n \geq 1} n^2 \mathbb{P}(N = n)\right]$$

$$= s^2 \mathbb{E}\, N + M^2 \mathbb{E}\,(N^2).$$

Since $\mathbb{E}\,(N^2) = \sigma^2 + m^2$, we get $\mathbb{V}\text{ar}\, S_N = s^2 m + M^2 \sigma^2$.

2. This follows by Point 2 of Theorem 3.45, because

$$\mathbb{P}(S_N \leq s) = \sum_{n \geq 1} \mathbb{P}(S_n \leq s, N = n) = \sum_{n \geq 1} \mathbb{P}(S_n \leq s)\mathbb{P}(N = n).$$

3. According to Theorem 3.49 for i.i.d. variables, $\psi_{S_n}(t) = \psi_P(t)^n$, so

$$\psi_{S_N}(t) = \sum_{n \geq 1} \mathbb{E}\,[e^{-tS_n}\mathbb{1}_{(N=n)}] = \sum_{n \geq 1} \mathbb{E}\,(e^{-tS_n})\mathbb{P}(N = n)$$

$$= \sum_{n \geq 1} \psi_P(t)^n \mathbb{P}(N = n) = g_N(\psi_P(t)).$$

Similarly, for the characteristic function, $\phi_{S_N}(t) = g_N(\phi_P(t))$.

4. We have

$$\mathbb{P}(S_N = k) = \sum_{n \geq 1} \mathbb{P}(N = n, S_n = k) = \sum_{n \geq 1} \mathbb{P}(N = n)\mathbb{P}(S_n = k).$$

According to Theorem 3.49 again, $g_{S_n}(s) = g_P(s)^n$, so

$$g_{S_N}(s) = \sum_{k \geq 1} \mathbb{P}(S_N = k)s^k = \sum_{n \geq 1} \mathbb{P}(N = n) \sum_{k \geq 1} \mathbb{P}(S_n = k)s^k$$

$$= \sum_{n \geq 1} \mathbb{P}(N = n)g_{S_n}(s) = \sum_{n \geq 1} \mathbb{P}(N = n)g_P(s)^n,$$

or $g_{S_N}(s) = g_N(g_P(s))$.

□

Definition 4.24 Let (X_n) be an i.i.d. random sequence with distribution P. Let $N \sim \mathcal{P}(\lambda)$ be independent of (X_n).

The randomly indexed sum $S_N = \sum_{n=1}^{N} X_n$ is said to have a compound Poisson distribution, and we will write $S_N \sim \mathcal{CP}(\lambda, P)$.

▷ *Example 4.25 (Poisson and Compound Poisson Distribution)* Suppose that the number N of accidents per year in a certain type of factory has a Poisson distribution with parameter λ so that $g_N(s) = e^{\lambda(s-1)}$; see Example 2.65. Suppose that the probability of zero or one casualties per accident is p, so that $X_n \sim \mathcal{B}(p)$, and $g_{X_n}(s) = ps + 1 - p$; see Example 2.70.

Let us determine the distribution of the number of casualties per year. According to Proposition 4.23, $g_{S_N}(s) = e^{\lambda p(s-1)}$, and hence $S_N \sim \mathcal{P}(p\lambda)$. Therefore, the compound Poisson distribution $\mathcal{CP}(\lambda, \mathcal{B}(p))$ reduces to the ordinary Poisson distribution with parameter $p\lambda$.   ◁

A random walk is a type of random sequences, which is not i.i.d. but has i.i.d. increments.

Definition 4.26 Let (X_n) be an i.i.d. random sequence taking values in $\mathbb{Z}$. Let S_0 be a random variable also taking values in $\mathbb{Z}$ and independent of (X_n). The sequence (S_n) of random variables

$$S_n = S_0 + X_1 + \cdots + X_n, \quad n \geq 1,$$

is called a random walk.

If $X_n \sim \mathcal{B}(p)$ for all n, it is a Bernoulli random walk. The random walk is said to be simple when X_n takes only the values -1 and 1, and symmetric if $\mathbb{P}(X_n = -1) = \mathbb{P}(X_n = 1)$.

An example of a random walk is given above in Exercise 3.3.

▷ *Example 4.27 (A Periodically Tossed Coin)* Let (ζ_n) be an i.i.d. random sequence with distribution $\mathbb{P}(\zeta_n = 1) = p = 1 - \mathbb{P}(\zeta_n = -1)$. Let $s \in \mathbb{R}_+^*$ and $T \in \mathbb{N}^*$ be fixed. Setting $S_n = s \sum_{j=1}^n \zeta_{jT}$ defines a random walk taking the values $(2k - n)s$ for $k \in [\![1, n]\!]$. A geometrical interpretation is shown in Fig. 4.1. We compute

$$\mathbb{P}(S_n = (2k - n)s) = \binom{n}{k} p^k (1 - p)^{n-k}.$$

For instance, for $p = 1/2$, we get $\mathbb{E}\, S_n = 0$ and $\mathbb{V}\text{ar}\, S_n = ns^2$. Another particular case, for $s = T = 1$, is studied in Exercise 4.5.   ◁

4.2   Stochastic Convergence

The main types of convergence of random sequences are presented. Relations between them and numerous practical criteria of convergence are investigated.

4.2.1   Different Types of Convergence

All the random variables considered in this section are defined on a probability space $(\Omega, \mathcal{F}, \mathbb{P})$.

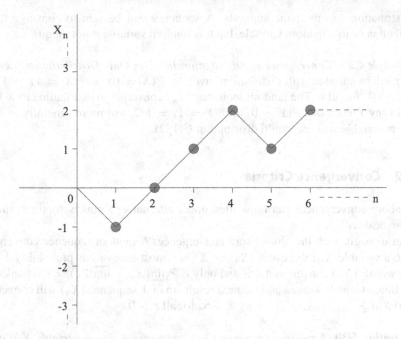

Fig. 4.1 A simple random walk trajectory

Definition 4.28 Let (X_n) be a random sequence, and let X be a random variable.

- $X_n \overset{\text{a.s.}}{\to} X$ (almost surely or with probability 1)
  if $\mathbb{P}[\{\omega \in \Omega : X_n(\omega) \to X(\omega)\}] = 1$
- $X_n \overset{\mathbb{P}}{\to} X$ (in probability)
  if for all $\varepsilon > 0$, $\mathbb{P}(|X_n - X| > \varepsilon)$ converges to zero when n tends to infinity
- $X_n \overset{\mathcal{D}}{\to} X$ (in distribution)
  if $\mathbb{E}[h(X_n)]$ converges to $\mathbb{E}[h(X)]$, when n tends to infinity, for any bounded continuous function $h : \mathbb{R} \to \mathbb{R}$
- $X_n \overset{L^p}{\to} X$ (in L^p-norm), where $0 < p \leq +\infty$,
  if $X_n \in L^p$ for all $n \in \mathbb{N}$, $X \in L^p$, and $\|X_n - X\|_p$ converges to zero when n tends to infinity

Note that all above convergences depend on the considered probability $\mathbb{P}$. The convergence in L^p-norm is the topological convergence of the space $L^p(\Omega, \mathcal{F}, \mathbb{P})$ equipped with the norm $\| \cdot \|_p$. The convergence in L^1-norm is also referred to as convergence in mean, and the convergence in L^2-norm as quadratic mean (or mean square) convergence.

The limit random variable X is unique for the a.s., probability, and L^p convergences. On the contrary, only distributions are involved for the convergence in distribution—as its name suggests. A sequence can be said to converge to a distribution or to a random variable, but this random variable is not unique.

▷ *Example 4.29 (Convergence in Distribution Involves Only Distributions)* Let X be a random variable with distribution given by $\mathbb{P}(X = 0) = \mathbb{P}(X = 1) = 1/2$. Set $X_n = X$ for all n. The random sequence (X_n) converges in distribution to X but also to any Y such that $\mathbb{P}(Y = 0) = \mathbb{P}(Y = 1) = 1/2$, and more generally to any random variable with Bernoulli distribution $\mathcal{B}(1/2)$. ◁

4.2.2 Convergence Criteria

The above convergences can be written under alternative forms, sometimes more easy to deal with.

Let us begin with the almost sure convergence. A random sequence converges a.s. to a variable X if the event $(X_n \to X)$ is almost sure or has probability 1. In other words, (X_n) converges a.s. if and only if $\mathbb{P}(\underline{\lim} X_n = \overline{\lim} X_n) = 1$. Thanks to both Borel–Cantelli lemma and the next result, an i.i.d. sequence (X_n) will converge a.s. to X if $\sum_{n \geq 0} \mathbb{P}(|X_n - X| > \varepsilon) < +\infty$ for all $\varepsilon > 0$.

Proposition 4.30 *A random sequence (X_n) converges a.s. to a variable X if and only if $\mathbb{P}(\overline{\lim} |X_n - X| > \varepsilon) = 0$ for all $\varepsilon > 0$.*

Proof We have

$$(X_n \longrightarrow X) =$$

$$= \{\omega \in \Omega : \forall N, \exists n \text{ such that if } k \geq n \text{ then } |X_k(\omega) - X(\omega)| < 1/N\}$$

$$= \bigcap_{N \geq 1} \bigcup_{n \geq 0} \bigcap_{k \geq n} (|X_k - X| < 1/N) = \bigcap_{N \geq 1} \underline{\lim_n} (|X_n - X| < 1/N).$$

The sequence of events $\underline{\lim}_n (|X_n - X| < 1/N)$ is decreasing in N. Their intersection has probability one if and only if these events have probability one for every N; in other words, if $\mathbb{P}(\underline{\lim}(|X_n - X| < \varepsilon) = 1$, or, by taking complements, if $\mathbb{P}(\overline{\lim} |X_n - X| > \varepsilon) = 0$, for all $\varepsilon > 0$. □

Convergence in L^p induces convergences of the sequences of moments.

Proposition 4.31 *Let (X_n) be an L^p random sequence and X an L^p random variable. If $X_n \overset{L^p}{\to} X$, then $\mathbb{E}(|X_n|^p) \to \mathbb{E}|X|^p$. Moreover, $\mathbb{E} X_n \to \mathbb{E} X$.*

Proof The first result is a consequence of Minkowski's inequality. The second is an immediate consequence of $X_n - X \leq |X_n - X|$ for all $n \geq 0$. □

As for other topological convergences, a Cauchy sequence is a convergent sequence for the almost sure, L^p, and in probability convergences. For instance, for the latter,

$$\text{if } \forall \varepsilon > 0, \forall a > 0, \exists N(a), \quad n, m > N(a) \Rightarrow \mathbb{P}(|X_n - X_m| > \varepsilon) < a,$$

then a random variable X exists such that $\mathbb{P}(|X_n - X| > \varepsilon)$ converges to zero.

Convergence in distribution for sequences of random variables appears as a property of strong convergence of their probability distributions.

Definition 4.32 Let (μ_n) be a sequence of finite measures on $\mathbb{R}^d$, and let μ be a finite measure on $\mathbb{R}^d$. The sequence (μ_n) is said to converge strongly to μ if, for all bounded continuous $h \colon \mathbb{R}^d \to \mathbb{R}$,

$$\int_{\mathbb{R}^d} h \, d\mu_n \to \int_{\mathbb{R}^d} h \, d\mu.$$

Note that the bounded continuous functions can be replaced by continuous or uniformly continuous functions with compact supports.

The following criteria of strong convergence of measures will lead to similar criteria in terms of convergence of random sequences.

Theorem 4.33 (Alexandrov) *Let (μ_n) be a sequence of finite measures on $\mathbb{R}^d$, and let μ be a finite measure on $\mathbb{R}^d$. Then (μ_n) converges strongly to μ if one of the four following equivalent conditions is satisfied:*

1. *$\int_{\mathbb{R}^d} h \, d\mu_n$ converges to $\int_{\mathbb{R}^d} h \, d\mu$ for all bounded continuous $h \colon \mathbb{R}^d \to \mathbb{R}$.*
2. *$\overline{\lim} \, \mu_n(F) \leq \mu(F)$ for all closed Borel sets F.*
3. *$\underline{\lim} \mu_n(O) \geq \mu(O)$ for all open Borel sets O.*
4. *$\mu_n(B)$ converges to $\mu(B)$ for all Borel sets B such that $\mu(\partial B) = 0$, where ∂B denotes the boundary of B in $\mathbb{R}^d$.*

Proof $1 \Rightarrow 2$ Let F be a closed Borel set, let $\varepsilon > 0$, and let $\delta > 0$. Set

$$F_\delta = \{x \in \mathbb{R}^d : d(x, F) \leq \delta\}. \tag{4.1}$$

This function is decreasing and converges to F when δ tends to zero. So for δ small enough, and since μ is finite, we have $\mu(F_\delta) \leq \mu(F) + \varepsilon$.

Let f be defined on $\mathbb{R}^d$ by $f(y) = g(d(y, F)/\varepsilon)$, where $g(t) = \mathbb{1}_{]-\infty,0[}(t) + (1-t)\mathbb{1}_{[0,1]}(t)$. Clearly, f is continuous, $f(y) = 1$ on F, and $f(y) = 0$ on $\overline{F}_\delta$, with $0 \leq f(y) \leq 1$ for all $y \in \mathbb{R}^d$. Moreover 1 says that $\int_{\mathbb{R}^d} f \, d\mu_n$ converges to $\int_{\mathbb{R}^d} f \, d\mu$. We have

$$\mu_n(F) = \int_F d\mu_n \leq \int_{\mathbb{R}^d} f \, d\mu_n$$

and

$$\int_{\mathbb{R}^d} f d\mu - \int_{F_\delta} f d\mu \le \mu(F_\delta) \le \mu(F) + \varepsilon.$$

Therefore,

$$\overline{\lim} \, \mu_n(F) \le \lim \int_{\mathbb{R}^d} f d\mu_n = \int_{\mathbb{R}^d} f d\mu \le \mu(F) + \varepsilon.$$

Since this holds for all $\varepsilon > 0$, 2 is satisfied.

$2 \Rightarrow 1$ Let f be a bounded continuous function on $\mathbb{R}^d$. Let us first show that $\overline{\lim} \int_{\mathbb{R}^d} f d\mu_n \le \int_{\mathbb{R}^d} f d\mu$.

We can suppose that $0 < f < 1$, by considering if necessary $\widetilde{f} = (M - f)/2M$, where $M = \sup |f|$.

Set $F_{i,k} = \{x \in \mathbb{R}^d : i/k \le f(x)\}$ and $G_{i,k} = \{x \in \mathbb{R}^d : (i-1)/k \le f(x) \le i/k\}$, for $i \in [\![1, k]\!]$ and $k \in \mathbb{N}^*$. We have

$$\sum_{i=1}^{k} \frac{i-1}{k} \mu(G_{i,k}) \le \int_{\mathbb{R}^d} f d\mu \le \sum_{i=1}^{k} \frac{i}{k} \mu(G_{i,k}).$$

Since

$$\sum_{i=1}^{k} i \mu(G_{i,k}) = \sum_{i=1}^{k} i [\mu(F_{i-1,k}) - \mu(F_{i,k})] = 1 + \sum_{i=1}^{k} \mu(F_{i,k})$$

and

$$\sum_{i=1}^{k} (i-1) \mu(G_{ik}) = \sum_{i=1}^{k} \mu(F_{i,k}),$$

we obtain

$$\frac{1}{k} \sum_{i=1}^{k} \mu(F_{i,k}) \le \int_{\mathbb{R}^d} f d\mu \le \frac{1}{k} \Big[1 + \sum_{i=1}^{k} \mu(F_{i,k}) \Big].$$

The above inequalities also hold for every μ_n. Since $F_{i,k}$ is a closed Borel set for $i \in [\![1, k]\!]$, 2 says that $\overline{\lim} \, \mu_n(F_{i,k}) \le \mu(F_{i,k})$, so

$$\overline{\lim} \int_{\mathbb{R}^d} f d\mu_n \le \frac{1}{k} + \frac{1}{k} \sum_{i=1}^{k} \overline{\lim} \, \mu_n(F_{i,k}) \le \frac{1}{k} + \frac{1}{k} \sum_{i=1}^{k} \mu(F_{i,k}) \le \frac{1}{k} + \int_{\mathbb{R}^d} f d\mu.$$

For k tending to infinity, we get $\overline{\lim} \int_{\mathbb{R}^d} f d\mu_n \leq \int_{\mathbb{R}^d} f d\mu$. Similarly, replacing f by $-f$, we obtain $\underline{\lim} \int_{\mathbb{R}^d} f d\mu_n \geq \int_{\mathbb{R}^d} f d\mu$, and 1 follows.

2 ⇔ 3 by taking complements.

2 ⇒ 4 Since 2 and 3 are equivalent, for all Borel sets B,

$$\mu(\widetilde{B}) \geq \overline{\lim} \mu_n(\widetilde{B}) \geq \overline{\lim} \mu_n(B) \geq \underline{\lim} \mu_n(B) \geq \underline{\lim} \mu_n(\overset{\circ}{B}) \geq \mu(\overset{\circ}{B}),$$

where $\widetilde{B}$ denotes the closure of B in $\mathbb{R}^d$, $\overset{\circ}{B}$ its interior, and ∂B its boundary. If $\mu(\widetilde{B} \setminus \overset{\circ}{B}) = 0$, then $\mu(\overset{\circ}{B}) = \mu(B) = \mu(\widetilde{B})$, from which 4 follows.

4 ⇒ 2 Let F be a closed Borel set, and let F_δ be defined by (4.1) for $\delta > 0$. Since each ∂F_δ is a subset of $\{x \in \mathbb{R}^d : d(x, F) = \delta\}$, these boundaries are disjoint for distinct δ. Moreover, $\mathbb{R}^d = \cup_{\delta \in \mathbb{Q}_+^*} F_\delta$, hence $\mu(\mathbb{R}^d) \leq \sum_{\delta \in \mathbb{Q}_+^*} \mu(\partial F_\delta) \leq 1$.

Therefore, only at most an enumerable quantity of these boundaries have non-null measures. Among the others, a sequence $(\delta_m)_{m \in \mathbb{N}}$ exists that tends to the empty set. Since $\mu(\partial F_{\delta_m}) = 0$, we know that $\mu_n(\partial F_{\delta_m})$ converges to $\mu(\partial F_{\delta_m})$ when n tends to infinity, for all m. We have $F \subset F_{\delta_m}$, so

$$\overline{\lim_n} \mu_n(F) \leq \lim_n \mu_n(F_{\delta_m}) = \mu(F_{\delta_m}).$$

The sequence (F_{δ_m}) is decreasing and converges to the closed set F. Therefore $\mu(F_{\delta_m})$ is decreasing and converges to $\mu(F)$, from which the result follows. □

We state without proofs the next convergence criteria.

Theorem 4.34 *Let (μ_n) be a sequence of finite measures on $\mathbb{R}^d$.*

Let μ be a finite measure on $\mathbb{R}^d$. The sequence (μ_n) converges strongly to μ if and only if the sequence of Fourier transforms of (μ_n) converges to the Fourier transform of μ.

If the sequence of Fourier transforms of (μ_n) converges to a function g that is continuous at 0, then g is the Fourier transform of some measure μ such that μ_n converges strongly to μ.

Finally, the convergence of a sequence of measures induces the convergence of the sequences of their image measures as follows.

Theorem 4.35 (Continuous Mapping) *Let (μ_n) be a sequence of finite measures on $(\mathbb{R}^d, \mathcal{B}(\mathbb{R}^d))$, and let $h: \mathbb{R}^d \to \mathbb{R}^{d'}$ be an a.e. continuous Borel function. If (μ_n) converges strongly to μ, then the sequence $(\mu_n \circ h^{-1})$ of image measures of μ_n by h converges strongly to the image measure $\mu \circ h^{-1}$.*

Proof We prove the theorem for a function that is continuous everywhere.

Let F be a closed Borel set of $\mathbb{R}^d$. Since the function h is continuous, the set $h^{-1}(F)$ is closed. Since μ_n converges strongly to μ, criterion 2 of Alexandrov's theorem yields $\overline{\lim} \mu_n[h^{-1}(F)] \le \mu[h^{-1}(F)]$, which proves the desired convergence.   □

All above results on strong convergence of measures induce the following criteria of convergence in distribution of random sequences.

Theorem 4.36 *Let (X_n) be a random sequence. The seven following statements are equivalent:*

1. *(X_n) converges in distribution to X.*
2. *$(\mathbb{P}_{X_n})$ converges strongly to $\mathbb{P}_X$.*
3. *$\overline{\lim} \mathbb{P}(X_n \in F) \le \mathbb{P}(X \in F)$ for all closed Borel sets F.*
4. *$\underline{\lim} \mathbb{P}(X_n \in O) \ge \mathbb{P}(X \in O)$ for all open Borel sets O.*
5. *$\mathbb{P}(X_n \in B)$ converges to $\mathbb{P}(X \in B)$ for all Borel sets B such that $\mathbb{P}(X \in \partial B) = 0$.*
6. *$F_{X_n}(x)$ converges to $F_X(x)$ at every continuity point $x \in \mathbb{R}$ of F_X.*
7. *$\phi_{X_n}(t)$ converges to $\phi_X(t)$ for all $t \in \mathbb{R}$.*

Note that for nonnegative variables, 7 also holds true for Laplace transform.

Proof Points 2–5 are direct consequences of Alexandrov's Theorem. Points 6 and 4 are equivalent because the rectangles generate the Borel σ-algebra; see Theorem 1.13. Finally, 7 is an immediate consequence of Theorem 4.34.   □

For absolutely continuous variables, the convergence of the associated sequence of densities gives a criterion of convergence in distribution. Note that in general the converse does not hold true.

Theorem 4.37 (Scheffé) *Let (X_n) be a sequence of random variables with densities (f_n), and let X be a random variable with density f. If f_n converges to f λ-a.e., then X_n converges in distribution to X.*

Proof Set $I_x =]-\infty, x]$ for $x \in \mathbb{R}$, and $\phi_n = f_n - f$. Since densities are integrable, with integral 1, we have $\int_{\mathbb{R}} \phi_n(t)dt = 0$, that is:

$$\int_{I_x} \phi_n(t)dt = -\int_{\mathbb{R}\backslash I_x} \phi_n(t)dt,$$

and hence

$$2\left|\int_{I_x} \phi_n(t)dt\right| = \left|\int_{I_x} \phi_n(t)dt - \int_{\mathbb{R}\backslash I_x} \phi_n(t)dt\right| \le \int_{I_x} |\phi_n(t)|dt + \int_{\mathbb{R}\backslash I_x} |\phi_n(t)|dt,$$

or

$$2 \left| \int_{I_x} \phi_n(t) dt \right| \leq \int_{\mathbb{R}} |\phi_n(t)| dt.$$

Since $\int_{\mathbb{R}} \phi_n(t) dt = 0$ and $|\phi_n| + \phi_n = 2\phi_n^+$, we have

$$\frac{1}{2} \int_{\mathbb{R}} |\phi_n(t)| dt = \int_{\mathbb{R}} \phi_n^+(t) dt,$$

which converges to zero when n tends to infinity, because (f_n) converges to f.

Moreover, $0 \leq \phi_n^+(t) \leq f(t)$ for $t \in \mathbb{R}$. The dominated convergence theorem applies to show that (f_n) converges to f in L^1. Therefore,

$$|F_{X_n}(x) - F_X(x)| = \left| \int_{I_x} (f_n - f)(t) dt \right|$$

converges to zero when n tends to infinity, and the conclusion follows by Point 6 of Theorem 4.36. □

For discrete variables, the convergence in distribution is reflected in the convergence of the sequences of distribution weights.

Theorem 4.38 *A discrete random sequence (X_n) converges in distribution to a discrete random variable X if and only if $\mathbb{P}(X_n = x)$ converges to $\mathbb{P}(X = x)$ when n tends to infinity, for all $x \in \mathbb{R}$.*

Proof We prove the theorem for sequences of variables taking integer values.

We have $F_{X_n}(t) = \sum_{k=0}^{[t]} \mathbb{P}(X_n = k)$ and $F_X(t) = \sum_{k=0}^{[t]} \mathbb{P}(X = k)$ for all $t \in \mathbb{R}$.

If $\mathbb{P}(X_n = k)$ converges to $\mathbb{P}(X = k)$ for all integers k, then $F_{X_n}(t)$ converges to $F_X(t)$, and hence (X_n) converges in distribution to X.

Conversely, let $k \in \mathbb{N}$.

- Suppose $\mathbb{P}(X = k) = 0$. Since $\partial\{k\} = \{k\}$, we get by criterion 5 of Theorem 4.36 that $\mathbb{P}(X_n \in \{k\})$ converges to $\mathbb{P}(X \in \{k\})$, or $\mathbb{P}(X_n = k)$ to $\mathbb{P}(X = k)$.
- Suppose $\mathbb{P}(X = k) \neq 0$. Let $f : \mathbb{R} \to \mathbb{R}$ be defined by:

$$f(x) = \begin{cases} 2(x - k) + 1 & \text{if } k - 1/2 < x \leq k, \\ -2(x - k) + 1 & \text{if } k \leq x < k + 1/2, \\ 0 & \text{otherwise.} \end{cases}$$

This function f is continuous and bounded and (X_n) converges in distribution to X, so, by definition, $\mathbb{E}[f(X_n)]$ converges to $\mathbb{E}[f(X)]$. Moreover, $f(m) = 0$ for all $m \in \mathbb{N} \setminus \{k\}$, and hence

$$\mathbb{E}[f(X)] = \sum_{m \geq 0} f(m)\mathbb{P}(X = m) = f(k)\mathbb{P}(X = k) = \mathbb{P}(X = k).$$

Similarly, $\mathbb{E}[f(X_n)] = \mathbb{P}(X_n = k)$ for all $n \in \mathbb{N}$, and the conclusion follows.

□

A sequence of discrete variables may converge in distribution to a random variable with a density, as shown in the next example.

▷ *Example 4.39 (A Discrete Random Sequence with a Continuous Limit)* Let X_n be a uniformly distributed random variable such that $\mathbb{P}(X_n = k/n) = 1/n$ for $k \in [\![1, n-1]\!]$. Using Point 6 of Theorem 4.36 easily shows that (X_n) converges in distribution to $X \sim \mathcal{U}([0, 1])$. ◁

The characteristic function gives a criterion of convergence in distribution, which requires no previous knowledge of the limit, an immediate consequence of Theorem 4.34. A similar criterion can be stated for the Laplace transform of nonnegative variables.

Theorem 4.40 (Lévy's Continuity) *Let (X_n) be a random sequence. If the sequence of characteristic functions of (X_n) converges to a function g continuous at 0, then g is the characteristic function of a random variable, say X, and (X_n) converges in distribution to X.*

4.2.3   Links Between Convergences

The relationships between the different convergences can be shown in a diagram.

$$
\begin{array}{c}
\text{a.s.} \\
\Downarrow \\
L^\infty \Rightarrow L^{p'} \Rightarrow L^p \Rightarrow L^1 \Rightarrow \mathbb{P}\ , \quad p' \geq p \geq 1. \\
\Downarrow \\
\mathcal{L}
\end{array}
$$

We will prove the implications between the different convergences and then give counter-examples for the converses.

Theorem 4.41 *Let (X_n) be a random sequence converging to a random variable X.*

1. *If convergence holds either in probability or in mean, a subsequence converges a.s. to X.*

2. *Convergence in probability induces convergence in distribution.*
3. *Convergence in L^p induces convergence in probability.*
4. *Almost sure convergence induces convergence in probability.*

Proof

1. If (X_n) converges in probability to X, then for any positive integer n, some $m_n \in \mathbb{N}^*$ exists such that $\mathbb{P}(|X_{m_n} - X| \geq 1/n) \leq 1/n^2$; the sequence (m_n) can be supposed to be non-decreasing. Since $\sum_{n \geq 1} \mathbb{P}(|X_{m_n} - X| \geq 1/n) < +\infty$, Borel–Cantelli lemma yields

$$\mathbb{P}\left(\overline{\lim_{m_n}} |X_{m_n} - X| < \frac{1}{n} \right) = 1.$$

This is true for all n, so the result is proven for convergence in probability. The proof for convergence in mean is omitted.

2. Let us show that $\mathbb{E}[h(X_n)]$ converges to $\mathbb{E}[h(X)]$ for all bounded continuous h. Otherwise, some non-decreasing sequence (n_k) and $\eta > 0$ would exist such that

$$\left| \mathbb{E}[h(X_{n_k})] - h(X) \right| \geq \eta. \tag{4.2}$$

Then (X_{n_k}) would converge in probability to X, and, thanks to 1, a subsequence $(X_{n_{kk'}})$ would converge a.s. to X. Moreover, h is continuous, so $(h(X_{n_{kk'}})$ would converge a.s. to $h(X)$. Thanks to the dominated convergence theorem, $\mathbb{E}[h(X_{n_{kk'}})]$ would then converge to $\mathbb{E}[h(X)]$. This contradicts (4.2) and the result is proven.

3. Markov's inequality yields, for all $\varepsilon > 0$,

$$\mathbb{P}(|X_n - X| \geq \varepsilon) \leq \frac{1}{\varepsilon^p} \mathbb{E}[|X_n - X|^p],$$

from which the conclusion follows.

4. Proposition 4.30 induces that if (X_n) converges to X a.s., then $\mathbb{P}(\overline{\lim} |X_n - X| > \varepsilon) = 0$ for all $\varepsilon > 0$. Let us set $E_n = \cup_{k \geq n}(|X_n - X| > \varepsilon)$. Since the sequence (E_n) is decreasing, we have

$$\lim_n \mathbb{P}(E_n) = \mathbb{P}\left(\bigcap_{n \geq 0} E_n \right) = 0.$$

Therefore, since $(|X_n - X| > \varepsilon) \subset E_n$ for all n, we have $\mathbb{P}(|X_n - X| > \varepsilon) \leq \mathbb{P}(E_n)$, so $\mathbb{P}(|X_n - X| > \varepsilon)$ converges to 0 when n tends to infinity.

□

The converse implications do not hold in general, as shown by the following examples.

▷ *Example 4.42 ($X_n \xrightarrow{\mathcal{D}} X \nRightarrow X_n \xrightarrow{\mathbb{P}} X$)* Let $X \sim \mathcal{N}(0, 1)$. Set $X_n = -X$ for all $n \in \mathbb{N}$. Since the distribution $\mathcal{N}(0, 1)$ is symmetric, the sequence (X_n) converges in distribution to X too. On the contrary, $\mathbb{P}(|X_n - X| > \varepsilon) = \mathbb{P}(|X| > \varepsilon/2) > 0$ for all $\varepsilon > 0$, so X_n does not converge in probability to X.   ◁

▷ *Example 4.43 ($X_n \xrightarrow{\mathbb{P}} X \nRightarrow X_n \xrightarrow{a.s.} X$ and $X_n \xrightarrow{L^1} X \nRightarrow X_n \xrightarrow{a.s.} X$)* Let (X_n) be an i.i.d. random sequence with distribution $\mathcal{B}(1/n)$. Since

$$\mathbb{P}(|X_n| > \varepsilon) = \mathbb{P}(X_n = 1) = \frac{1}{n} \quad \varepsilon \in]0, 1[,$$

the sequence (X_n) converges in probability to zero. Moreover, $\mathbb{E}(|X_n|) = 1/n$ converges to zero, so (X_n) also converges in mean to zero. Since a.s. convergence implies convergence in probability, if (X_n) converged a.s., the limit would necessarily be zero. The events $(X_n > \varepsilon)$ are independent and

$$\sum_{n \geq 1} \mathbb{P}(X_n > \varepsilon) = \sum_{n \geq 1} \mathbb{P}(X_n = 1) = \sum_{n \geq 1} \frac{1}{n}$$

is a divergent series; thus, Borel–Cantelli lemma yields $\mathbb{P}[\overline{\lim}(X_n > \varepsilon)] = 1$. Therefore, X_n does not converge a.s. to zero.   ◁

▷ *Example 4.44 ($X_n \xrightarrow{\mathbb{P}} X \nRightarrow X_n \xrightarrow{L^1} X$)* Let (u_n) be a sequence of positive real numbers, and let (X_n) be a random sequence such that $\mathbb{P}(X_n = u_n) = 1/n$, and $\mathbb{P}(X_n = 0) = 1 - 1/n$.

For all $\varepsilon > 0$, we have $\mathbb{P}(|X_n| > \varepsilon) = 1/n$, so (X_n) converges in probability to zero. If X_n converges in mean, its limit is necessarily zero.

If $u_n = \sqrt{n}$, then $\mathbb{E}(|X_n|) = 1/\sqrt{n}$ converges to zero and X_n indeed converges in mean to zero. On the contrary, if $u_n = n$, then $\mathbb{E}(|X_n|) = 1$, or if $u_n = n^2$, then $\mathbb{E}(|X_n|) = n$, so X_n cannot converge in mean to zero.   ◁

▷ *Example 4.45 ($X_n \xrightarrow{a.s.} X \nRightarrow X_n \xrightarrow{L^1} X$)* The random sequence $(n\mathbb{1}_{]0,1/n]})$, defined on the probability space $(]0, 1], \mathcal{B}(]0, 1]), \lambda)$, clearly converges a.s. to zero but not in mean.   ◁

Up to some strong assumptions, converse implications may become true.

Proposition 4.46 *If (X_n) converges in distribution to an a.s. constant variable C, then (X_n) converges in probability to C.*

Proof Let $\varepsilon > 0$. We have $(|X_n - C| > \varepsilon) = (X_n \in \mathbb{R}\backslash]C - \varepsilon, C + \varepsilon[)$. The Borel set $\mathbb{R}\backslash]C - \varepsilon, C + \varepsilon[$ is closed, and since (X_n) converges in distribution to C, criterion 3 of Theorem 4.36 yields

$$0 \leq \overline{\lim} \, \mathbb{P}(|X_n - C| > \varepsilon) \leq \mathbb{P}(C \in \mathbb{R}\backslash]C - \varepsilon, C + \varepsilon[).$$

Since the above right-hand side probability is null, clearly $\mathbb{P}(|X_n - C| > \varepsilon)$ converges to zero. □

The following extension of the dominated convergence theorem is proven in Exercise 4.10, together with other convergence criteria.

Proposition 4.47 *Let (X_n) be an L^p random sequence, and let $X \in L^p$, for some $p \geq 1$. If (X_n) converges in probability to X, and if $|X_n| < Y$ a.s. for all n, with $Y \in L^p$, then (X_n) converges to X in L^p.*

4.2.4 Convergence of Sequences of Random Vectors

The definitions of the convergence a.s. in probability or in L^p for sequences of random vectors are similar to Definition 4.28, by replacing the difference $X_n - X$ by a norm of this quantity, for example, the Euclidean norm

$$\|X_n - X\| = \sqrt{\sum_{i=1}^{d} [X_n(i) - X(i)]^2}.$$

The convergence of a sequence of vectors is thus equivalent to the convergence of the random sequences of its coordinates.

On the contrary, the convergence in distribution amounts to the strong convergence of the sequence of probability distributions, as in Theorem 4.36, thanks to Alexandrov's theorem. Since this is not a consequence of the convergence of the coordinates, we will focus on this type of convergence. For sake of simplicity, most results will be stated or proven for bidimensional vectors only.

Up to some strong conditions, the convergence in distribution of two sequences may induce the convergence in distribution of the pair.

Theorem 4.48 *Let (X_n) and (Y_n) be two random sequences such that (X_n) converges in distribution to X and (Y_n) converges in distribution to Y.*

1. *If $Y = C$ where C is an a.s. constant variable, then (X_n, Y_n) converges in distribution to (X, C).*
2. *If (X_n) and (Y_n) are independent, then (X_n, Y_n) converges in distribution to (X, Y).*

Proof

1. Let h be a bounded uniformly continuous function.

$$\mathbb{E}\left[h(X_n, Y_n)\right] = \mathbb{E}\left[h(X_n, C)\right] + \mathbb{E}\left[h(X_n, Y_n) - h(X_n, C)\right].$$

On the one hand, C does not depend on n, so $\mathbb{E}\left[h(X_n, C)\right]$ converges to $\mathbb{E}\left[h(X, C)\right]$.

On the other hand, since (Y_n) converges in distribution to C, according to Proposition 4.46, $\mathbb{P}(|Y_n - C| \geq \varepsilon)$ converges to zero for all $\varepsilon > 0$.
Since h is uniformly continuous, we obtain

$$\forall \alpha > 0, \ \exists \varepsilon_0 \text{ such that } \quad |Y_n - C| < \varepsilon_0 \Rightarrow |h(X_n, Y_n) - h(X_n, C)| < \alpha.$$

Finally,

$$\mathbb{E}\left[|h(X_n, Y_n) - h(X_n, C)|\right] \leq (i) + (ii),$$

where

$$(i) = \mathbb{E}\left[\mathbb{1}_{(|Y_n - C| \geq \varepsilon_0)}|h(X_n, Y_n) - h(X_n, C)|\right],$$
$$(ii) = \mathbb{E}\left[\mathbb{1}_{(|Y_n - C| < \varepsilon_0)}|h(X_n, Y_n) - h(X_n, C)|\right].$$

Since (i) converges to zero when n tends to infinity and $(ii) \leq \varepsilon_0 \alpha$ for any α, the proof is complete.   □

The following particular cases are often of use.

Corollary 4.49 (Slutsky) *Let (X_n) and (Y_n) be two random sequences. If (X_n) converges in distribution to X and if (Y_n) converges in distribution to C where C is a.s. constant, then*

$$X_n + Y_n \xrightarrow{\mathcal{D}} X + C, \quad Y_n X_n \xrightarrow{\mathcal{D}} CX \quad and \quad \frac{X_n}{Y_n} \xrightarrow{\mathcal{D}} \frac{X}{C} \ (if \ C \neq 0).$$

▷ *Example 4.50 (Convergence of a Quotient)* Let $\sigma \in \mathbb{R}^*$. If (X_n) converges in distribution to $X \sim \mathcal{N}(0, \sigma^2)$ and (Y_n) converges in distribution to σ, Slutsky's theorem applies to show that (X_n/Y_n) converges in distribution to $X/\sigma \sim \mathcal{N}(0, 1)$.
                                                                                        ◁

▷ *Example 4.51 (Convergence of Other Classical Quotients)* Let (X_n) be an i.i.d. sequence of random variables with distribution $\mathcal{N}(0, 1)$, and let (Z_n) be an independent sequence of random variables with distribution $\chi^2(n)$. Suppose (X_n)

and (Z_n) are independent. Then

$$\frac{X_n}{\sqrt{Z_n/n}} \xrightarrow{\mathcal{D}} \mathcal{N}(0, 1).$$

Indeed, since $\mathbb{E}\,(Z_n/n) = 1$ and $\mathbb{V}\mathrm{ar}\,(Z_n/n) = 2/n$, Chebyshev's inequality yields that (Z_n/n) converges in probability, and hence in distribution, to 1. The conclusion follows from Slutsky's theorem.   ◁

Note that it can also be proven that if (X_n) converges in probability to X and (Y_n) to Y, then

$$X_n + Y_n \xrightarrow{\mathbb{P}} X + Y, \quad \text{and} \quad Y_n X_n \xrightarrow{\mathbb{P}} XY. \tag{4.3}$$

Finally, the convergence in distribution of a sequence of vectors induces the convergence of all the sequences of its coordinates, but the convergence of all linear combinations is necessary for the converse to hold.

Proposition 4.52 *A sequence of d-dimensional random vectors (X_n) converges in distribution to a d-dimensional random vector X if and only if $(a' X_n)$ converges in distribution to $a' X$ for all $a \in \mathbb{R}^d$.*

Proof The direct implication derives from Slutsky's theorem and the converse from Lévy's continuity theorem.   ⊔

The convergence of a random sequence induces the convergence of a function of this sequence, as follows.

Theorem 4.53 *Let (X_n) be a random sequence of d-dimensional random vectors, and let X be a random vector. Let $h : \mathbb{R}^d \to \mathbb{R}$ be an a.s. continuous function. If (X_n) converges to X, then $(h(X_n))$ converges to $h(X)$ for the same type of convergence.*

Proof For sequences of random variables.

The theorem is a corollary of Theorem 4.35 for the convergence in distribution. Let us prove it for the convergence in probability. The other types of convergence are omitted.

Suppose (X_n) converges in probability to X. Let $\varepsilon > 0$. Set

$$A_\delta = \{x \in \mathbb{R} : \exists y, \ |x - y] < \delta \text{ and } |h(x) - h(y)| > \varepsilon\}, \quad \delta > 0.$$

If $X(\omega) \notin A_\delta$ and $|h(X_n(\omega)) - h(X(\omega))| > \varepsilon$, then $|X_n(\omega) - X(\omega)| \geq \delta$. Therefore,

$$
\begin{aligned}
\mathbb{P}[|h(X_n(\omega)) - h(X(\omega))| > \varepsilon] &= \mathbb{P}[|h(X_n(\omega)) - h(X(\omega))| > \varepsilon, X \in A_\delta] \\
&\quad + \mathbb{P}[|h(X_n(\omega)) - h(X(\omega))| > \varepsilon, X \notin A_\delta] \\
&\leq \mathbb{P}(X \in A_\delta) + \mathbb{P}(|X_n - X| \geq \delta).
\end{aligned}
$$

The function h is a.s. continuous, say on C_h with $\mathbb{P}(\mathbb{R} \backslash C_h) = 0$. Thus, when δ tends to zero, $A_\delta \cap C_h$ converges to the empty set, and $\mathbb{P}(X \in A_\delta)$ to zero. Finally, by assumption, $\mathbb{P}(|X_n - X| \geq \delta)$ converges to zero when n tends to infinity. □

The delta method plays an important part in statistics. For sake of simplicity, we state it for random variables, but the result holds true for sequences of vectors.

Theorem 4.54 (Delta Method) *Let (X_n) be a sequence of random variables, and let (a_n) be a sequence of real numbers converging to infinity. Suppose a real number a and a random variable X exist such that*

$$
a_n(X_n - a) \xrightarrow{\mathcal{D}} X, \quad n \to +\infty.
$$

Let $h : \mathbb{R} \longrightarrow \mathbb{R}$ be a function differentiable at a such that $h'(a) \neq 0$. Then

$$
a_n[h(X_n) - h(a)] \xrightarrow{\mathcal{D}} h'(a)X \quad n \to +\infty.
$$

Proof Since $(a_n(X_n - a))$ converges in distribution to X and (a_n) tends to infinity, necessarily $(|X_n - a|)$ converges in distribution to zero so also in probability; hence, for all $\delta > 0$, $\mathbb{P}(|X_n - a| \leq \delta) \to 1$.

By definition of derivatives, for all $\varepsilon > 0$ some $\delta > 0$ exists such that

$$
|x - y| \leq \delta \implies |h(y) - h(x) - h'(x)(y - x)| \leq \varepsilon |y - x|,
$$

so $\mathbb{P}[|h(X_n) - h(a) - h'(a)(X_n - a)| \leq \varepsilon |X_n - a|] \longrightarrow 1$, or

$$
\mathbb{P}[a_n|h(X_n) - h(a) - h'(a)(X_n - a)| \leq \varepsilon a_n |X_n - a|] \longrightarrow 1, \quad n \to \infty
$$

and hence $(a_n[h(X_n) - h(a)] - h'(a)a_n(X_n - a))$ converges in distribution to zero. Since $(a_n(X_n - a))$ converges in distribution to X, according to Theorem 4.53, $(h'(a)a_n(X_n - a))$ converges in distribution to $h'(a)X$. Then Slutsky's theorem yields the result. □

4.3 Limit Theorems

The convergence theorems to be stated here will mainly concern weighted sums of i.i.d. random variables. Nevertheless, we begin with results concerning sequences of discrete random variables with particular distributions—that are not i.i.d..

4.3.1 Asymptotics of Discrete Distributions

Let us begin by sequences of variables with binomial or hyper-geometric distributions. All of them are proven by elementary analytic calculus.

Proposition 4.55 (Poisson's Law of Large Numbers) *Let $X_n \sim \mathcal{B}(n, p_n)$. If np_n converges to λ when n tends to infinity, with $\lambda \in \mathbb{R}_+^*$, then*

$$\mathbb{P}(X_n = k) \longrightarrow \mathbb{P}(Z = k), \quad \text{where} \quad Z \sim \mathcal{P}(\lambda), \quad n \to +\infty.$$

Proof For $k = 0$, we compute

$$\mathbb{P}(X_n = 0) = \binom{n}{0} p_n^0 (1 - p_n)^n = (1 - p_n)^n = \exp\left(n \log\left[1 - \frac{\lambda}{n} + o(1/n)\right]\right),$$

which converges to $e^{-\lambda}$.

For $k \neq 0$, since p_n converges to zero, the quotient

$$\frac{\mathbb{P}(X_n = k)}{\mathbb{P}(X_n = k - 1)} = \frac{(n - k + 1)p_n}{k(1 - p_n)}$$

converges to λ/k when n tends to infinity. Setting $k = 1$, we obtain $\mathbb{P}(X_n = 1) \approx \lambda\mathbb{P}(X_n = 0)$, which converges to $\lambda e^{-\lambda}$. Similarly, $\mathbb{P}(X_n = 2) \approx \lambda\mathbb{P}(X_n = 1)/2$ converges to $\lambda^2 e^{-\lambda}/2$. The desired convergence follows by induction. □

In other words, if $p_n = O(1/n)$, the occurrences of the studied events become rare and the binomial distribution is asymptotically well approximated by a Poisson distribution. In practice, the Poisson's law of large numbers allows one to approximate the probability of an experiment with binomial distribution $\mathcal{B}(n, p)$ by a Poisson distribution $\mathcal{P}(np)$, as soon as p is small (the characteristics of the Poisson distributed experiments), and hence n is large enough for np to remain reasonable.

▷ *Example 4.56 (A Vaccination Process)* One over 1000 vaccinated individuals has a bad reaction. 2000 individuals are vaccinated. Let us compute the probability of observing k bad reactions. The associated random variable X has a binomial distribution $\mathcal{B}(2000, 0.001)$. Since $p = 0.001$ is small and $n = 2000$ is large, it can be approximated by a Poisson distribution with parameter np, that is, $\mathcal{P}(2)$. Hence

$$\mathbb{P}(X = k) \approx \frac{e^{-2}2^k}{k!}.$$

The exact expression,

$$\mathbb{P}(X = k) = \binom{2000}{k}(0.001)^k(1 - 0.001)^{2000-k},$$

would be far less easy to compute.   ◁

The next Laplace theorem states that the number of successes in n independent repetitions of any experiment with two outcomes is normally distributed for n large enough. This is a particular case of the central limit theorem stated in the next section. Still, it is of great historic significance in the theory of probability, and the presented proof uses only basic—yet tedious—analytic calculations.

Theorem 4.57 (Laplace) *Let $p \in]0, 1[$ and set $q = 1 - p$. If $X \sim \mathcal{B}(n, p)$, then, for $(k - np)/\sqrt{npq} \in]\lambda_1, \lambda_2]$,*

$$\mathbb{P}(X = k) = \frac{1}{\sqrt{2\pi npq}}e^{-(k-np)^2/2npq}[1 + O(1/\sqrt{n})] \qquad (4.4)$$

Moreover,

$$\mathbb{P}\left(\lambda_1 < \frac{X - np}{\sqrt{npq}} \leq \lambda_2\right) = \frac{1}{\sqrt{2\pi}}\left(\int_{\lambda_1}^{\lambda_2} e^{-z^2/2}dz\right)[1 + O(1/\sqrt{n})].$$

Proof Stirling's formula says that $n! = \sqrt{2\pi n}n^n e^{-n}e^{\theta_n}$, where $0 < \theta_n < 1/12n$. Since $\binom{n}{k} = n!/k!(n-k)!$, we get

$$\binom{n}{k}p^k q^{n-k} = \frac{1}{\sqrt{2\pi}}\sqrt{\frac{n}{k(n-k)}}\left(\frac{np}{k}\right)^k\left(\frac{nq}{n-k}\right)^{n-k}e^{\theta_n-\theta_k-\theta_{n-k}},$$

with $0 < \theta_n < 1/12n$, $0 < \theta_k < 1/12k$, and $0 < \theta_{n-k} < 1/12(n-k)$.

Set $z = (k - np)/\sqrt{npq}$. We have $np/k = (1 + z\sqrt{q/np})^{-1}$ and $k = z\sqrt{npq} + np$, so

$$\log\left[\left(\frac{np}{k}\right)^k\right] = -(np + z\sqrt{npq})\log\left(1 + z\sqrt{\frac{q}{np}}\right)$$

$$= -(np + z\sqrt{npq})\left[z\sqrt{\frac{q}{np}} - \frac{z^2}{2}\frac{q}{np} + \frac{z^3}{3}\left(\frac{q}{np}\right)^{3/2} + z^4\varepsilon_1(z)\right]$$

$$= -z\sqrt{npq} - \frac{z^2}{2}q + \frac{z^3}{6}q\sqrt{\frac{q}{np}} + z^4\frac{\varepsilon_2(z)}{n}.$$

In the same way, $n - k = nq - z\sqrt{npq}$, so

$$\log\left[\left(\frac{nq}{n-k}\right)^{n-k}\right] = z\sqrt{npq} - \frac{z^2}{2}p - \frac{z^3}{6}p\sqrt{\frac{p}{nq}} + z^4\varepsilon_3(z),$$

hence

$$\log\left[\left(\frac{np}{k}\right)^k\left(\frac{nq}{n-k}\right)^{n-k}\right] = \frac{-z^2}{2}(p+q) + \frac{z^3}{\sqrt{n}}\frac{\varepsilon_4(z)}{n},$$

where $\varepsilon_i(z)$ converges to zero when z tends to zero, for $i \in [\![1,4]\!]$. If z remains bounded, we get

$$\left(\frac{np}{k}\right)^k\left(\frac{nq}{n-k}\right)^{n-k} = e^{-z^2/2}e^{z^3\varepsilon_4(z)/\sqrt{n}} = e^{-z^2/2}\left[1 + \frac{\varepsilon_5(z)}{\sqrt{n}}\right].$$

Moreover,

$$\sqrt{\frac{n}{k(n-k)}} = \sqrt{n}\left[(np + z\sqrt{npq})(nq - z\sqrt{npq})\right]^{-1/2}$$

$$= \frac{1}{\sqrt{npq}}\left(1 + z\frac{q-p}{\sqrt{npq}} - \frac{z^2}{n}\right)^{-1/2} = \frac{1}{\sqrt{npq}}\left[1 + \frac{\varepsilon_6(z)}{\sqrt{n}}\right].$$

Setting $\theta = \theta_n - \theta_k - \theta_{n-k}$, we have $|\theta| \le [1/n + 1/k + 1/(n-k)]/12$, but

$$\frac{1}{k} + \frac{1}{n-k} = \frac{1}{npq + (q-p)z\sqrt{npq} - z^2pq} = \frac{1}{npq}\left(1 + z\frac{q-p}{\sqrt{npq}} - \frac{z^2}{n}\right)^{-1}$$

$$= \frac{1}{npq}\left[1 + \varepsilon_7(z)/\sqrt{n}\right],$$

hence

$$|\theta| \le \frac{1}{12}\left[\frac{1}{n} + \frac{1}{npq} + \frac{1}{npq}\frac{\varepsilon_7(z)}{\sqrt{n}}\right],$$

that is, $\theta = \varepsilon_8(z)/n$. Thus, if z remains bounded, we have

$$\binom{n}{k} p^k q^{n-k} = \frac{1}{\sqrt{2\pi npq}} \left[1 + \frac{\varepsilon_6(z)}{\sqrt{n}} \right] e^{-z^2/2} \left[1 + \frac{\varepsilon_5(z)}{\sqrt{n}} \right] e^{\varepsilon_8(z)/n}$$

$$= \frac{1}{\sqrt{2\pi npq}} e^{-z^2/2} \left[1 + \frac{\varepsilon_9(z)}{\sqrt{n}} \right], \qquad (4.5)$$

or (4.4). Finally,

$$\mathbb{P}\left(\lambda_1 < \frac{X - np}{\sqrt{npq}} \leq \lambda_2 \right) = \sum_{i=1}^{r} \mathbb{P}(X = k_i),$$

where $k_1, \ldots, k_r$ denote the r integers of the interval $]np + \lambda_1 \sqrt{npq}, np + \lambda_2 \sqrt{npq}]$. Setting $z_i = (k_i - np)/\sqrt{npq}$, we obtain by (4.5)

$$\sum_{i=1}^{r} \mathbb{P}(X = k_i) = \sum_{i=1}^{r} \binom{n}{k_i} p^{k_i} q^{n-k_i} = \frac{1}{\sqrt{2\pi npq}} \sum_{i=1}^{r} e^{-z_i^2/2} \left[1 + \frac{\varepsilon_8(z_i)}{\sqrt{n}} \right].$$

Therefore,

$$\frac{1}{\sqrt{npq}} \sum_{i=1}^{r} e^{-z_i^2/2} \longrightarrow \int_{\lambda_1}^{\lambda_2} e^{-z^2/2} dz$$

when n tends to infinity, and $\varepsilon_8(z_i)/\sqrt{n} = O(1/\sqrt{n})$.   □

In other words, Laplace theorem says that, when n is large enough, the binomial distribution $\mathcal{B}(n, p)$ is well approximated by the Gaussian distribution $\mathcal{N}(np, npq)$.

▷ *Example 4.58 (Coin Tossing)* Let us compare the probability of obtaining as many times heads as tails when tossing a fair coin, respectively, 100 times and 200 times.

The random variable X_n counting the number of tails in n tosses has a binomial distribution $\mathcal{B}(n, 1/2)$. Laplace theorem yields

$$\mathbb{P}(X_{100} = 50) \stackrel{\sim}{=} \frac{1}{\sqrt{50\pi}} \quad \text{and} \quad \mathbb{P}(X_{200} = 100) \stackrel{\sim}{=} \frac{1}{\sqrt{100\pi}},$$

so $\mathbb{P}(X_{100} = 50)/\mathbb{P}(X_{200} = 100) \stackrel{\sim}{=} \sqrt{2}$.   ◁

The approximation obtained by using for a fixed n the continuous Gaussian distribution instead of the discrete binomial one, specifically

$$\binom{n}{k} p^k (1-p)^{n-k} \approx \frac{1}{\sqrt{2\pi npq}} e^{-(k-np)^2/2npq},$$

includes a certain error. For minimizing it, the so-called correction of continuity methods have been developed. For example, if $X \sim \mathcal{B}(n, p)$, the error is reduced by considering

$$\mathbb{P}(X = k) \approx \mathbb{P}\left[\frac{\sqrt{n}}{\sigma}\left(\frac{k-1/2}{n} - p\right) < Z \leq \frac{\sqrt{n}}{\sigma}\left(\frac{k+1/2}{n} - p\right)\right],$$

$$\mathbb{P}(X \leq k) \approx \mathbb{P}\left[Z \leq \frac{\sqrt{n}}{\sigma}\left(\frac{k+1/2}{n} - p\right)\right],$$

where $Z \sim \mathcal{N}(0, 1)$ and $\sigma = \sqrt{p(1-p)}$.

Further, the hyper-geometric distribution can be asymptotically approximated by either a binomial or Poisson distribution.

Proposition 4.59 *Let $X_{N,M} \sim \mathcal{H}(N, M, n)$. Suppose that both M and N tend to infinity.*

1. *If M/N converges to p and if n is fixed, then $\mathbb{P}(X_{N,M} = k)$ converges to $\mathbb{P}(Y = k)$, where $Y \sim \mathcal{B}(n, p)$.*
2. *If nM/N converges to λ, then $\mathbb{P}(X_{N,M} = k)$ converges to $\mathbb{P}(Z = k)$, where $Z \sim \mathcal{P}(\lambda)$.*

Proof

1. We have

$$\mathbb{P}(X_{N,M} = k) = \frac{\binom{M}{k}\binom{N-M}{n-k}}{\binom{N}{n}}$$

$$= \frac{n!}{k!(n-k)!} \times \frac{M!}{(n-k)!} \times \frac{(N-M)!}{(N-M-n+k)!} \times \frac{(N-n)!}{N!}$$

$$= \binom{n}{k} \frac{M \ldots (M-k+1).(N-M)\ldots(N-M-n+k+1)}{N \ldots (N-n+1)}$$

$$= \binom{n}{k} \frac{M^k}{N^k} \frac{(N-M)^{n-k}}{N^{n-k}} \frac{\prod_{j=1}^{k-1}(1-j/M)[1-j/(N-M)]}{\prod_{j=1}^{n-1}(1-j/N)},$$

which converges to $\binom{n}{k} p^k (1-p)^{n-k}$.

2. We have

$$\mathbb{P}(X_{N,M} = 0) = \frac{\binom{M}{0}\binom{N-M}{n}}{\binom{N}{n}} = \frac{(N-M)\dots(N-M-n+1)}{N\dots(N-n+1)},$$

so $\log \mathbb{P}(X_{N,M} = 0) = \log(1 - M/N) + \dots + \log[1 - M/(N-n+1)]$. Hence

$$n \log\left(1 - \frac{M}{N}\right) \leq \log \mathbb{P}(X_{N,M} = 0) \leq n \log\left(1 - \frac{M}{N-n+1}\right),$$

meaning that $\mathbb{P}(X_{N,M} = 0)$ converges to $e^{-\lambda}$. Moreover,

$$\frac{\mathbb{P}(X_{N,M} = k+1)}{\mathbb{P}(X_{N,M} = k)} = \frac{n-k}{k+1} \times \frac{M-k}{N-M-n+k+1}$$

$$\approx \frac{n}{k+1} \times \frac{M}{N} \approx \frac{\lambda}{k},$$

and hence, by induction, $\mathbb{P}(X_{N,M} = k)$ converges to $e^{-\lambda}\lambda^k/k!$.

□

▷ *Example 4.60 (Urn Problem)* An urn contains N bowls among which M are white; n bowls are drawn at random. As shown in Example 2.43, if the drawing is done with replacement, the number of drawn white bowls X has a binomial distribution $\mathcal{B}(n, M/N)$; without replacement, X has a hyper-geometric distribution $\mathcal{H}(N, M, n)$.

Clearly, if n is fixed and much smaller than both M and N, replacing or not the bowls in the urn will not change much the distribution of X. On the contrary, if the proportion of white balls becomes small, drawing a white ball becomes a rare event, and the distribution of X becomes of the Poisson type.   ◁

4.3.2   Laws of Large Numbers

A large number law is a theorem giving conditions under which

$$\frac{1}{n}\sum_{i=1}^{n}(X_i - \mathbb{E}\,X_i) \xrightarrow[n\to\infty]{} 0.$$

It is said to be weak if the convergence holds in probability and strong if the convergence is almost sure. Let us begin with the weak law for i.i.d. variables even if it is a corollary of the strong law next stated.

Theorem 4.61 (Weak Law of Large Numbers) *Let (X_n) be an i.i.d. integrable random sequence. Then*

$$\frac{X_1 + \cdots + X_n}{n} \xrightarrow{\mathbb{P}} \mathbb{E} X_1.$$

The convergence also holds in mean.

Proof We prove the theorem for square integrable random variables and convergence in probability.

Set $S_n = X_1 + \cdots + X_n$. Let σ^2 denote the variance of X_n.

We have $\mathbb{E} S_n/n = m$, and $\mathbb{V}\text{ar} S_n = n\sigma^2$, so $\mathbb{V}\text{ar}(S_n/n) = \sigma^2/n$. Thanks to Chebyshev's inequality, $\mathbb{P}(|S_n/n - m| > \varepsilon) \leq \sigma^2/n\varepsilon^2$, for all $\varepsilon > 0$, and hence $\mathbb{P}(|S_n/n - m| > \varepsilon)$ converges to zero. $\qquad\qquad\square$

▷ *Example 4.62 (Convergence of a Quotient)* Let (X_n) be an i.i.d. random sequence with uniform distribution $\mathcal{U}(0, 1)$. For any fixed non null integers k_1 and k_2,

$$\frac{X_1^{k_1} + \cdots + X_n^{k_1}}{X_1^{k_2} + \cdots + X_n^{k_2}} \xrightarrow{\mathbb{P}} \frac{k_2 + 1}{k_1 + 1}.$$

Indeed, the law of large numbers and Example 2.58 together give

$$\frac{1}{n}(X_1^{k_i} + \cdots + X_n^{k_i}) \xrightarrow{\mathbb{P}} \mathbb{E}(X_1^{k_i}) = \frac{1}{k_i + 1}, \quad i = 1, 2.$$

The convergence of the quotient holds thanks to the properties of convergence in probability; see (4.3) Sect. 4.2.4. ◁

▷ *Example 4.63 (Frequencies and Probabilities)* Let (X_n) be an i.i.d. random sequence with distribution P. The random variables $\mathbb{1}_{(X_n \in B)}$ are i.i.d. too, with expected value of their common distribution $P(B)$ for all Borel sets B. The law of large numbers yields

$$P(B) = \lim_{n \to +\infty} \frac{1}{n} \sum_{i=1}^{n} \mathbb{1}_{(X_i \in B)} = \lim_{n \to +\infty} \frac{\text{number of } i \leq n \text{ such that } X_i \in B}{n}.$$

This justifies the estimation or definition of probabilities by frequencies. ◁

Numerous criteria of convergence of weighted sums of random variables exist. Let us present some of them, interesting in themselves and necessary for proving the strong law of large numbers to come.

Proposition 4.64 *Let (X_n) be an independent random sequence. If $\sum_{n \geq 1} \operatorname{Var} X_n$ is finite, then $\sum_{i=1}^{n} X_i$ converges a.s.*

Proof An a.s. Cauchy sequence converges a.s.; the series $\sum_{i=1}^{n} X_i$ is an a.s. Cauchy sequence if, for all $\varepsilon > 0$, $\mathbb{P}[\cup_{m \geq 1}(|X_{n+1} + \cdots + X_{n+m}| > \varepsilon)]$ converges to zero when n tends to $+\infty$. We compute

$$\mathbb{P}\left[\bigcup_{m=1}^{N} ((X_{n+1} + \cdots + X_{n+m}) > \varepsilon) \right] =$$

$$= \mathbb{P}\left[\max_{1 \leq m \leq N} (|X_{n+1} + \cdots + X_{n+m}|) > \varepsilon \right]$$

$$\overset{(1)}{\leq} \frac{1}{\varepsilon^2} \operatorname{Var}(X_{n+1} + \cdots + X_{n+N}) \leq \frac{1}{\varepsilon^2} \sum_{m \geq 1} \operatorname{Var} X_{n+m}.$$

(1) by Kolmogorov's inequality. Since $\sum_{n \geq 1} \operatorname{Var} X_n$ is finite, the rest $\sum_{m \geq 1} \operatorname{V} ar X_{n+m}$ converges to zero when n tends to infinity and the result is proven. □

The next result, whose proof is straightforward, is a key to the proof of the following Kolmogorov's criterion.

Lemma 4.65 (Stochastic Kronecker's Lemma) *Let (X_n) be a random sequence such that $\sum_{n \geq 1} X_n$ is a.s. finite. Let (u_n) be a non-decreasing sequence of real numbers converging to infinity. Then*

$$\frac{1}{u_n} \sum_{i=1}^{n} u_i X_i \xrightarrow{a.s.} 0.$$

Proposition 4.66 (Kolmogorov's Criterion) *Let (X_n) be an independent random sequence. If*

$$\sum_{n \geq 1} \operatorname{Var}\left(\frac{X_n}{n} \right) < +\infty, \tag{4.6}$$

then

$$\frac{1}{n} \sum_{i=1}^{n} (X_i - \mathbb{E} X_i) \xrightarrow{a.s.} 0.$$

Proof Set $Y_n = X_n/n$. Since $\sum_{n \geq 1} \operatorname{Var} Y_n$ is finite, thanks to Proposition 4.64, the series with general term $Y_n - \mathbb{E} Y_n$ converges a.s. to Y. Necessarily $\mathbb{E} Y = 0$, so Y is a.s. finite. Applying the stochastic Kronecker's lemma to the sequence $(Y_n - \mathbb{E} Y_n)$ for $u_n = n$ yields the result. □

If (4.6) is not satisfied, the conclusion cannot hold true. The independence of the variables X_n is not sufficient—even with equal expected values.

▷ *Example 4.67 (A Counter-Example)* Let (X_n) be an independent random sequence such that

$$\mathbb{P}(X_n = -n) = 1 - \frac{1}{n^2} = 1 - \mathbb{P}(X_n = n - n^3).$$

We have $\mathbb{E}\,X_n = 0$ for all $n \geq 1$, but (X_n/n) converges a.s. to -1, and hence $\sum_{i=1}^n X_i/n$ cannot converge a.s. to a finite random variable. ◁

On the contrary, if the sequence is i.i.d., Kolmogorov's criterion induces the a.s. convergence of arithmetic means.

Theorem 4.68 (Strong Law of Large Numbers) *Let (X_n) be an i.i.d. integrable random sequence. Then*

$$\frac{X_1 + \cdots + X_n}{n} \xrightarrow{\text{a.s.}} \mathbb{E}\,X_1.$$

If the sequence is square integrable, then the convergence also holds in quadratic mean.

Proof We can suppose without loss of generality that the variables are centered.

Set $X_n^* = X_n \mathbb{1}_{(X_n \leq n)}$. We have $\sum_{n \geq 1} \mathbb{P}(X_n^* \neq X_n) = \sum_{n \geq 1} \mathbb{P}(X_n > n)$. Since the sequence (X_n) is i.i.d., $\mathbb{P}(|X_n| \geq n) = \mathbb{P}(|X_1| \geq n)$ for all n, and using Point 3 of Exercise 2.6 yields that $\sum_{n \geq 1} \mathbb{P}(|X_n| \geq n) \leq \mathbb{E}\,X_1$. Borel–Cantelli lemma finally induces that $\mathbb{P}(X_n^* \neq X_n$ infinitely often$) = 0$.

Thus, in order to prove the searched a.s. convergence, it is sufficient to prove that $\sum_{i=1}^n X_i^*/n$ converges a.s. to zero. We have

$$\sum_{n \geq 1} \text{Var}\left(\frac{X_n^*}{n}\right) \leq \sum_{n \geq 1} \frac{1}{n^2} \mathbb{E}\,(X_n^{*2}),$$

and

$$\sum_{n \geq 1} \frac{1}{n^2} \mathbb{E}\,(X_n^{*2}) = \sum_{n \geq 1} \frac{1}{n^2} \mathbb{E}\,(X_1^{*2}) = \sum_{n \geq 1} \frac{1}{n^2} \sum_{m=1}^n \mathbb{E}\left[(X_1^*)^2 \mathbb{1}_{(m-1 \leq |X_1| < m)}\right]$$

$$= \sum_{m \geq 1} \mathbb{E}\left[(X_1^*)^2 \mathbb{1}_{(m-1 \leq |X_1| < m)}\right] \sum_{n \geq m} \frac{1}{n^2}.$$

Comparing $\sum_{n\geq 1}(1/n^2)$ to $\int_1^{+\infty} x^{-2}dx$ yields for some real constant K,

$$\sum_{n\geq 1}\mathrm{Var}\left(\frac{X_n^*}{n}\right) \leq \frac{K}{m}\sum_{m\geq 1}\mathbb{E}\left[|X_1^*||X_1^*|\mathbb{1}_{(m-1\leq |X_1|<m)}\right]$$

$$\leq K\sum_{m\geq 1}\frac{1}{m}\mathbb{E}\left[m|X_1^*|\mathbb{1}_{(m-1\leq |X_1|<m)}\right] = K\mathbb{E}|X_1^*| < +\infty.$$

Hence, according to Kolmogorov's criterion, the series $(\sum_{i=1}^n (X_i^* - \mathbb{E}\,X_i^*)/n)$ converges a.s. to zero. Moreover, due to the dominated convergence theorem, $\mathbb{E}\,X_n^* = \mathbb{E}[X_1\mathbb{1}_{(X_1\leq n)}]$ converges to zero, so, by Cesàro's lemma, the sequence $\sum_{i=1}^n \mathbb{E}\,X_i^*/n$ converges to zero too, and the conclusion follows.

For square integrable variables, it remains to prove that $\overline{X}_n = \sum_{i=1}^n X_i/n$ converges to zero in L^2. We compute

$$\mathbb{E}\left(|\overline{X}_n|^2\right) = \frac{1}{n^2}\sum_{i=1}^n \mathbb{E}\left(X_i^2\right) + \sum_{i\neq j}\mathbb{E}\left(X_iX_j\right) = \frac{1}{n^2}\sum_{i=1}^n \mathbb{E}\left(X_1^2\right) = \frac{1}{n}\mathbb{E}\left(X_1^2\right).$$

Therefore, $\mathbb{E}\left(|\overline{X}_n|^2\right)$ indeed converges to zero. $\square$

Moreover, if X_1 is not integrable, the sequence $(\sum_{i=1}^n X_i/n)$ is divergent.

Proposition 4.69 *Let (X_n) be an i.i.d. random sequence. If the sequence with general term X_i/n converges a.s., then (X_n) is integrable.*

Proof If $\sum_{i=1}^n X_i/n$ converges a.s., then (X_n/n) converges a.s. to zero.

Since the variables are independent, necessarily, according to Borel–Cantelli lemma, $\sum_{n\geq 1}\mathbb{P}(|X_n| > n) < +\infty$. Hence, since all the X_n have the same distribution, $\sum_{n\geq 1}\mathbb{P}(|X_1| > n) < +\infty$; so, thanks to Exercise 2.6, $\mathbb{E}|X_1| < +\infty$.
$\square$

The strong law of large numbers remains true for a random number of variables.

Theorem 4.70 (Anscombe's Theorem) *Let (X_n) be an integrable i.i.d. random sequence. Let (N_n) be a random sequence taking values in $\mathbb{N}^*$, a.s. finite for all n, independent of (X_n), and converging to infinity a.s. when n tends to infinity. Then*

$$\frac{X_1 + \cdots + X_{N_n}}{N_n} \xrightarrow{a.s.} \mathbb{E}\,X_1, \quad n \to +\infty.$$

Proof Set $S_n = X_1 + \cdots + X_n$. The sequence S_n/n converges to $\mathbb{E}\, X_1$ a.s. by the large number law, so

$$\forall \varepsilon > 0, \, \exists n_\varepsilon \in \mathbb{N}^*, \, n > n_\varepsilon \implies \left| \frac{S_n}{n} - \mathbb{E}\, X_1 \right| < \varepsilon \quad \text{a.s.}$$

Since N_n tends to infinity a.s.,

$$\exists n_\varepsilon > 0, \, n > n_\varepsilon \implies N_n > n_\varepsilon \quad \text{a.s.,}$$

and the conclusion follows.   □

Note that most of the above convergence results carry over to sequences of random vectors without change, coordinate by coordinate.

4.3.3 Central Limit Theorem

Below is a result of paramount importance as well in probability theory as in statistics.

Theorem 4.71 (Central Limit) *Let (X_n) be an i.i.d. square integrable random sequence. Then*

$$\sqrt{n} \left(\frac{X_1 + \cdots + X_n}{n} - \mathbb{E}\, X_1 \right) \xrightarrow{\mathcal{D}} \mathcal{N}(0, \mathbb{V}\mathrm{ar}\, X_1).$$

Proof Set $\overline{X}_n = \sum_{i=1}^n X_i/n$. The characteristic function of $\sqrt{n}\,\overline{X}_n$ is

$$\varphi_{\sqrt{n}\,\overline{X}_n}(t) = \mathbb{E}\left[\exp\left(\frac{it}{\sqrt{n}} \sum_{i=1}^n X_i \right) \right] = \varphi_{X_1}\left(\frac{t}{\sqrt{n}} \right)^n.$$

We can suppose that the variables are centered. Set $\mathbb{V}\mathrm{ar}\, X_1 = \sigma^2$. We have $\varphi'_{X_1}(0) = 0$ and $\varphi''_{X_1}(0) = -\sigma^2$, so, using Taylor's formula, $\varphi_{X_1}(t/\sqrt{n}) = 1 - \sigma^2 t^2/2n + O(1/n)$. Therefore,

$$\log \varphi_{\sqrt{n}\,\overline{X}_n}(t) = n \log \varphi_{X_1}(t/\sqrt{n}) = n \log[1 - \sigma^2 t^2/2n + O(1/n)],$$

so $\log \varphi_{\sqrt{n}\,\overline{X}_n}(t)$ converges to $-\sigma^2 t^2/2$ and $\varphi_{\sqrt{n}\,\overline{X}_n}(t)$ converges to $e^{-\sigma^2 t^2/2}$, for $t \in \mathbb{R}$. This limit is the characteristic function of the distribution $\mathcal{N}(0, \sigma^2)$, so the conclusion follows from Lévy's continuity theorem.   □

Using criterion 5 of Theorem 4.36, the central limit theorem induces in particular
that

$$\mathbb{P}\Big[\lambda_1 < \sqrt{n}\Big(\frac{X_1 + \cdots + X_n}{n} - \mathbb{E}\,X_1\Big) \leq \lambda_2\Big] \longrightarrow \frac{1}{\sqrt{2\pi}} \int_{\lambda_1}^{\lambda_2} e^{-\sigma^2 t^2/2} dt,$$

that is to say a generalization of Laplace theorem to all i.i.d. square integrable
random sequences.

▷ *Example 4.72 (Stirling's Formula)* Let (X_n) be an i.i.d. random sequence with
Poisson distribution $\mathcal{P}(1)$. Set $S_n = X_1 + \cdots + X_n$. On the one hand, we know that
$S_n \sim \mathcal{P}(n)$, so $\mathbb{P}(S_n = n) = e^{-n} n^n / n!$ for all $n \in \mathbb{N}$. On the other hand, by the
central limit theorem applied to (X_n), for n large enough, we have

$$\mathbb{P}(n - 1 < S_n \leq n) = \mathbb{P}(-1/\sqrt{n} < Z_n \leq 0) \approx \mathbb{P}(-1/\sqrt{n} < Z \leq 0),$$

where $Z_n = \sqrt{n}(S_n/n - 1) \xrightarrow{\mathcal{D}} Z \sim \mathcal{N}(0,1)$. Then, since $e^{-x^2/2} \sim_0 1$,

$$\mathbb{P}(-1/\sqrt{n} < Z \leq 0) = \frac{1}{\sqrt{2\pi}} \int_{-1/\sqrt{n}}^{0} e^{-x^2/2} dx \approx \frac{1}{\sqrt{2\pi n}}.$$

The well-known formula

$$n! \approx n^n e^{-n} \sqrt{2\pi n},$$

follows—referred to as Stirling's formula.   ◁

We state without proof the following four results. First, the extension of
the central limit theorem to random sequences with random indices completes
Anscombe's theorem.

Proposition 4.73 *Let $(X_n)_{n \in \mathbb{N}^*}$ be an integrable i.i.d. random sequence with
centered distribution with finite variance σ^2. Let (N_n) be a random sequence taking
values in $\mathbb{N}^*$, a.s. finite for all n, independent of (X_n), and converging to infinity a.s.
when n tends to infinity. Then*

$$\frac{X_1 + \cdots + X_{N_n}}{\sigma \sqrt{N_n}} \xrightarrow{\mathcal{D}} \mathcal{N}(0,1), \quad n \to +\infty.$$

The next central limit theorem for functions of random sequences, particularly
useful in statistics, is an immediate consequence of the central limit theorem and of
the delta method.

Proposition 4.74 *Let (X_n) be an i.i.d. square integrable random sequence. Let $h :$ $\mathbb{R} \longrightarrow \mathbb{R}$ be a function differentiable at $m = \mathbb{E} X_1$. Then*

$$\sqrt{n} \left[h \left(\frac{X_1 + \cdots + X_n}{n} \right) - h(m) \right] \xrightarrow{\mathcal{D}} \mathcal{N}(0, h'(m)^2 \text{Var} X_1).$$

The following result can also be seen as a central limit theorem, with a.s. convergence to the distribution function of the distribution $\mathcal{N}(0, 1)$.

Theorem 4.75 (Almost Sure Central Limit Theorem) *Let (X_n) be an i.i.d. random sequence such that $\mathbb{E} X_1 = 0$ and $\text{Var} X_1 = 1$. Then*

$$\frac{1}{\log n} \sum_{k=1}^{n} \frac{1}{k} \mathbb{1}_{\{k^{-1/2}(X_1 + \cdots + X_k) \leq x\}} \xrightarrow{a.s.} F_{\mathcal{N}(0,1)}(x), \quad x \in \mathbb{R}.$$

Finally, extreme bounds for the speed of convergence in the limit central theorem are given by a divergence result.

Theorem 4.76 (Law of Iterated Logarithm) *Let (X_n) be an i.i.d. square integrable random sequence. Then*

$$\overline{\lim} \frac{\sqrt{n}(\overline{X}_n - \mathbb{E} X_1)}{\sqrt{\text{Var} X_1} \sqrt{2 \log \log n}} = 1 \quad and \quad \underline{\lim} \frac{\sqrt{n}(\overline{X}_n - \mathbb{E} X_1)}{\sqrt{\text{Var} X_1} \sqrt{2 \log \log n}} = -1,$$

where $\overline{X}_n = \sum_{i=1}^{n} X_i / n$.

Let (X_n) be an i.i.d. square integrable random sequence and set $S_n = \sum_{i=1}^{n} X_i$. According to the law of large numbers, $S_n \approx n \mathbb{E} X_1$, and, according to the central limit theorem, $S_n - n\mathbb{E} X_1 = o(\sqrt{n})$ when n tends to infinity. Of course, the order of the deviation between S_n and $n\mathbb{E} X_1$ can be more than $\sqrt{n}$. The theory of large deviations investigates the asymptotic behavior of the quantity $\mathbb{P}(|S_n - n\mathbb{E} X_1| > n^\alpha)$ for $\alpha > 1/2$. The next result, for $\alpha = 1$, is the simplest of the so-called large deviations theorems.

Theorem 4.77 (Chernoff) *Let (X_n) be an i.i.d. square integrable random sequence whose moment generating function M is finite on an interval $I \neq \{0\}$. Let h denote its Cramér transform. Set $m = \mathbb{E} X_1$. Let a be a nonnegative real number. Then for all $n \in \mathbb{N}^*$,*

$$\mathbb{P}\left(\frac{X_1 + \cdots + X_n}{n} - m > a \right) \leq e^{-nh(m+a)}, \tag{4.7}$$

$$\mathbb{P}\left(\frac{X_1 + \cdots + X_n}{n} - m < a \right) \leq e^{-nh(m-a)}. \tag{4.8}$$

Proof For centered variables.

Set $S_n = X_1 + \cdots + X_n$. According to Chernoff's inequality, $\mathbb{P}(S_n \leq an) \leq e^{-h_{S_n}(na)}$. Moreover, by definition, since $M_{S_n}(u) = \mathbb{E}(e^{uS_n}) = \mathbb{E}(\prod_{i=1}^n e^{uX_i}) = M(u)^n$, we have

$$h_{S_n}(na) = \sup_{u \in I}[nau - \log M_{S_n}(u)] = \sup_{u \in I}[nau - n \log M(u)] = nh(a).$$

The inequality (4.7) follows. Then (4.8) can be proven in the same way. □

▷ *Example 4.78 (Large Deviations for a Gaussian Sequence)* If (X_n) is a standard Gaussian sequence, thanks to Example 2.84, $h(t) = t^2/2$. Hence

$$\mathbb{P}\left(\frac{X_1 + \cdots + X_n}{n} > a\right) \leq e^{-na^2/2}, \quad n \in \mathbb{N}^*,$$

and similarly for $\mathbb{P}[(X_1 + \cdots + X_n)/n < a]$. ◁

Many of these asymptotic results remain true for vectors. Let us state the most important one.

Theorem 4.79 (Multivariate Central Limit Theorem) *Let (X_n) be a sequence of independent square integrable d-dimensional random vectors. Let M denote the expected value vector and Γ the covariance matrix of their distribution. Then*

$$\frac{1}{\sqrt{n}}\left(\sum_{i=1}^n X_i - nM\right) \xrightarrow{\mathcal{D}} \mathcal{N}_d(0, \Gamma), \quad n \to +\infty.$$

Proof According to the central limit theorem, we have for all $u \in \mathbb{R}^d$,

$$Z_u = \frac{1}{\sqrt{n}}\left(\sum_{i=1}^n u_i X_i - n \sum_{i=1}^n u_i M_i\right) \longrightarrow \mathcal{N}_1(0, u'\Gamma u), \quad n \to +\infty$$

so $\phi_{Z_u}(1)$ converges to $e^{-u'\Gamma u/2}$, and the conclusion follows from criterion 7 of Theorem 4.36. □

4.4 Stochastic Simulation Methods

Stochastic simulation methods include technics for generating random variables with fixed discrete or continuous distributions, as well as the celebrated Monte Carlo method.

Simulating a given distribution requires a generator of random numbers, present in every computer. Let us first see how such a generator works. It seems to produce "at random" sequences of numbers between 0 and $m - 1$. Actually, randomness is involved only in the initial choice of a real number $x_0 \in [\![1, m - 1]\!]$; then, $x_{m+1} = ax_n + b$ modulo m is deterministically designed by induction. Dividing x_n by m produces a number in $[0, 1]$. Usual choices are $a = 7^5$ and $m = 2^{31} - 1$ or $a = 5^5$ and $m = 2^{35} - 31$, with $b = 0$. Thus, a generator of random numbers directly simulates only values for random variables with uniform distribution on $[0, 1]$. All others are calculated from these ones, through deterministic functions.

4.4.1 Generating Random Variables

Simulating any type of random variables with discrete distributions is very easy.

Let $U \sim \mathcal{U}(0, 1)$. Let P be any discrete distribution taking values $\{x_n \in \mathbb{R} : n \in \mathbb{N}\}$, and set $t_n = \sum_{j=0}^{n} P(\{x_j\})$. Let X be the random variable defined by:

$$X(\omega) = x_n \quad \text{if} \quad t_n \leq U(\omega) < t_{n+1}.$$

We compute $\mathbb{P}(X = x_n) = t_{n+1} - t_n = P(\{x_n\})$, meaning that $X \sim P$.

▷ *Example 4.80 (Simulation of Bernoulli Distributions)* For simulating a Bernoulli distribution with parameter $p \in]0, 1[$, it is enough to take $X = \mathbb{1}_{(U > p)}$.   ◁

The inverse function transformation is a general method for generating any type of continous distributions.

Theorem 4.81 *Let X be a random variable with continuous distribution function F_X, and let $U \sim \mathcal{U}(0, 1)$. Then:*

1. $F_X(X) \sim \mathcal{U}(0, 1)$.
2. $F_X^{-1}(U) \sim X$, where F_X^{-1} denotes either the inverse function of F_X if it exists or its generalized inverse defined by:

$$F_X^{-1}(x) = \inf\{t \in \mathbb{R} : F_X(t) > x\}, \quad t \in [0, 1]. \tag{4.9}$$

Thus, a random variable with any fixed continuous distribution function can be constructed from a uniformly distributed variable. This constitutes the inverse transformation method.

Proof

1. Since F_X is continuous, if it is increasing, then it has an inverse $F_X^{-1} : [0, 1] \to \mathbb{R}$, and we get

$$\mathbb{P}[F_X(X) \leq x] = \mathbb{P}[X \leq F_X^{-1}(x)] = F_X(F_X^{-1}(x)) = x, \quad x \in [0, 1]. \tag{4.10}$$

If F_X is not increasing, it nevertheless has a generalized inverse $F_X^{-1} : [0, 1] \rightarrow \mathbb{R}$ defined by (4.9). Both functions F_X and F_X^{-1} are increasing so $\{t \in \mathbb{R} : t < F_X^{-1}(x)\} = \{t \in \mathbb{R} : F_X(t) < x\}$.

In both cases, (4.10) defines the distribution function of the uniform distribution on $[0, 1]$.

2. Set $V = F_X^{-1}(U)$. We compute

$$\mathbb{P}(V \leq t) = \mathbb{P}[F_X^{-1}(U) \leq t] = \mathbb{P}[U \leq F_X(t)] = F_X(t),$$

so V has the same distribution as X.

□

▷ *Example 4.82 (Simulation of Erlang and Exponential Distributions)* If $Y \sim \mathcal{E}(\lambda)$, then $X = \exp(-\lambda Y) \sim \mathcal{U}(0, 1)$. Conversely, if $U \sim \mathcal{U}(0, 1)$, then $V = -(\log U)/\lambda \sim \mathcal{E}(\lambda)$. The simulation of a random variable with exponential distribution follows.

Thanks to Example 3.53, a random variable with Erlang distribution can then be simulated as the sum of i.i.d. exponentially distributed random variables.   ◁

We state the inverse transformation method extended to random vectors without proof.

Theorem 4.83 (Fundamental Theorem of Simulation) *For all random vector X with dimension $d \geq 1$, a d-dimensional random vector $Y = \phi(U_1, \ldots, U_n)$ exists such that $Y \sim X$, where ϕ is a λ-a.s. continuous function and $(U_1, \ldots, U_n)$ is an n-dimensional random vector with uniform distribution on $[0, 1]^n$.*

Further, Von Neumann's method is general for any type of distributions.

For a discrete distribution, suppose we are given the realizations $U(\omega_1)$, $U(\omega_2), \ldots$, of $U \sim \mathcal{U}(0, 1)$ and the realizations $Y(\omega_1)$, $Y(\omega_2) \ldots$, of a random variable Y taking integer values with $\mathbb{P}(Y = n) = q_n$ for $n \geq 0$. Let $P = (p_n)_{n \geq 0}$ be some distribution on $\mathbb{N}$. Take $c \geq 1$ satisfying $p_n/q_n \leq c$ for all n such that $q_n \neq 0$. The following algorithm then yields realizations of a random variable $X \sim P$.

1. Let $Y(\omega)$.
2. Let $U(\omega)$.
3. If $U(\omega) \leq \frac{p_{Y(\omega)}}{cq_{Y(\omega)}}$, then $X(\omega) = Y(\omega)$. If not, $Y(\omega)$ is rejected; return to 1.

Let us check that $X \sim P$. Indeed, if $A = \{\omega \in \Omega : U_n < p_n/cq_n\}$, then

$$\mathbb{P}(X = n) = \mathbb{P}(Y = n \mid A) = \frac{\mathbb{P}[(Y = n) \cap A]}{\mathbb{P}(A)}.$$

Fig. 4.2 An illustration of von Neuman's method

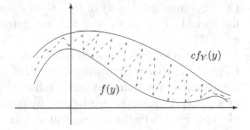

We compute

$$\mathbb{P}[(Y = n) \cap A] = \mathbb{P}(Y = n, U \le p_n/cq_n) = \mathbb{P}(Y = n)\mathbb{P}(U \le p_n/cq_n)$$

$$= q_n F_U(p_n/cq_n),$$

and hence $\mathbb{P}(X = n) = p_n/c\,\mathbb{P}(A)$. Necessarily, $\sum_{n \ge 0}\mathbb{P}(X = n) = 1$, so $c\,\mathbb{P}(A) = 1$ and $\mathbb{P}(X = n) = p_n$, meaning that indeed $X_n \sim P$.

Let now Y be a random variable with density f_Y on $\mathbb{R}$, and let f be some density on $\mathbb{R}$ such that $c \in \mathbb{R}_+^*$ exists satisfying $f(y) \le cf_Y(y)$ for all $y \in \mathbb{R}$. Let us construct from Y a random variable X with density f, as illustrated in Fig. 4.2.

Let $U \sim [0, 1]$. Another random variable is well defined by setting $X(\omega) = Y(\omega)$ on the event $(U \le f(Y)/cf_Y(Y))$, and $X(\omega)$ non-finite elsewhere; note that $\mathbb{P}(X = +\infty) = 0$. Let us check that the density of X is f. Indeed, for all real numbers x,

$$\mathbb{P}(X \le x) = \mathbb{P}\left[Y \le x \Big| U \le \frac{f(Y)}{cf_Y(Y)}\right] = \frac{1}{K}\mathbb{P}\left[Y \le x, U \le \frac{f(Y)}{cf_Y(Y)}\right]$$

$$= \frac{1}{K}\int_{-\infty}^{x}\int_{0}^{f(y)/cf_Y(y)} f_Y(y)\,du\,dy = \frac{1}{cK}\int_{-\infty}^{x} f(y)\,dy.$$

Since f is a density, when x tends to infinity, this quantity converges to 1, and hence $cK = 1$.

Direct methods, based on specific properties, exist for certain distributions. Here are some examples.

▷ *Example 4.84 (Simulation of Geometric Distributions)* A geometric distribution with parameter ρ it is the distribution of the waiting time of the first success in an experiment with two possible outcomes; see Chap. 2. Therefore, if (U_n) is an i.i.d. random sequence with distribution $\mathcal{U}(0, 1)$, then the variable Y defined by $Y(\omega) = \min\{n \in \mathbb{N}^* : U_n(\omega) > \rho\}$ has the distribution $\mathcal{G}(\rho)$. ◁

▷ *Example 4.85 (Box–Muller Method for Simulating Gaussian Distributions)* Let U_1 and U_2 be two independent variables with distribution $\mathcal{U}(0, 1)$.

Then $X_1 = \sqrt{-2\log U_1}\cos(2\pi U_2)$ and $X_2 = \sqrt{-2\log U_1}\sin(2\pi U_2)$ are independent standard Gaussian variables, as shown by computing the density of

(X_1, X_2) from the density of (U_1, U_2) by some change of variable method. Hence, the variable $Y = mX_1 + \sigma$ has the distribution $\mathcal{N}(m, \sigma^2)$ for all $m \in \mathbb{R}$ and $\sigma^2 \in \mathbb{R}_+^*$.

This also induces the simulation of a standard Gaussian d-dimensional vector $(X_1, \ldots, X_d)$ from simulating d standard Gaussian variables independently. The simulation of any Gaussian vector follows by using Theorem 3.70. The simulation of a $\chi^2(d)$ distribution also follows by considering the sum $\sum_{i=1}^{d} X_i^2$; see Example 3.58. Finally Fisher and Student distributions follow by using Proposition 3.59.

$\triangleleft$

4.4.2   Monte Carlo Simulation Method

We sill present the principle of the Monte Carlo method by showing how to compute—using the strong law of large numbers—an approximate value of the multiple integral

$$I = \int_0^1 \cdots \int_0^1 h(t_1, \ldots, t_d) dt_1 \ldots dt_d,$$

where $h : \mathbb{R}^d \to \mathbb{R}$ is a continuous function and $d \geq 1$.

We can write

$$I = \mathbb{E}\left[h(X_1, \ldots, X_d)\right],$$

where $X = (X_1, \ldots, X_d)'$ is a random vector with uniform distribution on $[0, 1]^d$. Consider N realizations $x^1 = X(\omega_1), \ldots, x^N = X(\omega_N)$ of X. For N large enough, the strong law of large numbers induces that

$$I \approx \frac{1}{N} \sum_{i=1}^{N} h(x^i).$$

More generally, the Monte Carlo method consists in using the strong law of large numbers for computing numerically such approximations as expected values. See Exercise 5.2 for an application to computing the value of π.

Proposition 4.86 *The speed of convergence of the Monte Carlo method is, when N tends to infinity, of order:*

1. $O(1/\sqrt{N})$ *in mean.*
2. $O(\sqrt{\log \log N / N})$ *at worst.*

Proof

1. Set $R_N = -I + \sum_{i=1}^{N} h(X_i)/N$. The central limit theorem yields

$$\mathbb{E}\,|R_N| \approx \frac{\sigma_{h(X_i)}}{\sqrt{N}}, \quad N \to +\infty,$$

and the result follows.
2. is a straightforward application of the law of iterated logarithm.

$\square$

The convergence of the Monte Carlo method is slow. Specifically, for an accuracy of ε, the order of the number of iterations is $1/\varepsilon^2$. The need for developing methods for accelerating the convergence—such as the methods of reduction of variance—naturally arises. Compared to deterministic methods, the Monte Carlo method is of practical interest mainly for either very irregular functions or when $d \geq 3$.

The precision of the method is given by $\alpha > 0$ and $0 < \theta < 1$ such that

$$\mathbb{P}(|R_N| \geq \alpha) \leq \theta.$$

A first approach to determine the minimal value of N necessary for reaching this precision is to use Chebyshev's inequality, that is:

$$P(|R_N| \geq \alpha) \leq \frac{\mathrm{Var}\,(h(X))}{N\alpha^2}.$$

Hence, the searched precision is obtained for

$$N \geq \left[\frac{\mathrm{Var}\,(h(X))}{\theta\alpha^2}\right] + 1,$$

where $[x]$ denotes the integer part of x.

4.5   Exercises and Complements

∇ **Exercise 4.1 (Weierstrass Theorem)** Let (X_n) be an i.i.d. random sequence with Bernoulli distribution with parameter x. Set $\overline{X}_n = \sum_{i=1}^{n} X_i/n$. Let $f : [0, 1] \to \mathbb{R}$ be a continuous function.

1. Show that for all $\varepsilon > 0$, there exists $\delta > 0$ such that

$$\mathbb{E}\,[|f(\overline{X}_n) - f(x)|\mathbb{1}_{(|\overline{X}_n - x| < \delta)}] \leq \varepsilon.$$

2. Show using Chebyshev's inequality that

$$\mathbb{E}\left[|f(\overline{X}_n) - f(x)|\mathbb{1}_{(|\overline{X}_n - x| \geq \delta)}\right] \leq \frac{M}{2n\delta^2},$$

for a real number M not depending on f.
3. Show that f is the uniform limit of the sequence of Bernstein's polynomials defined by:

$$P_n(x) = \sum_{k=0}^n f\left(\frac{k}{n}\right)\binom{n}{k}x^k(1-x)^{n-k}, \quad x \in [0, 1].$$

Solution
We can write, for any $\delta > 0$,

$$|\mathbb{E}[f(\overline{X}_n) - f(x)]| \leq$$
$$\leq \underbrace{\mathbb{E}[|f(\overline{X}_n) - f(x)|\mathbb{1}_{(|\overline{X}_n - x| < \delta)}]}_{(i)} + \underbrace{\mathbb{E}[|f(\overline{X}_n) - f(x)|\mathbb{1}_{(|\overline{X}_n - x| \geq \delta)}]}_{(ii)}.$$

1. Since f is uniformly continuous on $[0, 1]$, we have

$$\forall \varepsilon > 0, \exists \delta > 0, \quad |u - x| \leq \delta \Longrightarrow |f(u) - f(x)| \leq \varepsilon.$$

For such a δ, we have $(i) \leq \varepsilon$.
2. Setting $M = \sup_{x \in [0,1]} |f(x)|$, we have

$$(ii) \leq 2M\mathbb{P}(|\overline{X}_n - x| \geq \delta).$$

Moreover, $\mathbb{E}\,\overline{X}_n = x$ and $\mathrm{Var}\,\overline{X}_n = x(1-x)/n$, and $x(1-x) \leq 1/4$, so Chebyshev's inequality yields

$$(ii) \leq \frac{2Mx(1-x)}{n\delta^2} \leq \frac{M}{2n\delta^2}.$$

3. Therefore, for n large enough, $\sup_{x \in [0,1]} |\mathbb{E}(f(\overline{X}_n) - f(x))| \leq 2\varepsilon$, and the desired convergence follows because $P_n(x) = \mathbb{E}\,f(\overline{X}_n)$.

Note that if f is a continuous function defined on any other interval of $\mathbb{R}$, the result remains valid, up to an homotety and a translation. This constitutes an example of an analytical result shown by a probabilistic method. $\triangle$

∇ **Exercise 4.2 (Paradox of the Empty Box)**   Suppose an infinite number of numbered bowls is given. At time $t - \tau$, bowls 1–10 are put in a box, then one is drawn at random without replacement. At time $t - \tau/2$, bowls 11–20 are added in the box, then one is drawn. At time $t - \tau/2^2$, bowls 21–30 are added in the box, then one is drawn, and so on. Show that at time t, the box is a.s. empty.

Solution
At the first drawing, $\Omega_1 = [\![1, 10]\!]$, and one bowl $j_1 \in [\![1, 10]\!]$ is drawn. At the second, a bowl $j_2 \in \Omega_2 = [\![1, 20]\!] \setminus \{j_1\}$ is taken out, and so on. At the n-th drawing, $\Omega_n = [\![1, 10n]\!] \setminus \{j_1, \ldots, j_{n-1}\}$. For each n, the associated σ-algebra is the set of all subsets of Ω_n, and the probability is uniform. According to Kolmogorov's theorem, the probability space associated with the experiment is the infinite product of these spaces.

Set $A_n(i)$="the bowl i is in the box after the n-th drawing," $A(i) = $ "the bowl i is in the box at time t," for all positive integers i and n. Since $A(i)$ is the decreasing limit of the sequence $A_n(i)$, we have $\mathbb{P}(A(i)) = \lim_{n \to +\infty} \mathbb{P}(A_n(i))$. For $1 \le i \le 10$,

$$\mathbb{P}[A_n(i)] = \frac{9}{10} \frac{18}{19} \cdots \frac{9n}{9n+1}.$$

Since $\log \mathbb{P}[A_n(i)] = \sum_{i=1}^{n} \log[1 + 1/9i]$ and $\log[1 + 1/9i] \approx -1/9i$ when i tends to infinity, it follows that $A(i)$ is a null set. Moreover, $\mathbb{P}(A(i)) \ge \mathbb{P}(A(j))$ for all $i \in [\![1, 10]\!]$ and $j \ge 10$.

Finally, let $A = $ "the box is empty at time t." Clearly, $A = \cup_{i \ge 1} A(i)$, so by σ-subadditivity, $\mathbb{P}(A) \le \sum_{l \ge 1} \mathbb{P}(A(i))$, and hence A has probability zero. $\triangle$

∇ **Exercise 4.3 (Superior Limits of Events and of Real Numbers)** Let (X_n) be a random sequence.

1. Set $\overline{\lim} X_n = \inf_{n \ge 0} \sup_{k \ge n} X_k$. Show that $\overline{\lim}(X_n > x) = (\overline{\lim} X_n > x)$ for all $x \in \mathbb{R}$.
2. The sequence is supposed to be i.i.d.. Show that the distribution function F of $\overline{\lim} X_n$ takes values in $\{0, 1\}$, and then that $\overline{\lim} X_n$ is a.s. constant, equal to $x_0 = \inf\{x \in \mathbb{R} : F(x) = 1\}$.
3. Show that $\overline{\lim} X_n$ is measurable for the tail σ-algebra of the natural filtration of (X_n) and prove again that $\overline{\lim} X_n$ is a.s. constant.

Solution
1. By definition, $\overline{\lim}(X_n > x) = \cap_{n \ge 0} \cup_{k \ge n} (X_n > x)$, so is a measurable event. We have

$$\omega \in \overline{\lim}(X_n > x) \Leftrightarrow \forall n, \exists k \ge n, X_k(\omega) > x \Leftrightarrow \forall n, \sup_{k \ge n} X_k(\omega) > x$$
$$\Leftrightarrow \inf_n \sup_{k \ge n} X_k(\omega) > x \Leftrightarrow \overline{\lim} X_n(\omega) > x$$
$$\Leftrightarrow \omega \in (\overline{\lim} X_n > x).$$

2. Borel–Cantelli lemma implies that $\mathbb{P}(\overline{\lim}(X_n > x)) = 0$ or 1 for all $x \in \mathbb{R}$. According to 1, $\mathbb{P}(\overline{\lim} X_n > x) = 0$ or 1 too, that is, $F(x) = 0$ or 1. Finally

$$\mathbb{P}(\overline{\lim} X_n = x_0) = F(x_0) - \lim_{n \to +\infty} F(x_0 - 1/n) = 1 - 0 = 1.$$

3. Using 1 yields $(\overline{\lim} X_n > x) = \cap_{n \geq 0} A_n$, where $A_n = \cup_{k \geq n}(X_k > x)$. Clearly, $A_n \in \sigma(X_n, X_{n+1}, \dots)$. The sequence (A_n) is decreasing, so $\cap_{n \geq 0} A_n$ is a tail event, and the conclusion follows from Corollary 4.19. $\triangle$

▽ **Exercise 4.4 (Total Cost of a Component)**   An essential component for the continuous functioning of a machine is automatically and instantaneously replaced at each failure by a new identical one. The lifetime of the component is exponentially distributed with parameter λ.

1. Each replacement is assumed to cost b Euros and the rate of devaluation of the Euro to be constant and equal to $a > 0$. Compute the mean total cost of failures.
2. The updated cost of each failure at time $t \geq 0$ is now assumed to be given by a nonnegative function g defined on $\mathbb{R}_+$. Compute the mean total cost of failures.
3. Deduce 1 from 2.

Solution
1. Let $(T_n)_{n \in \mathbb{N}^*}$ denote the sequence of the failure times of the components. We have $T_n = X_1 + \cdots + X_n$, $n \geq 1$, where (X_n) is the sequence of the times between two successive failures, i.i.d. with distribution $\mathcal{E}(\lambda)$.
   The cost of the n-th failure is $b \exp(-aT_n)$ in updated Euros, so the total cost of the failures is $C = \sum_{n \geq 1} b \exp(-aT_n)$, and the mean total cost can be written as

$$\mathbb{E} C = \mathbb{E}\left(\sum_{n \geq 1} b e^{-aT_n} \right) = b \sum_{n \geq 0} \mathbb{E} e^{-aT_n}$$

$$= b \sum_{n \geq 1} (e^{-aX_1})^n = b \sum_{n \geq 1} \left(\frac{\lambda}{\lambda + a}\right)^n = \frac{\lambda b}{a}.$$

2. The cost of the n-th failure is $g(T_n)$, and the mean total cost is $\mathbb{E} C = \sum_{n \geq 1} \mathbb{E} g(T_n)$. Since $T_n \sim \gamma(n, \lambda)$, we compute

$$\mathbb{E} g(T_n) = \int_0^{+\infty} g(t) f_n(t) \, dt = \int_0^{+\infty} g(t) \frac{(\lambda t)^{n-1}}{(n-1)!} \lambda e^{-\lambda t} \, dt,$$

so $\mathbb{E} C = \lambda \int_0^{+\infty} g(t) \, dt$.
3. Setting $g(t) = b e^{-at}$, we get $\mathbb{E} C = \lambda b \int_0^t e^{-at} \, dt = \lambda b / a$. $\triangle$

▽ **Exercise 4.5 (A Simple Random Walk)** A coin is tossed indefinitely. Let $p \in$ $]0, 1[\setminus\{1/2\}$ be the probability of obtaining heads at one toss. Let S_n be the random variable equal to the number of heads less the number of tails in n tosses. We set $S_0 = 0$.

1. Show that S_n is a simple random walk.
2. Compute $\mathbb{P}(S_n = 0)$, depending on the parity of n.
3. Let $N = |\{n \in \mathbb{N}^* : S_n = 0\}|$ be the number of times when the trajectory cuts the time axis. Show that the expected value of N is finite.
4. Compute $\mathbb{P}(\overline{\lim}(S_n = 0))$.

Solution
1. It is enough to take, in Definition 4.26, X_n equal to 1 if the result of the n-th toss is heads and -1 otherwise.
2. Set $A_n = (S_n = 0)$. Clearly, $\mathbb{P}(A_{2n+1}) = 0$, and we can write $A_{2n} =$ "$X_i = 1$ for n indices i among the first n," so $\mathbb{P}(A_{2n}) = \binom{2n}{n} p^n (1 - p)^n$.
3. We can write $N = \sum_{n \geq 0} \mathbb{1}_{A_n}$, so $\mathbb{E}\, N = \sum_{n \geq 0} \mathbb{P}(A_n)$. Moreover,

$$\frac{\mathbb{P}(A_{2n+2})}{\mathbb{P}(A_{2n})} = 4p(1 - p).$$

   Since this quantity is less than one, the series converges.
4. By Definition 4.9, $\overline{\lim}\, A_n$ is the set of all the trajectories cutting the time axis an infinite number of times. The series with general term $\mathbb{P}(A_n)$ converges, so, using Borel–Cantelli lemma, we obtain $\mathbb{P}(\overline{\lim}\, A_n) = 0$, meaning that the trajectory will cut the time axis only a finite number of times.   △

▽ **Exercise 4.6 (Convergence of Non-integrable Variables)** Let (X_n) be an i.i.d. random sequence. Let P denote its distribution and φ its characteristic function.

1. Set $Y_n = (X_n + Y_{n-1})/n$ for $n \geq 1$, with $Y_0 = X_0/2$.
   (a) Write Y_n in terms of X_i with $0 \leq i \leq n$.
   (b) Write the characteristic function ϕ_n of Y_n in terms of ϕ and n.
   (c) Suppose P is the Cauchy distribution $\mathcal{C}(1)$. Show that (Y_n) converges in distribution to the same distribution.
2. Suppose again that P is the Cauchy distribution $\mathcal{C}(1)$. Study the convergence in distribution of the sequences $(\overline{X}_n)$ and (T_n) defined by $\overline{X}_n = \sum_{i=1}^n X_i/n$ and $T_n = \sum_{i=1}^n X_i/\sqrt{n}$.

Solution
1.(a) Clearly, $Y_n = \sum_{i=0}^n X_i/2^{n-i+1}$.
   (b) Since according to 1(a), X_n and Y_{n-1} are independent, Theorem 3.49 and property 5 of the characteristic functions jointly imply that

$$\phi_n(t) = \phi_{(Y_{n-1}+X_n)/2}(t) = \phi_{n-1}\left(\frac{t}{2}\right)\phi\left(\frac{t}{2}\right),$$

from which it follows that $\phi_n(t) = \prod_{i=0}^{n} \phi(t/2^{n-i+1}) = \prod_{j=1}^{n+1} \phi(t/2^j)$ for all t.

(c) We know that $\phi(t) = e^{-|t|}$ for all t; see Exercise 2.13. Therefore,

$$\phi_n(t) = \prod_{j=1}^{n+1} \phi\left(\frac{t}{2^j}\right) = \prod_{j=1}^{n+1} e^{-|t|/2^j} = \exp\left(-\sum_{j=1}^{n+1} \frac{|t|}{2^j}\right).$$

Finally, $\sum_{j=1}^{n+1}(1/2^j)$ is a geometrical series that converges to 1 when n tends to infinity. Hence, (Y_n) converges in distribution to a Cauchy distribution $\mathcal{C}(1)$.

2. The sequence (X_n) is not integrable, so neither the strong law of large numbers nor the limit central theorem apply. Nevertheless, we have

$$\phi_{\overline{X}_n}(t) = \phi\left(\frac{t}{n}\right)^n = \left(e^{-|t/n|}\right)^n = e^{-|t|}, \quad t \in \mathbb{R}.$$

The sequence $(\overline{X}_n)$ converges in distribution to the same Cauchy distribution.

On the contrary, $\phi_{T_n}(t) = \phi(t/\sqrt{n})^n = e^{-\sqrt{n}|t|}$ for all t. Therefore ϕ_{T_n} converges to the function $\mathbb{1}_{\{0\}}$, which is not continuous at zero so cannot be a characteristic function, and hence (T_n) cannot converge in distribution.   △

▽ Exercise 4.7 (Convergence of the Range—Continuation of Exercise 3.10)

Let (X_n) be an i.i.d. random sequence with uniform distribution on $[0, 1]$.

1. Show that $(X_{(n)})$ converges a.s. to 1.
2. Show that $(nX_{(1)})$ converges in distribution.
3. (a) Show that (E_n) converges a.s. to 1.
   (b) Show that $(n(1 - E_n))$ converges in distribution and give the limit distribution.

Solution

1. Let $0 < \varepsilon < 1$. We have

$$\mathbb{P}[|X_{(n)} - 1| > \varepsilon] = 1 - \mathbb{P}[1 - \varepsilon \leq X_{(n)} \leq 1 + \varepsilon] = 1 - F_{(n)}(1 + \varepsilon) + F_{(n)}(1 - \varepsilon).$$

Since $F_{(n)}(t) = t^n \mathbb{1}_{[0,1]}(t) + \mathbb{1}_{]1,+\infty[}(t)$, we obtain that $\mathbb{P}(|X_{(n)} - 1| > \varepsilon) = (1 - \varepsilon)^n$ converges to zero.

2. Since $F(x) = \mathbb{1}_{]1,+\infty[}(x)$, we obtain using (3.5) p. 116 that

$$F_{(1)}(x/n) = \mathbb{1}_{]1,+\infty[}(x) + [1 - (1 - x/n)^n]\mathbb{1}_{[0,1]}(x)$$

converges to $\mathbb{1}_{]1,+\infty[}(x) + (1 - e^{-x})\mathbb{1}_{[0,1]}(x)$ when n tends to infinity. This is the distribution function of an exponential distribution with parameter 1.

3. (a) Exercise 3.10 gives the density of E_n, that is, $f(x) = n(n-1)x^{n-2}(1-x)$ on $[0, 1]$. The sequence (E_n) is increasing and up-bounded, so converges a.s. Moreover $\mathbb{P}(1 - E_n > a) = F_{E_n}(1-a) = (1-a)^{n-1}[(n-1)a+1]$ converges to zero for all $a \in [0, 1]$, so (E_n) converges in probability to 1, and hence also a.s.

(b) We obtain by using Example 2.59,

$$f_{n(1-E_n)}(u) = \frac{n-1}{n}\left(1 - \frac{u}{n}\right)^{n-2} u \mathbb{1}_{[0,n]}(u),$$

which converges to $ue^{-u}\mathbb{1}_{\mathbb{R}_+}(u)$. Hence, according to Scheffé's theorem, the sequence converges in distribution to the Erlang distribution $\mathcal{E}(2, 1)$. $\triangle$

▽ **Exercise 4.8 (Convergence of the Maximum)** Let F be a distribution function. Set $S = \inf\{x \in \mathbb{R} : F(x) = 1\}$, with the convention $\inf \emptyset = +\infty$.

1. Let (X_n) be an i.i.d. random sequence with distribution $\mathcal{E}(\lambda)$. Compute S and the distribution function $F_{(n)}$ of $X_{(n)}$. Show that $X_{(n)}$ converges a.s. to S.
2. Same questions for the convergence in distribution of an i.i.d. random sequence (Y_n) with density $f(y) = \alpha(1-y)^{\alpha-1}\mathbb{1}_{[0,1]}(y)$, for $\alpha \in \mathbb{N}^*$.

Solution
1. Since $F(x) = 1 - e^{-\lambda x}$, we have $S = +\infty$, and we know that $F_{(n)}(x) = (F(x)^n$ by using (3.5) p. 116. For all $\varepsilon > 0$, we have

$$\sum_{n \geq 1} \mathbb{P}[X_{(n)} < \varepsilon] = \sum_{n \geq 1} F(\varepsilon)^n = \frac{F(\varepsilon)}{1 - F(\varepsilon)} = \frac{1}{e^{-\lambda \varepsilon}} - 1,$$

Borel–Cantelli lemma thus yields $\mathbb{P}[\overline{\lim}(X_{(n)} \leq \varepsilon)] = 0$, meaning that $(X_{(n)})$ converges a.s. to $S = +\infty$.
2. We have $F(y) = [1 - (1-y)^\alpha]\mathbb{1}_{[0,1]}(y) + \mathbb{1}_{]1,+\infty[}(y)$. Hence $F_{X_{(n)}} = F^n$ converges to $\mathbb{1}_{[1,+\infty[}$, the distribution function of the Dirac distribution δ_1. Finally, we check that $S = 1$. $\triangle$

▽ **Exercise 4.9 (Convergence in Probability and in Mean)** Let (X_n) be an integrable random sequence converging in probability to a random variable X. Set $X'_n = \min(X_n, X)$.

1. Show that (X'_n) converges in probability to X.
2. Suppose X is integrable. Show that X'_n converges to X in mean.
3. Suppose moreover that $\mathbb{E} X_n$ converges to $\mathbb{E} X$. Show that X_n converges to X in mean.

Solution

1. We have

$$\mathbb{P}(|X_n' - X| > \varepsilon) = \mathbb{P}(|X_n - X| > \varepsilon, X_n \leq X) \leq \mathbb{P}(|X_n - X| > \varepsilon),$$

and the latter quantity converges to zero by assumption.

2. We have $0 \leq X_n' \leq X$, so $|X_n' - X| = X - X_n'$ is integrable.

$$\mathbb{E}\,|X_n' - X| = \mathbb{E}\left[(X - X_n')\mathbb{1}_{(X - X_n' \leq \varepsilon)}\right] + \mathbb{E}\left[(X - X_n')\mathbb{1}_{(X - X_n' \geq \varepsilon)}\right]$$
$$= \qquad\qquad (i) \qquad\qquad + \qquad\qquad (ii).$$

On the one hand, (i) converges to zero. On the other hand, set $A_n = (X - X_n' \geq \varepsilon)$. Then $\mathbb{P}(A_n)$ converges to zero by 1, so $\mathbb{1}_{A_n}$ converges a.s. to zero, and $X\mathbb{1}_{A_n}$ too. Since $X\mathbb{1}_{A_n} \leq X$ and X is integrable, the dominated convergence theorem yields that (ii) converges to zero. Finally, X_n' converges to X in mean.

3. Let us set $X_n'' = \max(X_n, X)$. We have $\mathbb{E}\,X_n' + \mathbb{E}\,X_n'' = \mathbb{E}\,X_n + \mathbb{E}\,X$. Since $\mathbb{E}\,X_n'$ converges to $\mathbb{E}\,X$ by 2, and $\mathbb{E}\,X_n$ converges to $\mathbb{E}\,X$ by assumption, $\mathbb{E}\,X_n''$ converges to $\mathbb{E}\,X$. The conclusion follows because $\mathbb{E}\,|X_n - X| = \mathbb{E}\,(X_n'' - X_n')$.

Note that the dominated convergence theorem cannot be directly applied to the sequence $(X_n - X)$, hence the necessity to prove the result first for (X_n'). △

∇ **Exercise 4.10 (Equi-integrability and Convergence)** Let (X_n) be an integrable random sequence. It is said to be uniformly integrable if

$$\mathbb{E}\,(\mathbb{1}_{A_N}|X_n|) \longrightarrow 0 \text{ uniformly in } n, \quad N \to +\infty,$$

for all sequences of events (A_N) such that $\mathbb{P}(A_N)$ converges to zero when N tends to infinity.

1.(a) Show that (X_n) is equi-integrable—see Definition 4.5—if and only if it is bounded in L^1 and uniformly integrable.
  (b) **(Extension of Fatou's lemma)** Show that if (X_n) is equi-integrable, then

$$\mathbb{E}\,(\underline{\lim}\,X_n) \leq \underline{\lim}\,\mathbb{E}\,X_n \leq \overline{\lim}\,\mathbb{E}\,X_n \leq \mathbb{E}\,(\overline{\lim}\,X_n).$$

2.(a) Show that if (X_n) is equi-integrable and converges a.s. to X, then X is integrable and $\mathbb{E}\,X_n$ converges to $\mathbb{E}\,X$.
  (b) Show that if (X_n) is equi-integrable and converges in probability to X, then X is integrable and $\mathbb{E}\,X_n$ converges to $\mathbb{E}\,X$. Prove then Proposition 4.47 for $p = 1$.

(c) Show that if $(|X_n|^p)$ is equi-integrable and converges in probability to X, then X_n converges to X in L^p for all $p \geq 1$. Hint: use the inequality $(x + y)^p \leq 2^{p-1}(x^p + y^p)$, true for all $(x, y) \in \mathbb{R}_+^2$ and $p \geq 1$.

(d) Prove Proposition 4.47 for $p > 1$.

Solution

1.(a) Suppose that (X_n) is bounded in L^1 and uniformly integrable. Markov's inequality yields that for all $N \in \mathbb{N}^*$,

$$\mathbb{P}(|X_n| > N) \leq \frac{1}{N}\mathbb{E}|X_n| \leq \frac{1}{N}\sup_{n \geq 0}\mathbb{E}|X_n| \longrightarrow 0, \quad N \to +\infty,$$

so, using the uniform integrability, (X_n) is equi-integrable.

Conversely, let N_0 be such that if $N \geq N_0$, then $\mathbb{E}[\mathbb{1}_{(X_n > N)}|X_n|] \leq 1$. We have

$$\mathbb{E}(|X_n|) = \mathbb{E}[\mathbb{1}_{(X_n > N_0)}|X_n|] + \mathbb{E}[\mathbb{1}_{(X_n \leq N_0)}|X_n|] \leq 1 + N_0,$$

so the sequence $(\mathbb{E}(|X_n|))$ is bounded in L^1. Let $A_N \in \mathcal{F}$ and $\varepsilon > 0$. We have

$$\mathbb{E}[\mathbb{1}_{A_N}|X_n|] = \mathbb{E}[\mathbb{1}_{A_N \cap (X_n > N)}|X_n|] + \mathbb{E}[\mathbb{1}_{A_N \cap (X_n \leq N)}|X_n|]$$

$$\leq \mathbb{E}[\mathbb{1}_{(X_n > N)}|X_n|] + N_1\mathbb{P}(A_N) \leq \varepsilon,$$

if N_1 satisfies $\mathbb{E}[\mathbb{1}_{(X_n > N_1)}|X_n|] \leq \varepsilon/2$ and $\mathbb{P}(A_N) \leq \varepsilon/2N_1$.

(b) Fatou's lemma yields

$$\mathbb{E}(\underline{\lim} X_n) \leq \mathbb{E}[\underline{\lim}\, \mathbb{1}_{(X_n \geq -N)}X_n] \leq \underline{\lim}\, \mathbb{E}[\mathbb{1}_{(X_n \geq -N)}X_n].$$

We can write $\mathbb{E}X_n = \mathbb{E}[\mathbb{1}_{(X_n < -N)}X_n] + \mathbb{E}[\mathbb{1}_{(X_n \geq -N)}X_n]$. Thanks to equi-integrability, $|\mathbb{E}[\mathbb{1}_{(X_n < -N)}X_n]|$ converges to zero, so $\underline{\lim}_n \mathbb{E}[\mathbb{1}_{(X_n \geq -N)}X_n]$ converges to $\underline{\lim}\, \mathbb{E}X_n$ when N tends to infinity.

Therefore, $\mathbb{E}(\underline{\lim} X_n) \leq \underline{\lim}\, \mathbb{E}X_n$. The second inequality can be proven in the same way.

2.(a) According to 1(a), the sequence is bounded in L^1, say by M. Hence, Fatou's lemma yields that

$$\mathbb{E}(|X|) \leq \underline{\lim}\, \mathbb{E}(|X_n|) \leq M < +\infty.$$

The convergence of $\mathbb{E}X_n$ to $\mathbb{E}X$ is an immediate consequence of 1(b).

(b) If $(\mathbb{E}X_n)$ did not converge to $\mathbb{E}X$, some $\varepsilon > 0$ would exist such that $|EX_n - \mathbb{E}X| > \varepsilon$ for an infinite number of n, that is, for a subsequence of (X_n). This subsequence converging in probability to X would have a subsequence converging a.s. to X. This contradicts 2(a).

According to Example 4.6, a sequence bounded by an integrable variable is equi-integrable, so Proposition 4.47 is proven for $p = 1$.

(c) According to 2(b), we know that X^p is integrable and that $\mathbb{E}\,(|X_n|^p)$ converges to $\mathbb{E}\,(|X|^p)$. Moreover, $(|X_n|^p)$ is equi-integrable, hence, according to 1(a), is bounded in L^1. Since

$$|X_n - X|^p \leq (|X_n| + |X|)^p \leq 2^{p-1}(|X_n|^p + |X|^p),$$

the sequence $(|X_n - X|^p)$ is bounded in L^1. Let $A \in \mathcal{F}$. We have

$$\mathbb{E}\,(\mathbb{1}_A|X_n - X|^p) \leq 2^{p-1}[\mathbb{E}\,(\mathbb{1}_A|X_n|^p) + \mathbb{E}\,(\mathbb{1}_A|X|^p)].$$

Since (X_n) is equi-integrable, it is uniformly integrable. Therefore, $(|X_n - X|^p)$ is uniformly integrable too, and hence equi-integrable. Moreover, it converges in probability to zero. According to 2(b), $\mathbb{E}\,(|X_n - X|^p)$ converges to zero when n tends to infinity.

(d) For deducing Proposition 4.47 for $p > 1$ from 2(c), it is sufficient to prove that $(|X_n|^p)$ is equi-integrable. Set $M = \sup_{n \geq 0} \mathbb{E}\,[|X_n|^p]$. For all $\varepsilon > 0$, some N exists such that $|x|^{p-1} \geq M/\varepsilon$ for all real number $x \geq N$. For this N, we have

$$\mathbb{E}\,[\mathbb{1}_{(|X_n|>N)}|X_n|] \leq \mathbb{E}\left[\frac{\varepsilon}{M}\mathbb{1}_{(|X_n|>N)}|X_n|^p\right] \leq \frac{\mathbb{E}\,(|X_n|^p)}{M}\varepsilon \leq \varepsilon,$$

and the conclusion follows.   △

Introduction to Statistics

5

This chapter presents the basics of parametric and non- parametric statistics: point estimation, confidence intervals estimation, statistical tests. A section is especially dedicated to the linear model.

In statistics, unknown distributions are investigated from observations of a random phenomenon, contrary to probability theory, in which random variables with known distributions are considered.

Definition 5.1 A collection of independent random variables $X_1, \ldots, X_n$, each with the same distribution P on $(\mathbb{R}^d, \mathcal{B}(\mathbb{R}^d))$, is called an n-sample of P. The observed values (or observations) $x_1, \ldots, x_n$ are realizations of $X_1(\omega), \ldots, X_n(\omega)$.

If P is the distribution of a random variable X, then $X_1^n = (X_1, \ldots, X_n)$, such that $X_i \sim X$ for all i, is also called an n-sample of X.

Obviously, X_1^n is a real random vector with distribution $P^{\otimes n}$ defined on some probability space $(\Omega, \mathcal{F}, \mathbb{P})$. A possible choice is the identity function on $\mathbb{R}^d$, together with $(\Omega, \mathcal{F}, \mathbb{P}) = (\mathbb{R}^d, \mathcal{B}(\mathbb{R}^d), P)$. Therefore, the probability space is not often specified.

Two main aspects arise. In the parametric statistics setting, the statistical model is $(\mathbb{R}^d, \mathcal{B}(\mathbb{R}^d), (P_\theta)_{\theta \in \Theta})$. In other words, P is supposed to belong to an identified family of distributions $\{P_\theta : \theta \in \Theta\}$, and hence only its parameter $\theta \in \Theta \subset \mathbb{R}^r$ is to be determined. Otherwise—in the so-called non-parametric statistics setting— the statistical model is $(\mathbb{R}^d, \mathcal{B}(\mathbb{R}^d), \mathbf{P})$, where $\mathbf{P}$ may be any set of probabilities.

$\triangleright$ *Example 5.2 (Control of Quality)* A machine produces items from which a proportion p is defective. The value of p is estimated by controlling n of these items taken at random. Here, the distribution P is the Bernoulli distribution $\mathcal{B}(p)$. $\triangleleft$

$\triangleright$ *Example 5.3 (Error of Measure)* An instrument measures a certain physical quantity m, up to some error. The measure is assumed to be correct in mean, but

V. Girardin, N. Limnios, *Applied Probability*,
https://doi.org/10.1007/978-3-030-97963-8_5

to have a positive standard deviation. In other words, we observe the realizations $x_i = m + e_i$ of an n-sample of $X = m + \varepsilon$, with $\mathbb{E}\,\varepsilon = 0$ and $\mathbb{V}\text{ar}\,\varepsilon = \sigma^2$.

On the one hand, the instrument may have been calibrated prior to the experiment, while m is to be determined—for instance the weight or height of an individual.

On the other hand, the measured quantity m can be known in advance—for instance the speed of sound or light, and we aim to identify the error of measure of the instrument. If nothing is known on the noise ε, non-parametric methods will be used. Alternatively, a Gaussian noise may be assumed, $\varepsilon \sim \mathcal{N}(0, \sigma^2)$, of which σ^2 is the unknown.

Finally, both the measured quantity and the error of measure can be unknown. For instance the distribution P may be a normal distribution $\mathcal{N}(m, \sigma^2)$—the statistical model is said to be Gaussian, with parameter $(m, \sigma^2) \in \mathbb{R} \times \mathbb{R}_+^*$. $\triangleleft$

The notions presented here for random variables carry over to random vectors with only minor changes.

5.1   Non-parametric Statistics

Non-parametric methods are of use when no prior knowledge is available on the distribution of the observed sample.

5.1.1   Empirical Distribution Function

Definition 5.4 Let X_1^n be an n-sample of a distribution P on $(\mathbb{R}, \mathcal{B}(\mathbb{R}))$. Its empirical distribution is the function $\widehat{P}_n$ defined on $\Omega \times \mathcal{B}(\mathbb{R})$ by

$$\widehat{P}_n(\omega, A) = \frac{1}{n} \sum_{i=1}^{n} \delta_{X_i(\omega)}(A).$$

The probability $\sum_{i=1}^{n} \delta_{x_i}/n$ is associated with each set of observations $x_1, \ldots, x_n$. For any given A, the function $\widehat{P}_n(\cdot, A)$ is a random variable, and for any given ω, the function $\widehat{P}_n(\omega, \cdot)$ is a probability; precisely, $\widehat{P}_n$ is a random distribution.

We will denote by $\widehat{F}_n(t)$ the random variable and by $\widehat{F}_n(\omega, .)$ the (empirical) distribution function associated with the empirical distribution $\widehat{P}_n$, both defined by

$$\widehat{F}_n(\omega, t) = \frac{1}{n} \sum_{i=1}^{n} \mathbb{1}_{(X_i(\omega) \leq t)}.$$

Clearly, $\widehat{F}_n(\omega, .)$ is a step function. If F is the distribution function of P, then $n\widehat{F}_n(t) \sim \mathcal{B}(n, F(t))$, with

$$\widehat{F}_n(\omega, t) = \frac{k}{n} \quad \text{if } X_{(k)} \leq t < X_{(k+1)}, \quad 0 \leq k \leq n,$$

where $(X_{(1)}, \ldots, X_{(n)})$ is the order statistics of $(X_1, \ldots, X_n)$, and with the convention $X_{(0)} = -\infty$ and $X_{(n+1)} = +\infty$.

The strong law of large numbers takes the following form for empirical distribution functions, as induced by the ordinary one applied to the sequence $(\mathbb{1}_{(X_i \leq t)})$.

Proposition 5.5 (Empirical Law of Large Numbers) $\widehat{F}_n(t)$ *converges a.s. to* $F(t)$ *when n tends to infinity.*

The next result is stronger.

Theorem 5.6 (Glivenko–Cantelli) *The convergence of the empirical distribution functions is a.s. uniform, that is*

$$\sup_{t \in \mathbb{R}} |\widehat{F}_n(t) - F(t)| \xrightarrow{a.s.} 0, \quad n \to +\infty.$$

Proof Let $\varepsilon > 0$. Set $t_k = \inf\{t \in \mathbb{R} : F(t) \geq k\varepsilon\}$ for $k \in [\![1, N-1]\!]$ and $N \leq 1/\varepsilon$, where $t_0 = -\infty$, $t_N = +\infty$, with $F(t_N) = 1$ and $F(t_0) = 0$.

The set $\{t_0, \ldots, t_N\}$ is a finite partition of $\mathbb{R}$ such that $F(t_{k+1}^-) - F(t_k) \leq \varepsilon$, for $0 \leq k \leq N - 1$, where $F(t^-) = \lim_{h \to 0^+} F(t - h)$. This subdivision also contains the discontinuity points of F with jumps higher than ε.

For all $t \in [t_k, t_{k+1}[$, we have

$$\widehat{F}_n(\omega, t_k) - F(t_k) - \varepsilon \leq \widehat{F}_n(\omega, t) - F(t) \leq \widehat{F}_n(\omega, t_{k+1}^-) - F(t_{k+1}^-) + \varepsilon. \quad (5.1)$$

By the empirical law of large numbers, a null set E exists such that for all $\omega \notin E$, the sequences $\widehat{F}_n(\omega, t_k)$ and $\widehat{F}_n(\omega, t_k^-)$ converge respectively to $F(t_k)$ and $F(t_k^-)$, for all $k \in [\![1, N - 1]\!]$. Therefore, for n large enough and for all $\omega \notin E$,

$$|\widehat{F}_n(\omega, t_k) - F(t_k)| \leq \varepsilon \quad \text{and} \quad |\widehat{F}_n(\omega, t_k^-) - F(t_k^-)| \leq \varepsilon. \quad (5.2)$$

Inequalities (5.1) and (5.2) yield $\sup_{t \in \mathbb{R}} |\widehat{F}_n(\omega, t) - F(t)| < 2\varepsilon$ a.s., and the proof is finished. □

Corollary 5.7 *The sequence of empirical distributions* $\widehat{P}_n(\omega, .)$ *converges strongly to* P *when n tends to infinity.*

Proof Since the semi-open intervals generate the Borel σ-algebra $\mathcal{B}(\mathbb{R})$, this result derives from the inequality

$$|\widehat{P}_n(\omega,]a, b]) - P(]a, b])| \leq |\widehat{F}_n(\omega, b) - F(b)| + |\widehat{F}_n(\omega, a) - F(a)|,$$

true for $a < b$.   □

The next result derives straightforwardly from the central limit theorem.

Proposition 5.8 (Empirical Central Limit Theorem) *The following convergence in distribution holds true.*

$$\sqrt{n}\,\frac{\widehat{F}_n(t) - F(t)}{\sqrt{F(t)(1 - F(t))}} \xrightarrow{\mathcal{D}} \mathcal{N}(0, 1).$$

Note that the (worst) convergence speed can be specified by the iterated logarithm law, which of course is a divergence result.

Functions of the sample—called statistics—of which estimators are particular cases are the basic tools of mathematical statistics.

Definition 5.9 Let $X_1^n = (X_1, \ldots, X_n)$ be an n-sample of a distribution P. A (deterministic) function of the sample, say $T_n = T_n(X_1, \ldots X_n)$, is called a statistic.

If T_n can be written as a function T of the empirical distribution function $\widehat{F}_n$, say $T_n = T(\widehat{F}_n)$ where T does not depend on n, then T is called a statistical functional.

▷ *Example 5.10 (Linear Functionals)* Let φ be a Borel function. Then

$$T_n = \frac{1}{n} \sum_{i=1}^{n} \varphi(X_i)$$

is a statistical functional. Indeed, $T_n(X_1, \ldots, X_n) = T(\widehat{F}_n)$, where T is the functional defined by

$$T(F) = \int \varphi(x) dF(x).$$

Such functionals are of the linear type.   ◁

Definition 5.11 Let X_1^n be an n-sample of P and let g denote a real functional defined on a set of probabilities $\mathbf{P}$ containing P. An estimator of $g(P)$ is a statistic T_n taking values in $g(\mathbf{P})$.

Let T be a statistical functional defined on a set $\mathbf{F}$ containing the distribution function F and the empirical distribution functions F_n, for $n \geq 1$. An estimator of $T(F)$ is a statistic T_n taking values in $T(\mathbf{F})$.

If $x_1^n = (x_1, \ldots, x_n)$ is a set of observations of X_1^n, then $t_n = T_n(x_1 \ldots, x_n)$ is said to be an estimate of $g(P)$, itself called the estimand.

▷ *Example 5.12 (Statistical Functionals)* The mean, the variance, and more generally the moments of a distribution are statistical functionals. ◁

We will denote by $\mathbb{E}_P T_n$ the mean of any estimator T_n when the common distribution of the n-sample is P, or simply $\mathbb{E}\, T_n$ where no ambiguity may arise.

Two ideas important in statistics appear here, the asymptotic properties—for large sample sizes—and the mean properties of estimators.

Definition 5.13 An estimator T_n of $g(P)$ for $P \in \mathbf{P}$ is said to be

- Unbiased if $\mathbb{E}_P T_n = g(P)$.
- Asymptotically unbiased if $\mathbb{E}_P T_n$ converges to $g(P)$.
- Convergent (or consistent) if T_n converges in probability to $g(P)$.
- Strongly convergent if T_n converges a.s. to $g(P)$.

All above convergences are meant for n tending to infinity.

Empirical estimators can be deduced from the continuous function theorem and the empirical law of large numbers.

Definition 5.14 Let g be a functional defined on a set of probabilities containing both P and the empirical distribution $\widehat{P}_n$ for all integer n. Then $g(\widehat{P}_n)$ is called an empirical estimator of $g(P)$.

▷ *Example 5.15 (Estimators of Quantiles)* Let X_1^n be an n-sample of a continuous distribution P. The empirical median is

$$\widehat{q}_n(1/2) = \begin{cases} X_{(m+1)} & \text{if } n = 2m + 1, \\ (X_{(m)} + X_{(m+1)})/2 & \text{if } n = 2m, \end{cases}$$

where $(X_{(1)}, \ldots, X_{(n)})$ denotes the order statistics of the sample.

For $\alpha \neq 1/2$, the empirical estimator of the quantile q_α of order α, that is q_α such that $P(] - \infty, q_\alpha]) = \alpha$, is

$$\widehat{q}_n(\alpha) = \begin{cases} X_{(n\alpha)} & \text{if } n\alpha \in \mathbb{N}, \\ X_{([n\alpha]+1)} & \text{otherwise.} \end{cases}$$

Moreover, if F is non-decreasing, then q_α is known to be unique, and one can show that $\widehat{q}_n(\alpha)$ converges a.s. to q_α. Finally, if F is continuously differentiable, then $\sqrt{n}(\widehat{q}_n(\alpha) - q_\alpha)$ converges in distribution to a centered Gaussian distribution whose variance is a function of α.   ◁

In   the method of moments, if X_1^n is an n-sample of a random variable X, the empirical estimator of $\mathbb{E}[h(X)^p]$ is $\sum_{i=1}^{n} h(X_i)^p/n$, for any Borel function h such that $h(X) \in L^p$. The estimators $\overline{X}_n$ and S_n^{*2} below are obtained by the method of moments and called the empirical (or sampling) mean and variance of the n-sample.

Theorem 5.16 *If X_1^n is an n-sample of a distribution P with mean m and variance $\sigma^2 \neq 0$, then*

$$\overline{X}_n = \frac{1}{n} \sum_{i=1}^{n} X_i \quad and \quad S_n^{*2} = \frac{1}{n-1} \sum_{i=1}^{n} (X_i - \overline{X}_n)^2$$

are unbiased and convergent estimators of m and σ^2, respectively. Moreover,

$$\sqrt{n}\frac{(\overline{X}_n - m)}{\sigma} \xrightarrow{\mathcal{D}} \mathcal{N}(0,1) \quad and \quad \sqrt{n}\frac{(\overline{X}_n - m)}{S_n^*} \xrightarrow{\mathcal{D}} \mathcal{N}(0,1).$$

Proof For the mean,

$$\mathbb{E}\,\overline{X}_n = \frac{1}{n} \sum_{i=1}^{n} \mathbb{E}\,X_i = \frac{nm}{n} = m.$$

By the strong law of large numbers applied to (X_n), we know that $\overline{X}_n$ converges a.s. to m. The convergence in distribution is a restatement of the central limit theorem.
  For the variance,

$$\sum_{i=1}^{n} (X_i - m)^2 = \sum_{i=1}^{n} (X_i - \overline{X}_n)^2 + n(\overline{X}_n - m)^2 + 2\sum_{i=1}^{n} (X_i - \overline{X}_n)(\overline{X}_n - m).$$

Since $\sum_{i=1}^{n}(X_i - \overline{X}_n)(\overline{X}_n - m) = (\overline{X}_n - m)(\sum_{i=1}^{n} X_i - n\overline{X}_n) = 0$, we have

$$\sum_{i=1}^{n} \mathbb{V}\mathrm{ar}\,X_i = \mathbb{E}\left[\sum_{i=1}^{n}(X_i - m)^2\right] = \mathbb{E}\left[\sum_{i=1}^{n}(X_i - \overline{X}_n)^2\right] + n\mathbb{E}\left[(\overline{X}_n - m)^2\right].$$

Using $\mathbb{E}\left[(\overline{X}_n - m)^2\right] = \operatorname{Var} \overline{X}_n = \sigma^2/n$, we get $\mathbb{E}\left[\sum_{i=1}^n (X_i - \overline{X}_n)^2\right] = (n-1)\sigma^2$. Therefore, this estimator is biased, but S_n^* is unbiased. Let us show that it converges to σ^2. First,

$$(X_i - \overline{X}_n)^2 = X_i^2 + \frac{1}{n^2}\left(\sum_{j=1}^n X_j\right)^2 - \frac{2}{n}\sum_{j=1}^n X_i X_j,$$

so

$$S_n^{*2} = \frac{1}{n-1}\sum_{i=1}^n X_i^2 + \frac{n}{(n-1)n^2}\left(\sum_{i=1}^n X_j\right)^2 - \frac{2}{n(n-1)}\sum_{i=1}^n\sum_{j=1}^n X_i X_j$$

$$= \frac{1}{n-1}\left(\sum_{j=1}^n X_i^2\right) - \frac{n}{n-1}\overline{X}_n^2.$$

Since $\mathbb{E}(X_i^2) = \operatorname{Var} X_i + (\mathbb{E} X_i)^2 = \sigma^2 + m^2$, we obtain by the strong law of large numbers applied to (X_i^2) that

$$\frac{1}{n-1}\sum_{i=1}^n X_i^2 \overset{\text{a.s.}}{\to} \sigma^2 + m^2.$$

But $\overline{X}_n$ converges a.s. to m, so $\frac{n}{n-1}\overline{X}_n^2$ converges a.s. to m^2, and hence S_n^{*2} converges a.s. to σ^2.

Finally, $\sqrt{n}(\overline{X}_n - m)$ converges in distribution to $\mathcal{N}(0, \sigma^2)$ by the central limit theorem, and hence, thanks to Slutsky's theorem, $\sqrt{n}(\overline{X}_n - m)/S_n^*$ converges in distribution to $\mathcal{N}(0, 1)$. $\square$

Note that $\tilde{S}_n^2 = \sum_{i=1}^n (X_i - \overline{X}_n)^2/n$ is a biased estimator of the variance σ^2. If m is known, $S_n^2 = \sum_{i=1}^n (X_i - m)^2/n$ is an unbiased estimator of σ^2 because $\mathbb{E}(X_i - m)^2 = \operatorname{Var} X_i = \sigma^2$ for all i.

5.1.2 Confidence Intervals

Let X_1^n be an n-sample of a distribution P, and let g be a real functional defined on a set of probabilities $\mathbf{P}$ containing P. An interval whose bounds are functions of the sample is called a random interval.

Let $T_n = T_n(X_1, \dots X_n)$ be an estimator of $g(P)$. If the distribution of T_n is available, one can determine, from a set of observations $x_1^n = (x_1, \dots, x_n)$, a random interval containing $g(P)$ with a large probability.

Definition 5.17 Let I_n be a Borel function defined on $\mathbb{R}^n$ taking values in the set of intervals of $\mathbb{R}$. Let $\alpha \in]0, 1[$. Then $I_n = I_n(X_1, \ldots, X_n)$ is called a confidence interval for $g(P)$ at significance level $1 - \alpha$ if

$$P[g(P) \in I_n] \geq 1 - \alpha.$$

In view of the observations, $g(P)$ belongs to I_n with probability $1 - \alpha$. Of course, it is necessary to arbitrate between the size of the interval and the level.

▷ *Example 5.18 (Chebyshev's Inequality and Confidence Intervals)* By Chebyshev's inequality, if $T_n = T_n(X_1, \ldots X_n)$ is an unbiased estimator of $g(P)$, then

$$P[|T_n - g(P)| \geq c\sqrt{k}] \leq \frac{1}{c^2 k}\mathbb{E}_P([T_n - g(P)]^2), \quad c > 0, \ k > 0.$$

In particular, if $k_0 = \sup_{P \in \mathbf{P}} \mathbb{E}_P([T_n - g(P)]^2)$ is finite, then $P[|T_n - g(P)| \geq c\sqrt{k_0}] \leq 1/c^2$. For $c = 1/\sqrt{\alpha}$, we obtain the confidence interval

$$I_n = \left]T_n(X_1, \ldots, X_n) - \sqrt{k_0/\alpha}, \ T_n(X_1, \ldots, X_n) + \sqrt{k_0/\alpha}\right[$$

at level $1 - \alpha'$, with $\alpha' \leq \alpha$. The intervals obtained in this way are very large in general, because α' (whose exact value is unknown) may be much smaller than α. ◁

▷ *Example 5.19 (Confidence Interval for the Mean)* If n is large enough, Theorem 5.16 yields confidence intervals for the mean, for any distribution. For example, if the variance σ^2 of the distribution is known, then

$$I_n = \left]\overline{X}_n - r_{1-\alpha/2}\sigma/\sqrt{n}, \ \overline{X}_n + r_{1-\alpha/2}\sigma/\sqrt{n}\right[$$

is a confidence interval for the mean at level $1 - \alpha$, where $r_{1-\alpha/2}$ is the quantile of order $1 - \alpha/2$ of the distribution $\mathcal{N}(0, 1)$. ◁

The next example is more elaborated.

▷ *Example 5.20 (Confidence Tube for the Distribution Function)* Let F denote the distribution function of P. Let $r_{1-\alpha/2}$ be the quantile of order $1 - \alpha/2$ of the distribution $\mathcal{N}(0, 1)$. Thanks to the empirical law of large numbers, for n large enough,

$$I_n(t) = \left]\widehat{F}_n(t) - r_{1-\alpha/2}\sqrt{\frac{\widehat{F}_n(t)[1 - \widehat{F}_n(t)]}{n}}, \ \widehat{F}_n(t) + r_{1-\alpha/2}\sqrt{\frac{\widehat{F}_n(t)[1 - \widehat{F}_n(t)]}{n}}\right[$$

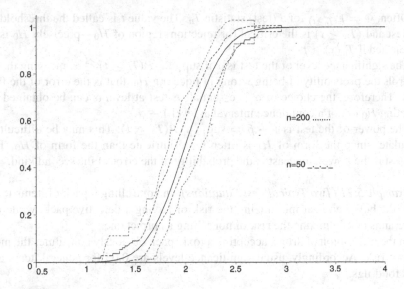

Fig. 5.1 Confidence tubes for F of $\mathcal{N}(2, 1/3)$—Example 5.20

is a confidence interval for $F(t)$ at the level $1-\alpha$, for any fixed t. Glivenko–Cantelli theorem ensures that the size of $I_n(t)$ is uniformly bounded in t. All these intervals constitute a confidence tube for F at level $1-\alpha$, as shown in Fig. 5.1.   ◁

5.1.3 Non-parametric Testing

Assume we study an unknown distribution P, belonging to a given set of distributions **P**, in order to validate some—statistical—hypothesis, say H_0:"$P \in \mathbf{P}_0$," against an alternative H_1: "$P \in \mathbf{P}_1$," where $\mathbf{P}_0$ and $\mathbf{P}_1$ are disjoint subsets of **P**.

A classical case is $\mathbf{P}_0 = \{P_0\}$, in order to decide if P is equal—or not—to a known distribution P_0, for instance encountered previously in the study of a likely phenomenon. This is a simple hypothesis in goodness-of-fit tests.

Another classical case is to decide whether or not two independently observed phenomena follow the same distribution.

All in all, we have to take a decision depending on observations of an n-sample through a binary test function Φ. The choice involves two possible errors: either rejecting the so-called null hypothesis H_0 when it is true, called the error of the first kind; or not rejecting it when the alternative is true, the error of the second kind. The two hypotheses are not symmetrical—H_0 is favored, and thus the induced cost of the error of the first kind is the worst.

Often, $\Phi = \mathbb{1}_{(T_n \geq t)}$ for a (test) statistic T_n. The value t is called the threshold of the test and $(T_n \geq t)$ is the critical (or rejection) region of H_0—precisely H_0 is to be rejected if $T_n(x) \geq t$.

The significance level of the test is α if $\sup_{P \in \mathbf{P}_0} P(T_n \geq t) \leq \alpha$, meaning that α controls the probability of being wrong in rejecting H_0, that is the error of the first kind. Therefore, the choice of α is essential. A test at level α can be obtained by rejecting H_0 out of a confidence interval at level $1 - \alpha$.

The power of the test is $1 - \beta = \sup_{P \in \mathbf{P}_1} P(T_n \geq t)$. This may be difficult to calculate, since the form of H_1 is often more intricate than the form of H_0. The greatest is the power, the least is the probability of the error of the second kind.

▷ *Example 5.21 (Two Typical Test Situations)* In controlling a pack of items to be sold, the buyer aims at minimizing the risk of buying a defective pack, while the seller aims at minimizing the risk of not selling a correct one.

In the production of drugs, accepting a toxic pack obviously constitutes the most serious risk. Accordingly, usual significance levels are 0.05 for general items and 0.01 for drugs.   ◁

The following results are linked to a typical non-parametric goodness-of-fit test. The second one is stated without proof.

Theorem 5.22 (Kolmogorov–Smirnov) *If X_1^n is an n-sample of an unknown distribution P with continuous distribution function F, then the distribution of the random variable $D_n = \sup_{t \in \mathbb{R}} |\widehat{F}_n(t) - F(t)|$ does not depend on P. Moreover,*

$$F_{D_n}\left(\frac{x}{\sqrt{n}}\right) \longrightarrow \sum_{k \in \mathbb{Z}} (-1)^k e^{-2k^2 x^2} \mathbb{1}_{\mathbb{R}_+}(x), \quad n \to +\infty.$$

Proof Set $U_i = F(X_i)$ for $i \in [\![1, n]\!]$. From Theorem 4.81, these random variables are known to be i.i.d. with distribution $\mathcal{U}(0, 1)$. Then

$$D_n = \sup_{1 \leq i \leq n} (|U_{(i)} - i/n|, |U_{(i)} - (i-1)/n|),$$

where $(U_{(1)}, \ldots, U_{(n)})$ is the order statistics of the sample U_1^n, and hence the distribution of D_n does not depend on P.   □

The Kolmogorov–Smirnov goodness-of-fit test derives straightforwardly from the above theorem. Let X_1^n be an n-sample of a distribution P. Suppose we have to compare P to a known distribution P_0 with continuous distribution function F, that is to test H_0: "$P = P_0$" against H_1: "$P \neq P_0$."

Under H_0, the distribution $\mathcal{D}_n$ of the above statistic D_n does not depend on F and, by the Glivenko–Cantelli theorem, D_n converges a.s. to 0. Under the alternative H_1, it does not converge to 0. Thus, the hypothesis H_0 is to be rejected if $D_n(x) \geq d_{n,1-\alpha}$, quantile of order $1 - \alpha$ of $\mathcal{D}_n$.

The convergence of the distribution function shows that for n large enough, the distribution of $\sqrt{n}D_n$ does not depend on n either. The limit function is the distribution function of a distribution $\mathcal{D}$ whose quantiles can be computed. So, when n is large, H_0 is to be rejected if $\sqrt{n}D_n(x) \geq d_{1-\alpha}$, quantile of order $1 - \alpha$ of $\mathcal{D}$.

Similarly, the distributions of two samples can be compared through a test, thanks to the following extension of Kolmogorov–Smirnov theorem.

Theorem 5.23 *Let $X_1^m = (X_1, \ldots, X_m)$ and $Y_1^n = (Y_1, \ldots, Y_n)$ be two samples of a continuous distribution P. Let $\widehat{F}_m^X$ and $\widehat{F}_n^Y$ denote the respective empirical distribution functions of the two samples. The distribution $\mathcal{D}_{m,n}$ of the random variable defined by $D_{m,n}(\omega) = \sup_{x \in \mathbb{R}}[\widehat{F}_m^X(\omega, x) - \widehat{F}_n^Y(\omega, x)]$ does not depend on P.*

Proof Set $U_i = F(X_i)$ and $V_j = F(Y_j)$, where F is the distribution function of P. All these random variables have the same distribution $\mathcal{U}(0, 1)$. Denoting as usual the order statistics of X_1^n, Y_1^n, U_1^n, and V_1^n, we can write

$$D_{m,n} = \sup\left\{\frac{i}{m} - \frac{j}{n} \,:\, Y_{(j)} \leq X_{(i)} < Y_{(j+1)}, \ i \in [\![1, m]\!], \ j \in [\![0, n-1]\!]\right\}$$

$$= \sup\left\{\frac{i}{m} - \frac{j}{n} \,:\, V_{(j)} \leq U_{(i)} < V_{(j+1)}, \ i \in [\![1, m]\!], \ j \in [\![0, n-1]\!]\right\},$$

and the result follows.   □

The Kolmogorov–Smirnov test for comparing two unknown distributions derives straightforwardly from the above theorem. Let $X_1^m = (X_1, \ldots, X_m)$ be an m-sample of a continuous distribution P with distribution function F, and let $Y_1^n = (Y_1, \ldots, Y_n)$ be an n-sample of a continuous distribution Q with distribution function $\widetilde{F}$. Let us test H_0: "$F = \widetilde{F}$" against H_1: "$F(x) \leq \widetilde{F}(x)$, $x \in \mathbb{R}$." Under H_1, the values taken by X are smaller than those taken by Y, so H_0 is to be rejected if $D_{m,n}(x, y) \geq d_{m,n,1-\alpha}$, the quantile of order $1 - \alpha$ of $\mathcal{D}_{m,n}$.

Another set of tests involves the chi-squared distance.

Definition 5.24 Let P and Q be two distributions on $[\![1, d]\!]$. Set $p_i = P(\{i\})$ and $q_i = Q(\{i\})$. The quantity

$$\chi^2(P, Q) = \sum_{i=1}^{d} \frac{(p_i - q_i)^2}{q_i},$$

is called the chi-squared distance between P and Q.

Mathematically speaking, the chi-squared distance is not a distance since it is not symmetric: It is an entropy functional, precisely an alternative to Kullback–Leibler divergence.

The next result is a straightforward consequence of Cochran's theorem. A direct proof might be derived from noticing that $N_n = (N_n(1), \ldots, N_n(d))$ has a multinomial distribution $\mathcal{M}(n; p_1, \ldots, p_d)$.

Theorem 5.25 *Let* $(X_1, \ldots, X_n)$ *be an n-sample of a distribution* P *defined on* $[\![1, d]\!]$. *Set* $p_i = P(\{i\})$ *and* $N_n(i) = \sum_{j=1}^n \mathbb{1}_{(X_j=i)}$. *Then*

$$T_n = \sum_{i=1}^d \frac{(N_n(i) - np_i)^2}{np_i} \xrightarrow{\mathcal{D}} \chi^2(d-1), \quad n \to +\infty.$$

The chi-squared goodness-of-fit test derives straightforwardly. Let $(X_1, \ldots, X_n)$ be an n-sample of a distribution P. Let us compare P to a known distribution Q, that is test H_0: "$P = Q$" against H_1: "$P \neq Q$."

For a discrete distribution on $[\![1, d]\!]$, the empirical estimator of P is given by $\widehat{P}_n(i) = N_n(i)/n$ for $i \in [\![1, d]\!]$. Under H_0, thanks to Theorem 5.25, $\chi^2(\widehat{P}_n, P)$ converges in distribution to $\chi^2(d-1)$; under H_1, $\chi^2(\widehat{P}_n, P)/n$ is known to converge a.s. to $\chi^2(P, Q)$ by the law of large numbers, so that $\chi^2(\widehat{P}_n, P)$ converges to infinity. Therefore, H_0 is to be rejected if $\chi^2(\widehat{P}_n(x), P) \geq \chi^2_{d-1,1-\alpha}$, the quantile of order $1 - \alpha$ of the distribution $\chi^2(d-1)$.

For a general distribution, we come back to the discrete case by considering $p_i = P(X \in I_i)$ where $\{I_i : i \in [\![1, d]\!]\}$ is a partition of $\mathbb{R}$. Let us set $Y_i = \mathbb{1}_{(X \in I_i)}$ and $q_i = Q(X \in I_i)$ and test H_0: "$p_i = q_i, i \in [\![1, d]\!]$" against H_1: "$\exists i, p_i \neq q_i$." Nevertheless, information on P is lost by this ranking by intervals, which does not happen with Kolmogorov–Smirnov test. Note that each of the classes has to contain enough elements—say at least 5—for the test to be somewhat meaningful.

A test of independence also derives from the properties of the chi-squared distance, the so-called chi-squared independence test.

Let X_1^n and Y_1^n be n-samples of two random variables X and Y taking respective values in $[\![1, d]\!]$ and $[\![1, d']\!]$. Let us test the independence of X and Y. If they are independent, then the distribution of the pair (X, Y) is the product of their marginal distributions, so H_0: "$P_{ij} = P_{i.}P_{.j}, i \in [\![1, d]\!], j \in [\![1, d']\!]$" and H_1: "$\exists(i, j), P_{ij} \neq P_{i.}P_{.j}$", where $P_{i.} = \sum_{j=1}^{d'} P_{ij}$ and $P_{.j} = \sum_{i=1}^d P_{ij}$.

The empirical distribution of the pair is

$$\widehat{P}_n(i, j) = \frac{1}{n} N_n(i, j) = \frac{1}{n} \sum_{l=1}^n \sum_{m=1}^n \mathbb{1}_{(X_l=i, Y_m=j)}.$$

Under H_0, this is $\overline{P}_n(i, j) = N_n(i, .)N_n(., j)/n^2$. The sequence of random variables $\chi^2(\widehat{P}_n, \overline{P}_n)$ can be shown to converge to a $\chi^2((d-1)(d'-1))$ distribution under H_0, and to infinity under H_1. The chi-squared independence test follows:

The null hypothesis H_0 is to be rejected if $\chi^2(\widehat{P}_n(x, y), \overline{P}_n(x, y)) \geq \chi^2_{(d-1)(d'-1),1-\alpha}$, the quantile of order $1 - \alpha$ of the distribution $\chi^2((d-1)(d'-1))$.

5.2 Parametric Statistics

Most of the notions developed above for real non-parametric statistical models have obvious equivalents in real parametric statistics. Therefore, we will rather insist on the differences.

Definition 5.26 A triple $(\mathbb{R}^d, \mathcal{B}(\mathbb{R}^d), \{P_\theta : \theta \in \Theta\})$ is called a parametric statistical model if P_θ is a probability on $(\mathbb{R}^d, \mathcal{B}(\mathbb{R}^d))$ for all $\theta \in \Theta$.

The set Θ is called the set of parameters. We will suppose that $\Theta \subset \mathbb{R}^r$ for some $r \geq 1$. The model will be assumed to be identifiable: if $\theta_1 \neq \theta_2$, then $P_{\theta_1} \neq P_{\theta_2}$.

▷ *Example 5.27 (Control of Quality)* Clearly, in Example 5.2, for a family of Bernoulli distributions, $P = \mathcal{B}(p)$, $\theta = p$ and $\Theta = [0, 1]$, and the model is identifiable.   ◁

▷ *Example 5.28 (Gaussian Statistical Model)* In Example 5.3, where $P = \mathcal{N}(m, \sigma^2)$, we have $\theta = (m, \sigma^2)$ and $\Theta = \mathbb{R} \times \mathbb{R}_+$, and the model is identifiable.◁

Furthermore, we will assume that for all $\theta \in \Theta$, P_θ is absolutely continuous with respect to a σ-finite reference measure μ, with density f_θ, where $f_\theta(x) > 0$ on $S_\theta \subset \mathbb{R}^d$. Then P_θ is said to have the support S_θ and the model to be dominated by μ. In the discrete case μ is the counting measure and we will set $f_\theta(x) = \mathbb{P}_\theta(\{x\})$.

Definition 5.29 The likelihood of an n-sample of a distribution $\mathbb{P}_\theta$ is the function L defined on $\Theta \times \mathbb{R}^n$ by $L(\theta, x) = L(\theta, x_1, \ldots, x_n) = \prod_{i=1}^n f_\theta(x_i)$.

Precisely, we will study the partial function $\theta \to L(\theta, x)$, for $x \in \mathbb{R}^d$, also called likelihood. Note that the definition of the likelihood depends on the measure μ, which is not unique in general.

▷ *Example 5.30 (Continuation of Example 5.27)* For a 4-sample of a Bernoulli distribution $\mathcal{B}(\theta)$, we have

$$L(\theta, x_1, x_2, x_3, x_4) = \prod_{i=1}^4 \theta^{\sum_{i=1}^4 x_i}(1 - \theta)^{\sum_{i=1}^4 (1-x_i)},$$

with $\theta \in [0, 1]$ and $x_i \in \{0, 1\}$ for $i = 1, \ldots, 4$.   ◁

Definition 5.31 Let $(X_1, \ldots, X_n)$ be an n-sample of the distribution P_θ, with $\theta = (\theta_1, \ldots, \theta_r)$. A statistic T_n is said to be

- Free for the parameter θ_i if its distribution does not depend on θ_i.
- Sufficient (or comprehensive) for the parameter θ_i if the conditional distribution of X given $(T_n = t)$ does not depend on θ_i.

Note that if T_n is sufficient for θ_i and if φ is a strictly monotonous Borel function, then $\varphi(T_n)$ is also a sufficient statistic for θ_i.

A sufficient statistic contains all the information in the sample relevant to the estimation of the specified parameter. In other words, the knowledge of T_n and its distribution is sufficient to provide that information. This property is checked in the likelihood.

Theorem 5.32 (Factorization) *Let X_1^n be an n-sample of the distribution P_θ and let T_n be a statistic taking values in $\mathbb{R}^r$, with density h_θ. Then T_n is sufficient for θ if and only if a Borel function $g : S_\theta \to \mathbb{R}_+$ exists such that*

$$L(\theta, x) = h_\theta(T_n(x))g(x), \quad \theta \in \Theta, \; n \in \mathbb{N}^*. \tag{5.3}$$

One can show that if (5.3) holds true for one $n > 1$, then it holds true for all integers.

Proof For discrete variables.

Set $E_n = \{x \in \mathbb{R}^n : T_n(x) = t\}$. We have $(T_n = t) = \cup_{x \in E_n}(X = x)$, so $P_\theta(T_n = t) = \sum_{x \in E_n} h_\theta(t)g(x)$. Therefore, if (5.3) holds true, then

$$P_\theta(X = x \mid T_n = t) = \frac{P_\theta(X = x, T_n = t)}{P_\theta(T_n = t)}$$

$$= \begin{cases} 0 & \text{if } x \notin E_n, \\ \dfrac{h_\theta(t)g(x)}{P_\theta(T_n = t)} = \dfrac{g(x)}{\sum_{x \in E_n} g(x)} & \text{if } x \in E_n. \end{cases}$$

Conversely, the functions $h_\theta(t) = P_\theta(T_n = t)$ and $g(x) = P_\theta(X = x \mid T_n = t)$ fulfil the required conditions. $\square$

▷ *Example 5.33 (Sufficient Statistics for the Poisson Distribution)* The sum S_n of an n-sample of a Poisson distribution $\mathcal{P}(\lambda)$ is a sufficient statistic for λ. Indeed, $S_n \sim \mathcal{P}(n\lambda)$, and

$$L(\lambda, x) = \prod_{i=1}^n \frac{e^{-\lambda x_i} \lambda^{x_i}}{x_i!} = h_\lambda(s) \frac{s! \, n^s}{\prod_{i=1}^n (x_i)!},$$

where $h_\lambda(s) = \mathbb{P}(S_n = s) = e^{-n\lambda}(\lambda n)^s/s!$ ◁

5.2.1 Point Estimation

Definitions of estimators in parametric statistics derive straightforwardly from the general case.

Definition 5.34 Let X_1^n be an n-sample of P_θ and let $g : \Theta \to \mathbb{R}^s$, with $s \leq r$. An estimator of $g(\theta)$ is a statistic $T_n = T_n(X_1, \ldots, X_n)$ taking values in $g(\Theta)$.

Since the distribution of T_n depends on the value of θ, we set

$$\mathbb{E}_\theta T_n = \int_{\mathbb{R}^n} T_n \, dP_\theta.$$

If $(x_1, \ldots, x_n)$ are the observations of X_1^n, then the estimate of $g(\theta)$ will be the value $T_n(x_1 \ldots, x_n)$ taken by the estimator. Note that if T_n is an estimator of θ, then $g(T_n)$ is called a plug-in estimator of $g(\theta)$.

For an estimator to be of use, some knowledge of its distribution is necessary. Asymptotic results are given by Theorem 5.16, whichever be the estimated distribution. In parametric statistics, the family of the distribution is known, and specific results can be proven for any finite n, such as the following one for a real valued Gaussian model.

Theorem 5.35 (Fisher) *Let X_1^n be an n-sample of a Gaussian distribution $\mathcal{N}(m, \sigma^2)$ with $\sigma \neq 0$. With the notation of Theorem 5.16,*

1. $\overline{X}_n \sim \mathcal{N}(m, \sigma^2/n)$, 2. $nS_n^2 \sim \sigma^2\chi^2(n)$,
3. $(n-1)S_n^{*2} \sim \sigma^2\chi^2(n-1)$, 4. $\overline{X}_n$ *and* S_n^{*2} *are independent,*
5. $\sqrt{n}(\overline{X}_n - m)/S_n^{*2} \sim \tau(n-1)$.

Conversely, if X_1^n is an n-sample of a distribution P on $(\mathbb{R}, \mathcal{B}(\mathbb{R}))$ such that $\overline{X}_n$ and S_n^{*2} are independent, then P is a normal distribution, as shown in Exercise 5.5 below.

Note that the χ^2-distribution loses one degree of freedom between S_n^2 and S_n^{*2}, corresponding to the estimation of one parameter, the mean m.

Proof Point 1 is clear, and Point 2 derives directly from $(X_i - m)/\sigma \sim \mathcal{N}(0, 1)$ for $i \in [\![1, n]\!]$.

For proving 3 and 4, we will use Cochran's theorem. Suppose that $m = 0$ and $\sigma^2 = 1$. Then

$$(n-1)S_n^{*2} = \sum_{i=1}^n X_i^2 + n\overline{X}_n^2 - 2\overline{X}_n \sum_{i=1}^n X_i = \sum_{i=1}^n X_i^2 - n\overline{X}_n^2.$$

Let A be an $n \times n$ matrix whose last line is $(1/\sqrt{n}, \ldots, 1/\sqrt{n})$ and such that $AA' = I_n$ (A is a unitary matrix). Setting $Y = AX$, we have $Y \sim \mathcal{N}_n(0, I)$, $\sum_{i=1}^{n} X_i^2 = \sum_{i=1}^{n} Y_i^2$ and $Y_n^2 = n\overline{X}_n^2$, and hence, $(n-1)S_n^{*2} = \sum_{i=1}^{n-1} Y_i^2$. Then, 3 and 4 indeed follow from applying Cochran's theorem.

If $m \neq 0$ or $\sigma \neq 1$, set $Z_i = (X_i - m)/\sigma$. Then $\overline{X}_n = m + \sigma\overline{Z}_n$ and $\sum_{i=1}^{n}(X_i - \overline{X}_n)^2 = \sum_{i=1}^{n}(Z_i - \overline{Z}_n)^2$, from which the conclusion follows.

Thanks to 3, $(n-1)S_n^{*2}/\sigma^2 \sim \chi^2(n-1)$, and thanks to 1, $\sqrt{n}(\overline{X}_n - m)/\sigma \sim \mathcal{N}(0, 1)$. By 4, these random variables are independent, and 5 follows.   □

5.2.2   Maximum Likelihood Method

The maximum likelihood estimation method is classical for determining estimators. Let us begin by an illustrative example.

▷ *Example 5.36 (Continuation of Example 5.30)* Assume that $P_\theta = \mathcal{B}(\theta)$, with $\theta \in \{1/4, 3/4\}$. Then $\mathbb{P}_{1/4}(\{1\}) = \mathbb{P}_{3/4}(\{0\}) = 1/4$ and $\mathbb{P}_{3/4}(\{1\}) = \mathbb{P}_{1/4}(\{0\}) = 3/4$.

If $(1, 1, 0, 1)$ is observed for 4 realizations of the phenomenon, then one will say that $\theta = 3/4$ is more likely to be true than $\theta = 1/4$.   ◁

The generalization of this intuitive idea leads to the method of maximum likelihood estimation.

Definition 5.37 Let X_1^n be an n-sample of a distribution P_θ, with $\Theta \subset \mathbb{R}^r$. An estimator $T_n = T_n(X_1, \ldots, X_n)$ is called a maximum likelihood estimator of θ if

$$L(T_n(x), x) = \sup_{\theta \in \Theta} L(\theta, x), \quad x \in \mathbb{R}^n.$$

If T_n is a maximum likelihood estimator of θ, then $G_n = g(T_n)$ is a maximum likelihood estimator of $g(\theta)$ for all smooth function g, a property of invariance of the maximum likelihood.

▷ *Example 5.38 (Continuation of Example 5.36)* We compute $L(3/4; 1, 1, 1, 0) = (3/4)^3 \times 1/4$ and $L(1/4; 1, 1, 1, 0) = (1/4)^3 \times 3/4$. The maximum likelihood estimate clearly takes the value $3/4$.   ◁

▷ *Example 5.39 (Maximum Likelihood Estimation for the Uniform Distribution)* Assume that P is the uniform distribution on an interval $[0, \theta]$, where $\theta \in \mathbb{R}_+$. Observations $x_1, \ldots, x_n$ are available, from which x_k is the largest. The likelihood of this model is

$$L(\theta, x_1, \ldots, x_n) = \frac{1}{\theta^n} \prod_{i=1}^{n} \mathbb{1}_{[0,\theta]}(x_i).$$

Necessarily $x_k \leq \theta$, otherwise it could not have been observed. Moreover, L decreases with respect to θ. Therefore, the maximum likelihood estimator of θ is $\widehat{\theta} = X_{(n)}$, and θ will be estimated by x_k.   ◁

▷ *Example 5.40 (Capture–Recapture Sampling)* In order to identify the size N of a herd of stags without census (counting all the stags), the following method is used. First, M stags are captured at random, marked, and freed. Subsequently, n stags are captured at random and the number of marked stags among them is counted. All stags are assumed to be captured with the same probability.

Let X be the number of marked recaptured stags. Clearly, $X \sim \mathcal{H}(N, M, n)$. The unknown parameter of this distribution is N. Thus, the likelihood of the model is

$$L(N, k) = \mathbb{P}(X = k) = \frac{\binom{M}{k}\binom{N-M}{n-k}}{\binom{N}{n}},$$

and hence

$$\frac{L(N, k)}{L(N+1, k)} = \frac{(N-n)(N-M)}{N(N-n-M+k)}.$$

The function L increases and then decreases in N. It is larger than 1 for $N < nM/k$, then it is less than 1. The maximum likelihood estimator of the model is $\widehat{N} = [nM/X]$ or $\widehat{N} = [nM/X] + 1$.

Note that the same estimator would have been obtained by applying the method of moments.   ◁

The next proposition is a direct consequence of the factorization theorem.

Proposition 5.41 *If T_n is a sufficient statistic of θ, then any maximum likelihood estimator of θ is a function of T_n.*

For many purposes—including the practical determination of maximum likelihood estimators, the logarithm of the likelihood or log-likelihood is more easy to cope with than the likelihood. For an n-sample,

$$\log L(\theta, x_1, \ldots, x_n) = \sum_{i=1}^{n} \log f_\theta(x_i).$$

If L is differentiable with respect to θ, and if $T_n = T_n(x) = T_n(x_1, \ldots, x_n)$ is a maximum likelihood estimator of the parameter θ of a distribution having a density, then T_n satisfies

$$\frac{\partial}{\partial \theta_i} \log L(\theta, x)\Big|_{\theta = T_n} = 0, \quad i = 1, \ldots, r. \tag{5.4}$$

Solving the above equation allows one to identify possible values for T_n.

Further, suppose that L is two times differentiable with respect to θ, and set $\Delta_1 = \frac{\partial^2}{\partial \theta_1^2} \log L(\theta, x)$, and let Δ_k for $k \in [\![2, r]\!]$ be the determinant of the matrix

$$\left(\frac{\partial^2}{\partial \theta_i \partial \theta_j} \log L(\theta, x) \right)_{1 \le i, j \le k}.$$

Then for a solution (θ, x) of (5.4) is known to be a maximum of $\log L(\theta, x)$, if $(-1)^k \Delta_k > 0$ for $k \in [\![1, r]\!]$.

$\triangleright$ *Example 5.42 (Maximum Likelihood Estimators for the Gaussian Model)* If $P_\theta = \mathcal{N}(m, \sigma^2)$ with $\theta = (m, \sigma^2)$, we have

$$L((m, \sigma^2), x) = \frac{1}{(\sqrt{2\pi}\sigma)^n} \exp\left[-\sum_{i=1}^{n} (x_i - m)^2 / 2\sigma^2 \right],$$

so

$$\frac{\partial \log L((m, \sigma^2), x)}{\partial m} = \frac{\sum_{i=1}^{n} (x_i - m)}{\sigma^2}$$

and

$$\frac{\partial \log L((m, \sigma^2), x)}{\partial (\sigma^2)} = \frac{-n}{2\sigma^2} + \frac{\sum_{i=1}^{n} (x_i - m)^2}{2\sigma^4}.$$

Therefore, $\widehat{m} = \overline{X}_n$ and $\widehat{\sigma^2} = \sum_{i=1}^{n} (X_i - \overline{X}_n)^2 / n$ are possible maximum likelihood estimators. We check by computing Δ_1 and Δ_2 that

$$T_n(x_1, \ldots, x_n) = \left(\overline{x}_n, \frac{1}{n} \sum_{i=1}^{n} (x_i - \overline{x}_n)^2 \right)$$

indeed gives the maximum of likelihood. $\triangleleft$

5.2.3 Precision of the Estimators

We have presented different methods for constructing estimators of parameters. Various criteria exist for measuring their precision. We state most of the results without proofs.

Definition 5.43 A sequence of estimators (T_n) of $g(\theta)$ is said to be:

- Unbiased if $\mathbb{E}_\theta T_n = g(\theta)$, for all $n \in \mathbb{N}^*$.
- Asymptotically unbiased if $\mathbb{E}_\theta T_n$ converges to $g(\theta)$.
- Convergent (or consistent) if T_n converges in P_θ-probability to $g(\theta)$.
- Convergent in mean square if $\mathbb{E}_\theta([T_n - g(\theta)]^2)$ converges to zero.
- Strongly convergent if T_n converges P_θ−a.s. to $g(\theta)$.
- Asymptotically normal if $\sqrt{n}[T_n - g(\theta)]$ converges in distribution to a Gaussian variable.

All above convergences are understood for n tending to infinity.

The maximum likelihood method yields estimators that are convergent but may be biased (such as σ^2 in Example 5.42). It may also yield several estimators, showing the necessity of comparison. The notion of risk is introduced for this purpose. We assume here for simplification that the statistical model is $(\mathbb{R}, \mathcal{B}(\mathbb{R}), \{P_\theta : \theta \in \Theta\})$, with $\Theta \subset \mathbb{R}$.

Definition 5.44 Let X_1^n be an n-sample of P_θ.

- If $T_n - T_n(X_1, \ldots, X_n)$ is an estimator of $g(\theta)$, its risk is defined by

$$R(T_n, \theta) = \int_{\mathbb{R}^n} l(g(\theta), T_n(x_1, \ldots, x_n)) dP_\theta(x_1) \ldots dP_\theta(x_n),$$

where l is a loss function, that is a nonnegative Borel function defined on $g(\Theta) \times \mathbb{R}^d$.
- An estimator T_n^* is said to be admissible if $R(T_n^*, g(\theta)) \leq R(T_n, g(\theta))$ for all $\theta \in \Theta$ and all estimators T_n of $g(\theta)$.

An important question is whether the risk has an inferior limit and how this limit can be determined. We will give only one example, for estimators of θ and the quadratic loss function $l(\theta, t) = (\theta - t)^2$. The associated risk R_2 is called the quadratic risk and the celebrated bias-variance analysis

$$R_2(T_n, \theta) = [\mathbb{E}(T_n - \theta)^2] + \mathbb{V}\text{ar } T_n$$

holds true, with simply $R_2(T_n, \theta) = \text{Var}_\theta T_n$ if T_n is unbiased. The unbiased estimators of θ can thus be ordered, the one with minimum variance being the closest to the parameter in the sense of least squares.

Definition 5.45 Let X_1^n be an n-sample of a distribution P_θ in a dominated statistical model. The Fisher information of the model—with possibly infinite coefficients—is defined as

$$\mathcal{I}_n(\theta) = \mathbb{E}_\theta\left(\left[\frac{\partial}{\partial\theta}\log L(\theta, X_1^n)\right]^2\right).$$

Computing Fisher information is easier for regular models.

Definition 5.46 Let $(\mathbb{R}, \mathcal{B}(\mathbb{R}), \{P_\theta : \theta \in \Theta\})$ be a dominated statistical model, with Θ an open subset of $\mathbb{R}$. The model is said to be regular if the function $L(\cdot, x)$ is differentiable over Θ, if $\frac{\partial}{\partial\theta}\log L(\theta, X_1^n)$ is square integrable and centered, and if

$$\frac{\partial}{\partial\theta}\mathbb{E}_\theta[YL(\theta, X_1^n)] = \mathbb{E}_\theta\left[Y\frac{\partial}{\partial\theta}\log L(\theta, X_1^n)\right], \quad Y \in L^2(\mathbb{R}, \mathcal{B}(\mathbb{R}), P_\theta).$$

Proposition 5.47 *Let X_1^n be an n-sample of a distribution P_θ with support S_θ in a regular dominated statistical model.*

The Fisher information of the model is

$$\mathcal{I}_n(\theta) = \text{Var}_\theta\left[\frac{\partial}{\partial\theta}\log L(\theta, X_1^n)\right] = \mathbb{E}_\theta\left(\left[\frac{\partial}{\partial\theta}\log L(\theta, X_1^n)\right]^2\right).$$

If $S_\theta = S$ does not depend on θ, then $\mathcal{I}_n(\theta) = n\mathcal{I}_1(\theta)$. If moreover $L(\cdot, x)$ is two times differentiable over Θ with

$$\mathbb{E}_\theta\left[\frac{\partial^2}{\partial\theta^2}L(\theta, X_1^n)\right] = \frac{\partial^2}{\partial\theta^2}\mathbb{E}_\theta[L(\theta, X_1^n)],$$

then

$$\mathcal{I}_n(\theta) = -\mathbb{E}_\theta\left[\frac{\partial^2}{\partial\theta^2}\log L(\theta, X_1^n)\right].$$

▷ *Example 5.48 (Continuation of Example 5.33)* If P is the Poisson distribution $\mathcal{P}(\lambda)$, the model is regular, with $\Theta = \mathbb{R}_+^*$, and

$$\frac{\partial^2}{\partial\lambda^2}\log L(\lambda, x) = \frac{\partial^2}{\partial\lambda^2}[-\lambda + x\log\lambda - \log(x!)] = \frac{-x}{\lambda^2}.$$

Therefore, $\mathcal{I}_1(\lambda) = \mathbb{E}_\lambda(X)/\lambda^2 = 1/\lambda$.   ◁

Theorem 5.49 (Cramér–Rao Inequality) *Let X_1^n be an n-sample of a distribution P_θ, with $S_\theta = S$ for $\theta \in \Theta \subset \mathbb{R}$. Let $T_n = T_n(X_1, \ldots, X_n)$ be a statistic such that $\mathbb{E}_\theta(T_n(X))$ is a differentiable function of θ. If*

$$\frac{d}{d\theta} \int_{\mathbb{R}^n} h(x_1, \ldots, x_n) L(\theta, x_1, \ldots, x_n) dx_1 \ldots dx_n =$$

$$\int_{\mathbb{R}^n} h(x_1, \ldots, x_n) \frac{\partial}{\partial\theta} L(\theta, x_1, \ldots, x_n) dx_1 \ldots dx_n$$

for any measurable function h such that $\mathbb{E}_\theta|h(X)| < +\infty$, then

$$\mathbb{V}\mathrm{ar}_\theta T_n \geq \frac{\left(\frac{d}{d\theta} \mathbb{E}_\theta[T_n(X)]\right)^2}{\mathbb{E}_\theta\left([\frac{\partial}{\partial\theta} \log L(\theta, X)]^2\right)}.$$

If T_n is an unbiased estimator of $g(\theta)$, we get

$$\mathbb{V}\mathrm{ar}_\theta T_n \geq \frac{[g'(\theta)]^2}{n\mathcal{I}_1(\theta)}.$$

The estimator T_n is said to be efficient if its variance is equal to the lower bound of the inequality, called Cramér–Rao bound. Then, its precision is maximum.

Theorem 5.50 (Efficiency) *Under the hypothesis of Theorem 5.49, an unbiased estimator T_n of $g(\theta)$ is efficient if and only if, for all n, a function $\varphi_n : \Theta \to \mathbb{R}$ exists such that*

$$\frac{\partial}{\partial\theta} \log L(\theta, x) = \varphi_n(\theta)[T_n(x) - g(\theta)]. \tag{5.5}$$

Then $\mathbb{V}\mathrm{ar}_\theta T_n = g'(\theta)/\varphi_n(\theta)$.

Obviously, all estimators satisfying (5.5) are unbiased. Note that maximum likelihood estimators are generally asymptotically efficient, in other words $\sqrt{n}(T_n - g(\theta))$ converges in distribution to $T \sim \mathcal{N}(0, g'(\theta)^2/I(\theta))$.

▷ *Example 5.51 (Continuation of Example 5.33)* If $X \sim \mathcal{P}(\lambda)$, the empirical mean $\overline{X}_n$ is efficient, since, according to Example 5.33,

$$\frac{\partial}{\partial\lambda} \log L(\lambda, x) = -n + \frac{\overline{x}_n}{\lambda} = \frac{n}{\lambda}(\overline{x}_n - \lambda).$$

We check that $\mathbb{V}\mathrm{ar}_\lambda \overline{X}_n = 1/\varphi_n(\lambda) = \lambda/n$. ◁

Proposition 5.52 (Rao–Blackwell) *If T_n is a sufficient statistic for θ and if S_n is an unbiased estimator of $g(\theta)$ with finite variance, then*

$$\mathrm{Var}_\theta\,[\mathbb{E}_\theta\,(S_n \mid T_n)] \leq \mathrm{Var}_\theta\, S_n, \quad \theta \in \Theta,$$

where the equality holds if and only if $S_n = \mathbb{E}_\theta(S_n \mid T_n)$.

Therefore, for identifying the estimators with the smallest variance, it is sufficient to consider the functions of a sufficient statistic.

Most of these notions carry on multi-dimensional parameters by considering the variance matrix of the vector $\mathrm{grad}\log L(\theta, X_1, \ldots, X_n)$, called the Fisher information matrix of the model.

5.2.4   Parametric Confidence Intervals

The construction of confidence intervals for general distribution functions is of use for parametric models. Still, specific methods yield better results for parameters of particular families of distributions.

Let $(\mathbb{R}, \mathcal{B}(\mathbb{R}), \{P_\theta : \theta \in \Theta\})$ be a parametric statistical model. Let $g : \Theta \to \mathbb{R}$ be a Borel function. Suppose $g(\theta)$ is to be estimated from observations $(x_1, \ldots, x_n)$ of an n-sample X_1^n of P_θ. In the above section, a point value was obtained, while now a confidence interval is searched for.

Definition 5.53 Let $\alpha \in]0, 1[$. Let I_n be a Borel function defined on $\mathbb{R}^n$ taking values in the set of intervals of $\mathbb{R}$. Then $I_n = I_n(X_1, \ldots, X_n)$ is called a confidence interval for $g(\theta)$ at level $1 - \alpha$ if

$$P_\theta[g(\theta) \in I_n] = 1 - \alpha, \quad \theta \in \Theta.$$

Precise results can be obtained especially in a Gaussian statistical model.

$\triangleright$ *Example 5.54 (Confidence Interval for the Mean in a Gaussian Model)* The family of distributions is $\mathcal{N}(m, \sigma^2)$ for $(m, \sigma^2) \in \mathbb{R} \times \mathbb{R}_+^*$.

If σ^2 is known, then $Z_n = \sqrt{n}(\overline{X}_n - m)/\sigma \sim \mathcal{N}(0, 1)$ for all n. Hence, with probability $1 - \alpha$,

$$m \in \,]\overline{X}_n - \sigma r_{\alpha/2}/\sqrt{n}, \ \overline{X}_n + \sigma r_{1-\alpha/2}/\sqrt{n}[\,,$$

where $\overline{X}_n = \sum_{i=1}^n X_i/n$ and $r_{1-\alpha/2}$ is the quantile of order $1 - \alpha/2$ of $\mathcal{N}(0, 1)$.

If σ^2 is unknown, thanks to Fisher's theorem, $T_n = \sqrt{n}(\overline{X}_n - m)/S_n^* \sim \tau(n-1)$. Hence, with probability $1 - \alpha$,

$$m \in \,]\overline{X}_n - t_{n,\alpha/2}S_n^*/\sqrt{n},\; \overline{X}_n + t_{n,1-\alpha/2}S_n^*/\sqrt{n}[\,,$$

where $t_{n,1-\alpha/2}$ is the quantile of order $1 - \alpha/2$ of $\tau(n-1)$.   ◁

▷ *Example 5.55 (Confidence Interval for the Variance in a Gaussian Model)*
Assume again that the distribution is $\mathcal{N}(m, \sigma^2)$.

If m is known, then $nS_n^2/\sigma^2 = \sum_{i=1}^{n}(X_i - m)^2/\sigma^2 \sim \chi^2(n)$ and, with probability $1 - \alpha$,

$$\sigma^2 \in \,]\sum_{i=1}^{n}(X_i - m)^2/\pi_{n,1-\alpha/2},\; \sum_{i=1}^{n}(X_i - m)^2/\pi_{n,\alpha/2}[\,,$$

where $\pi_{n,\alpha/2}$ is the quantile of order $\alpha/2$ and $\pi_{n,1-\alpha/2}$ the quantile of order $1 - \alpha/2$ of $\chi^2(n)$; see Fig. 5.2.

If m is unknown, then

$$(n-1)S_n^{*2}/\sigma^2 = \sum_{i=1}^{n}(X_i - m)^2/\sigma^2 \sim \chi^2(n-1)$$

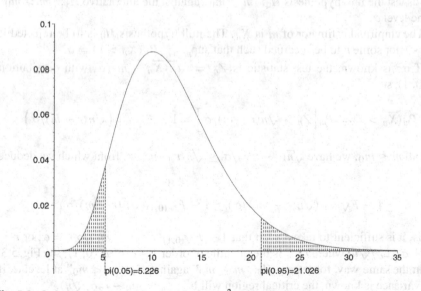

Fig. 5.2 Quantiles of order 0.05 and 0.95 of the $\chi^2(6)$ distribution

so $\mathbb{P}[\pi_{n-1,\alpha/2} < (n-1)S_n^{*2}/\sigma^2 < \pi_{n-1,1-\alpha/2}] = 1 - \alpha$, and, with probability $1 - \alpha$,

$$\sigma^2 \in \Big] \sum_{i=1}^n (X_i - \overline{X}_n)^2/\pi_{n-1,2}, \ \sum_{i=1}^n (X_i - \overline{X}_n)^2/\pi_{n-1,1} \Big[,$$

where $\pi_{n-1,\alpha/2}$ is the quantile of order $\alpha/2$ and $\pi_{n-1,1-\alpha/2}$ the quantile of order $1 - \alpha/2$ of the distribution $\chi^2(n-1)$. ◁

5.2.5 Testing in a Parametric Model

The observed distribution is known to depend on a parameter θ belonging to a set Θ, disjoint union of Θ_0 and Θ_1. Reasons—for instance previous studies—exist to believe that $\theta \in \Theta_0$. As in the non-parametric case, the question is to determine to which extent we are right in making this hypothesis, in view of observations $x_1, \ldots, x_n$ of an n-sample X_1^n of the phenomenon. A decision is taken depending on the value of observations through a binary test function Φ, for example $\Phi = \mathbb{1}_{(T_n \geq t)}$ for some test statistic $T_n(X_1, \ldots, X_n)$.

We will only consider the typical tests of the mean and the variance in a Gaussian model.

▷ *Example 5.56 (Test for the Mean in a Gaussian Model with Known Variance)* Let us test the null hypothesis H_0: "$m \leq m_0$" against the alternative H_1: "$m > m_0$" at the level α.

The empirical estimator of m is $\overline{X}_n$. The null hypothesis H_0 is to be rejected if $\overline{x}_n > r$ for some r to be specified such that $\sup_{m \leq m_0} P_m(\overline{X}_n > r) = \alpha$.

If σ^2 is known, the test statistic is $Z_n = \sqrt{n}(\overline{X}_n - m)/\sigma$, with distribution $\mathcal{N}(0, 1)$, so

$$P_m(\overline{X}_n > r) = P_m\Big[Z_n > \sqrt{n}(r - m)/\sigma\Big] = 1 - F_{\mathcal{N}(0,1)}\Big(\sqrt{n}(r - m)/\sigma\Big).$$

For all $m \leq m_0$, we have $\sqrt{n}(r - m_0)/\sigma \leq \sqrt{n}(r - m)/\sigma$, from which we deduce that

$$1 - F_{\mathcal{N}(0,1)}(\sqrt{n}(r - m_0)/\sigma) \geq 1 - F_{\mathcal{N}(0,1)}(\sqrt{n}(r - m)/\sigma).$$

Thus, it is sufficient to take r such that $1 - F_{\mathcal{N}(0,1)}(\sqrt{n}(r - m_0)/\sigma) = \alpha$, or $r = m_0 + \sigma r_{1-\alpha}/\sqrt{n}$, where $r_{1-\alpha}$ is the quantile of order $1 - \alpha$ of $\mathcal{N}(0, 1)$; see Fig. 5.3.

In the same way, for testing H_0: "$m \geq m_0$" against H_1: "$m < m_0$" at level α, if the variance is known, the critical region will be $(\overline{X}_n < m_0 - r_\alpha \sigma/\sqrt{n})$.

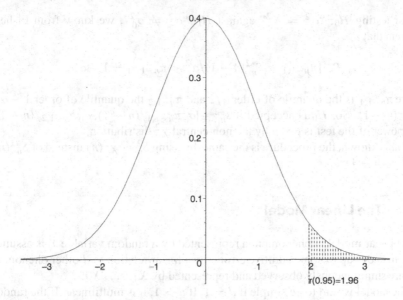

Fig. 5.3 Quantile of order 0.95 of the standard normal distribution

Finally, in testing H_0: "$m = m_0$" against H_1: "$m \neq m_0$" at level α, the null hypothesis H_0 is rejected if $\overline{X}_n$ is far from m_0, and the critical region will be

$$(\overline{X}_n < m_0 - r_{\alpha/2}\sigma/\sqrt{n}) \cup (\overline{X}_n > m_0 + r_{\alpha/2}\sigma/\sqrt{n}),$$

if the variance is known.   ◁

▷ *Example 5.57 (Student Test for the Mean)* Let again H_0: "$m \leq m_0$" and H_1: "$m > m_0$" in a Gaussian model.

If σ^2 is unknown, the statistic $Z_n = \sqrt{n}(\overline{X}_n - m)/S_n^*$ has the Student distribution $\tau(n)$. Hence $\overline{X}_n \in]-\infty, m + t_{n,1-\alpha}s_n/\sqrt{n}]$ with probability $1 - \alpha$, where $t_{n,1-\alpha}$ is the quantile of order $1 - \alpha$ of $\tau(n)$. If $m \leq m_0$ ($m > m_0$) we have $\overline{X}_n \in]-\infty, m_0 + t_{n,\alpha}s_n/\sqrt{n}]$ with a probability greater (smaller) than $1 - \alpha$.

The Student test follows: H_0 is not rejected if $\overline{x}_n \in]-\infty, m_0 + t_{n,\alpha}s_n/\sqrt{n}]$. One can show using a non-centered Student distribution that the probability of accepting H_0 is a decreasing function of $|m - m_0|$, which yields the power of the test.   ◁

▷ *Example 5.58 (Fisher Test for the Variance)* Let us test H_0: "$\sigma^2 \leq \sigma_0^2$" against H_1: "$\sigma^2 > \sigma_0^2$" in a Gaussian model.

If m is unknown, H_0 is accepted if s_n^{*2} belongs to the interval $[0, \sigma_0^2\pi_{n-1,1-\alpha}/(n-1)]$, where $\pi_{n-1,1-\alpha}$ is the quantile of order $1 - \alpha$ of the distribution $\chi^2(n - 1)$. One can show using the non-centered Fisher distribution that the probability of accepting H_0 is a non-decreasing explicit function of σ^2, which yields the power of the test.

For testing H_0: "$\sigma^2 = \sigma_0^2$" against H_1: "$\sigma^2 \neq \sigma_0^2$", we know from Fisher's theorem that

$$P_{\sigma^2}[\pi_{n-1,1} < S_n^{*2}(n-1)/\sigma^2 < \pi_{n-1,2}] = 1 - \alpha,$$

where $\pi_{n-1,1}$ is the quantile of order $\alpha/2$ and $\pi_{n-1,2}$ the quantile of order $1 - \alpha/2$ of $\chi^2(n-1)$. So, H_0 is accepted if $s_n^{*2} \in]\sigma_0^2\pi_{n-1,1}/(n-1), \sigma_0^2\pi_{n-1,2}/(n-1)[$. The power of the test is given by the non-central χ^2 distribution.

If m is known, the procedure is the same, by using $S_n^2 \sim \chi^2(n)$ instead of $S_n^{*2}(n-1) \sim \chi^2(n-1)$.   ◁

5.3   The Linear Model

In the linear model, a phenomenon represented by a random variable Y is assumed to be linked—up to some additive error—by a linear relation to d other phenomena that are simultaneously observed and represented by $X_1, \ldots, X_d$.

The model is said to be simple if $d = 1$. If $d > 1$, it is multilinear. If the random variables X_j are continuous, the model is called a regression model. If they take only a finite number of values, the model is an analysis of variance, or ANOVA. If the observations are grouped, a more thorough study is possible.

5.3.1   Linear and Quadratic Approximations

Let Y be a square integrable random variable. The expectation of Y is its orthogonal projection in L^2 on the subspace of constant random variables—the best approximation of Y by a constant in the sense of the L^2-norm; see Exercise 2.7. Note that the standard deviation σ_Y is the length of $Y - \mathbb{E} Y$ in L^2.

Let now $X_1, \ldots, X_d$ be square integrable random variables with covariance matrix Γ_X. The random variable Y is to be approximated by a linear function of $X_1, \ldots, X_d$; see Fig. 5.4, in which $d = 1$.

Fig. 5.4 Illustration of the linear model

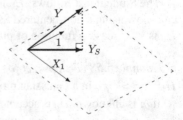

The projection $Y_S = M + \sum_{j=1}^{d} \alpha_j X_j$ has to satisfy the so-called normal equations:

$$\langle Y - Y_S, 1 \rangle = 0, \tag{5.6}$$

$$\langle Y - Y_S, X_j \rangle = 0, \quad \text{for } j = 1, \ldots, d.$$

Since the scalar product is that of centered variables in L^2, defined by

$$< X, Y >= \mathbb{E}(X - \mathbb{E}X)(Y - \mathbb{E}Y),$$

Since $M = \mathbb{E}Y - \sum_{j=1}^{d} \alpha_j \mathbb{E}X_j$, we get

$$\langle Y - Y_S, X_j \rangle = \langle Y - \mathbb{E}Y - \sum_{i=1}^{d} \alpha_i (X_i - \mathbb{E}X_i), X_j \rangle.$$

Due to (5.6), $\langle Y - \mathbb{E}Y - \sum_{i=1}^{d} \alpha_i (X_i - \mathbb{E}X_i), \mathbb{E}X_j \rangle = 0$, so we also have

$$\langle Y - \mathbb{E}Y - \sum_{i=1}^{d} \alpha_i (X_i - \mathbb{E}X_i), X_j - \mathbb{E}X_j \rangle = 0,$$

or, in matrix form,

$$\Gamma_X \begin{pmatrix} \alpha_1 \\ \vdots \\ \alpha_d \end{pmatrix} = \begin{pmatrix} \mathbb{C}\text{ov}(Y, X_1) \\ \vdots \\ \mathbb{C}\text{ov}(Y, X_d) \end{pmatrix}.$$

If X is not degenerated —that is if Γ_X is invertible, then

$$(\alpha_1, \ldots, \alpha_d) = (\mathbb{C}\text{ov}(Y, X_1), \ldots, \mathbb{C}\text{ov}(Y, X_d)) \, \Gamma_X^{-1}.$$

This defines the best linear approximation of Y by $(X_1, \ldots X_d)$ in the sense of the L^2-norm or least squares.

Note that if $(Y, X_1, \ldots, X_d)$ is a Gaussian vector, then $Y - Y_S$ and X_j are independent for $j = 1, \ldots, d$, and

$$\|Y\|_2^2 = \|Y - \sum_{i=1}^{d} \alpha_i X_i\|_2^2 + \|\sum_{i=1}^{d} \alpha_i X_i\|_2^2,$$

so the Pythagorean theorem is satisfied.

5.3.2   The Simple Linear Model

The simple linear model involves only two random variables. Let Y and X be two
square integrable random variables. The best linear approximation of Y by X is
$Y_S = \alpha X + m$, where

$$\alpha = \frac{\text{Cov}(X, Y)}{\text{Var}\, X} \quad \text{and} \quad m = \mathbb{E}\, Y - \alpha \mathbb{E}\, X. \tag{5.7}$$

The line with equation

$$y = \mathbb{E}\, Y + \frac{\text{Cov}(X, Y)}{\text{Var}\, X}(x - \mathbb{E}\, X)$$

is called the regression line of Y given X. This is the line around which the
distribution of (X, Y) is concentrated; in other words,

$$\int_{\mathbb{R}^2} (y - \alpha x - m)^2 d P_{(X,Y)}(x, y)$$

is minimum on $\mathbb{R}^2$ for α and m given by (5.7); see Fig. 5.4.

Let us consider estimation issues in the simple linear model

$$Y = m + \alpha X + \varepsilon,$$

where m and α are unknown. An n-sample of (X, Y) is observed. Does Y depends
linearly on X up to an additive error with zero mean ? In order to use the least
squares method, it is necessary to assume that X and ε are independent, that is to
say that the error does not depend on the observations of X.

Then, replacing the variance and covariance of X and Y by their empirical
estimators, we get

$$\widehat{\alpha} = \frac{\sum_{j=1}^{n}(X_j - \overline{X}_n)(Y_j - \overline{Y}_n)}{\sum_{j=1}^{n}(X_j - \overline{X}_n)^2} \quad \text{and} \quad \widehat{m} = \overline{Y}_n - \widehat{\alpha}\overline{X}_n.$$

An example is illustrated in Fig. 5.5, where the 13 points (x, y) of the scatterplot are
represented by +, and the vertical lines are the lengths $|y - \alpha x - \overline{y}_{13} - \widehat{\alpha}\overline{x}_{13}|$.

The variance of ε, the so-called residual variance, is estimated by

$$\widehat{\text{Var}\,\varepsilon} = \frac{1 - \widehat{\rho}_{X,Y}^2}{n - 2} \sum_{j=1}^{n}(Y_j - \overline{Y}_n)^2,$$

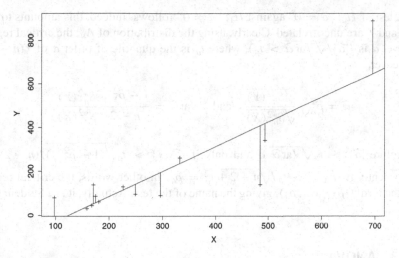

Fig. 5.5 A regression line

where the empirical estimator of the correlation coefficient of X and Y is

$$\widehat{\rho}_{X,Y} = \frac{\sum_{j=1}^{n}(X_j - \overline{X}_n)(Y_j - \overline{Y}_n)}{\sqrt{\sum_{j=1}^{n}(X_j - \overline{X}_n)^2 \sum_{j=1}^{n}(Y_j - \overline{Y}_n)^2}}.$$

Hence, the variances of $\widehat{\alpha}$ and $\widehat{m}$ are estimated by

$$\mathbb{V}\mathrm{ar}\,\widehat{\alpha} = \frac{\widehat{\mathbb{V}\mathrm{ar}\,\varepsilon}}{\sum_{j=1}^{n}(X_j - \overline{X}_n)^2} \quad \text{and} \quad \mathbb{V}\mathrm{ar}\,\widehat{m} = \left(\frac{1}{n} + \frac{\overline{X}_n^{\,2}}{\sum_{j=1}^{n}(X_j - \overline{X}_n)^2}\right)\widehat{\mathbb{V}\mathrm{ar}\,\varepsilon}.$$

These empirical estimators are convergent and unbiased for all distributions. In a Gaussian model, their distributions can be specified as follows.

▷ *Example 5.59 (Test of the Correlation Coefficient)* Assume that the considered simple linear model is Gaussian. Since the variance of the error is unknown, we have

$$A_n = \frac{\widehat{\alpha} - \alpha}{\sqrt{\mathbb{V}\mathrm{ar}\,\widehat{\alpha}}} \sim \tau(n-2).$$

A test of H_0: "$\alpha = 0$" against H_1: "$\alpha \neq 0$" follows. Indeed, this amounts to test if X and Y are uncorrelated. Clearly, using the distribution of A_n, the critical region at level a is $(|\widehat{\alpha}|/\sqrt{\mathrm{Var}\,\widehat{\alpha}} > t_a)$, where t_a is the quantile of order a of $\tau(n-2)$. Moreover,

$$\widehat{\alpha} = \widehat{\rho}_{X,Y}^2 \frac{\sqrt{S_n^{*2}(Y)}}{\sqrt{S_n^{*2}(X)}} \quad \text{and} \quad \mathrm{Var}\,\widehat{\alpha} = \frac{1 - \widehat{\rho}_{X,Y}^2}{n-2} \frac{S_n^{*2}(Y)}{S_n^{*2}(X)}.$$

Therefore $|\widehat{\alpha}| > t_a\sqrt{\mathrm{Var}\,\widehat{\alpha}}$ if and only if $|\widehat{\rho}_{X,Y}| > t_a\sqrt{(1 - \widehat{\rho}_{X,Y}^2)/(n-2)}$, or equivalently if $|\widehat{\rho}_{X,Y}| > t_a/\sqrt{n-2+t_a^2} = \rho_a$. In other words, the critical region has the form $(|\widehat{\rho}_{X,Y}| > \rho_a)$, giving the name of the test. Actually, it is a Student test.

◁

5.3.3 ANOVA

Samples of Gaussian distributions of any finite size can be compared through different methods. Thanks to the central limit theorem, for large enough samples, the results apply to all other distributions.

For Two Samples

Let a d-sample of a distribution $\mathcal{N}(m, \sigma^2)$ and a d'-sample of a distribution $\mathcal{N}(m', \sigma^2)$ be observed. The two samples are assumed to be independent. The first issue is to compare the means m and m'.

Let $X \sim \mathcal{N}_d(m\mathbf{1}_d, \sigma^2 I_d)$ and $Y \sim \mathcal{N}_{d'}(m'\mathbf{1}_{d'}, \sigma^2 I_{d'})$. Set $X_+ = \sum_{i=1}^d X_i/d$, $Y_+ = \sum_{j=1}^{d'} Y_j/d'$ and $S^2 = \sum_{i=1}^d (X_i - X_+)^2 + \sum_{j=1}^d (Y_j - Y_+)^2$. Thanks to Fisher's theorem, $X_+ - Y_+$ and S^2 are independent, with

$$X_+ - Y_+ \sim \mathcal{N}\left(m - m', \sigma^2\left(\frac{1}{d} + \frac{1}{d}'\right)\right) \quad \text{and} \quad S^2 \sim \chi^2(d + d' - 2).$$

Therefore, thanks to Proposition 3.59,

$$\frac{(X_+ - Y_+ - m + m')/\sqrt{1/d + 1/d'}}{S^2/\sqrt{d + d' - 2}} \sim \tau(d + d' - 2).$$

Confidence intervals for $m - m'$ follow as usual. Moreover, the ratio

$$T_{d,d'} = \frac{(X_+ - Y_+)/\sqrt{1/d + 1/d'}}{S^2/\sqrt{d + d' - 2}}$$

can be shown to have a non-central Student distribution, so that $\mathbb{P}(T_{d,d'} \geq t)$ increases with $m - m'$, at fixed t.

In testing H_0: "$m \leq m'''$" against H_1: "$m > m'''$" at level α, the null hypothesis H_0 is not rejected when $T_{d,d'} \leq t_{d+d'-2,\alpha}$. In testing H_0: "$m = m'''$" against H_1: "$m \neq m'''$", at level α, H_0 is not rejected when $t_{d+d'-2,1-\alpha/2} \leq T_{d,d'} \leq t_{d+d'-2,\alpha/2}$, where $t_{d+d'-2,1-\alpha/2}$ is the quantile of order $1 - \alpha/2$ of the distribution $\tau(d+d'-2)$.

One Way Model

For $d = d'$, grouping the samples makes this method amount to testing $m = 0$. The model if then said to be a one way model.

Let us consider d samples of the standard normal distribution, say $Y_k \sim \mathcal{N}_{n_k}(0, I_k)$ for $k \in [\![1, d]\!]$. Set $Y_{k+} = \sum_{l=1}^{n_k} Y_{kl}/n_k$ and $Y_{++} = \sum_{k=1}^{d} \sum_{l=1}^{n_k} Y_{kl}/n$.

The total sum of the squares is equal to the residual sum of squares plus the inter-lines sum of squares; in mathematical words,

$$\sum_{k=1}^{d} \sum_{l=1}^{n_k} (Y_{kl} - Y_{++})^2 = \sum_{k=1}^{d} \sum_{l=1}^{n_k} (Y_{kl} - Y_{k+})^2 + \sum_{k=1}^{d} n_k (Y_{k+} - Y_{++})^2,$$

or $T^2 = R^2 + L^2$. The following result is a consequence of Cochran's theorem or of specific results on the empirical variance.

Theorem 5.60 *The random variables R^2 and L^2 are independent, and*

$$R^2 \sim \chi^2(n - d) \text{ and } L^2 \sim \chi^2(d - 1),$$

and hence $T^2 \sim \chi^2(n - 1)$, where $n = \sum_{k=1}^{d} n_k$.

▷ *Example 5.61 (Equality of Means)* Testing the equality of the means of d samples $Y_k \sim \mathcal{N}_{n_k}(m_k \mathbf{1}_{n_k}, I_{n_k})$ for $k \in [\![1, d]\!]$ leads to test H_0: "$m_k = m$" against H_1: "$\exists (k_1, k_2)$ such that $m_{k_1} \neq m_{k_2}$." This can always be reduced to testing $m = 0$.

According to the above theorem, $(n - d)L^2/(d - 1)R^2 \sim \mathcal{F}(d - 1, n - d)$ under H_0. Under H_1, it can be shown to have a non-central Fisher distribution. A statistical test follows straightforwardly. ◁

Two Way Model

Let us consider a one way model with the additional assumption that all n_k are equal, say $n_k = d'$ for $k \in [\![1, d]\!]$. We can write

$$\sum_{k,l} (Y_{kl} - Y_{++})^2 = \sum_{k,l} (Y_{kl} - Y_{k+} - Y_{+l} + Y_{++})^2 + d' \sum_{k} (Y_{k+} - Y_{++})^2$$

$$+ d \sum_{l} (Y_{+l} - Y_{++})^2,$$

or $T^2 = R^2 + L^2 + C^2$, saying that the total sum of squares is equal to the residual sum of squares plus the inter-lines sum of squares plus the inter-columns sum of squares.

The following result is a consequence of Cochran's theorem too.

Theorem 5.62 *The random variables R^2, L^2, and C^2 are independent, with distributions $R^2 \sim \chi^2((d-1)(d'-1))$, $L^2 \sim \chi^2(d-1)$, and $C^2 \sim \chi^2(d'-1)$.*

▷ *Example 5.63 (Test of Equality of Factors)* Let $Y_{kl} = \alpha_k + \beta_l + \varepsilon_{kl}$, where $(\varepsilon_{kl}) \sim \mathcal{N}_{dd'}(0, I_{dd'})$, for $k \in [\![1, d]\!]$ and $l \in [\![1, d']\!]$. The above theorem yields statistical tests for the equality of both the factors α_k and the factors β_l. ◁

5.4   Exercises and Complements

∇ **Exercise 5.1 (Polling)** The proportion p of votes generally obtained by a certain political party is known to be comprised between 20 and 30% at the European elections. What is the number n of individuals to be questioned in a poll for estimating p with a precision greater than 3% and a probability of error less than 0.1,

1. Using Chebyshev's inequality?
2. Using the central limit theorem?

Solution
Let X_i, for $i \in [\![1, n]\!]$, be the random variable equal to 1 if the questioned individual i votes for this party and to 0 otherwise. Clearly, $X_i \sim \mathcal{B}(p)$. We know from Theorem 5.16 that $\overline{X}_n = (\sum_{i=1}^{n} X_i)/n$ is an unbiased estimator of p. We are looking for n such that

$$\mathbb{P}(|\overline{X}_n - p| \geq 0.03) \leq 0.1.$$

1. Chebyshev's inequality yields

$$\mathbb{P}(|\overline{X}_n - p| \geq 0.03) \leq \frac{\mathbb{V}\mathrm{ar}\,\overline{X}_n}{(0.03)^2}.$$

Thus, n such that $p(1-p)/n(0.03)^2 \leq 0.1$, or $n \geq p(1-p)/0.1(0.03)^2$, is convenient.

Since $0.2 < p < 0.3$, we have $0.16 < p(1-p) < 0.21$, and hence, $n > 0.21/[(0.1) \cdot (0.03)^2]$, or $n \geq 2334$.

2. For n large enough, thanks to the central limit theorem, we can consider that $Z_n = \sqrt{n}(\overline{X}_n - p)/\sqrt{p(1-p)} \sim \mathcal{N}(0, 1)$. So, n has to satisfy

$$\mathbb{P}\left(\frac{-0.03\sqrt{n}}{\sqrt{p(1-p)}} \leq Z_n \leq \frac{0.03\sqrt{n}}{\sqrt{p(1-p)}}\right) \geq 0.9,$$

or, since the standard normal distribution is symmetric,

$$2\mathbb{P}\left(Z_n \leq \frac{0.03\sqrt{n}}{\sqrt{p(1-p)}}\right) - 1 \geq 0.9,$$

that is $\mathbb{P}\left(Z_n \leq 0.03\sqrt{n}/\sqrt{p(1-p)}\right) \geq 0.95$.

We know that $\mathbb{P}(Z_n \leq 1.62) \cong 0.95$, so $0.03\sqrt{n}/\sqrt{p(1-p)} \geq 1.62$ is convenient, or $n \geq (1.62)^2 p(1 - p)/(0.03)^2$. Since $0.16 \leq p(1 - p) \leq 0.21$, it follows that $n \geq 0.21(1.62/0.03)^2$ or $n \geq 613$. △

∇ **Exercise 5.2 (Monte Carlo Estimation of π)** Consider the square with summits $A(0, 0)$, $B(0, 1)$, $C(0, 1)$, $D(1, 1)$, and the circle with center $M(1/2, 1/2)$ and radius $1/2$. We observe the realizations $(x, y) = (X(\omega), Y(\omega))$ of two i.i.d. variables X and Y with uniform distribution on $[0, 1]$. These variables can be regarded as the coordinates of a point P of the square $ABCD$. The value of π is estimated by 4 times the number of points inside the circle divided by the total; see Fig. 5.6.

1. Justify mathematically the approximation of π by this method, and write the associated algorithm.
2. Compute the minimum number of drawings N necessary for the probability of the estimate to be distant from more than 10^{-2} of the true value of π to be less than 10^{-3}, using:
   (a) Chebyshev's inequality;
   (b) The central limit theorem; note that $\mathbb{P}(|Z| > z) = 10^{-3}$ for $z = 3, 27$.

Fig. 5.6 Monte Carlo estimation of the value of π—Exercise 5.2

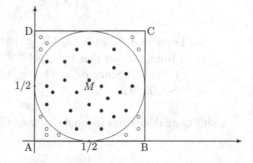

Solution

1. Let A_i = "the i-th drawn point is inside the circle." Let N be the total number of generated points and let N_A the number of these points inside the circle. Then $X_i = \mathbb{1}_{A_i} \sim \mathcal{B}(\pi/4)$, with $0 \leq \pi/4 \leq 1$, and $N_A = \sum_{i=1}^{N} \mathbb{1}_{A_i}$. Therefore, according to the strong law of large numbers,

$$\frac{N_A}{N} \xrightarrow{\text{a.s.}} \frac{\text{area of the circle}}{\text{area of the square}} = \frac{\pi}{4}, \quad N \to +\infty,$$

and hence, the approximate value of π is $4N_A/N$. The algorithm is as follows:

(a) $N := 0; N_A := 0$.

(b) $N := N + 1$; draw a random number $u = U(\omega)$ for $U \sim \mathcal{U}([0, 1])$.

(c) Draw another random number $v = V(\omega)$ for $V \sim \mathcal{U}([0, 1])$.

(d) If $(x - 1/2)^2 + (y - 1/2)^2 \leq 1/4$, then $N_A := N_A + 1$.

(e) Repeat from b. for the required number of iterations.

2. Since $\overline{X}_N$ is an unbiased estimator of $\pi/4$, the variable $T_N = 4\overline{X}_N$ is an unbiased estimator of π, with $\mathbb{V}\mathrm{ar}\, T_N = 16\mathbb{V}\mathrm{ar}\,\overline{X}_N = 4(1 - \pi/4)\pi/N$.

(a) Chebyshev's inequality yields

$$\mathbb{P}(|T_N - \pi| \geq 10^{-2}) \leq \frac{\mathbb{V}\mathrm{ar}\, T_N}{(10^{-2})^2}.$$

Since $0 \leq \pi/4 \leq 1$, we have $(1-\pi/4)\pi/4 \leq 1/4$, and hence $\mathbb{V}\mathrm{ar}\, T_N < 4/N$. Thus $\mathbb{P}(|T_N - \pi| \geq 10^{-2}) \leq 10^{-3}$ at least for all $N \geq 4.10^7$.

(b) Thanks to the central limit theorem applied to (X_N),

$$Z_N = \frac{\sqrt{N}}{\sqrt{(1 - \pi/4)\pi/4}} \left(\overline{X}_N - \pi/4\right) \xrightarrow{\mathcal{D}} Z \sim \mathcal{N}(0, 1).$$

Setting $(1 - \pi/4)\pi/4 = \sigma^2$, we have

$$\mathbb{P}(|T_N - \pi| \geq 10^{-2}) = \mathbb{P}\left(\left|\overline{X}_N - \frac{\pi}{4}\right| \geq \frac{10^{-2}}{4}\right) = \mathbb{P}\left(\frac{\sqrt{N}}{\sigma}\left|\overline{X}_N - \frac{\pi}{4}\right| \geq \frac{\sqrt{N}10^{-2}}{4\sigma}\right)$$

$$= \mathbb{P}\left(|Z_N| \geq \frac{\sqrt{N}10^{-2}}{4\sigma}\right).$$

We know that $\mathbb{P}(|Z| > 3, 27) \leq 10^{-3}$. Identifying the distribution of Z_N to its limit, we get that for $\sqrt{N}10^{-2}/4\sigma > 3.27$, we have $\mathbb{P}(|T_N - \pi| \geq 10^{-2}) \leq 10^{-3}$. Since $\sigma^2 = (1 - \pi/4)\pi/4 \leq 1/4$, we get $N > (3, 27)^2 10^4$, or $N > 106, 929$.

Using large deviations inequalities would induce even smaller bounds.   ◁

∇ **Exercise 5.3 (Estimation of the Size of a Population)** An urn contains N bowls numbered from 1 to N, from which n bowls are drawn at random with replacement. Let T_n be the largest drawn number.

1. Compute $\mathbb{P}(T_n \leq k)$ for $k \in \mathbb{N}$. Compute the distribution of T_n, and its mean.
2. Give an approximation of this mean when N is large enough, by using Riemann's sums.
3. Give an asymptotically unbiased estimator of the size of the population.

Solution
1. The probability space is here $(\Omega = \{1, \ldots, N\}^n, \mathcal{P}(\Omega), \mathbb{P})$, where $\mathbb{P}$ is uniform. We have $\mathbb{P}(X \leq k) = (k/N)^n$ for $1 \leq k \leq N$, and $\mathbb{P}(X \leq 0) = \mathbb{P}(X \geq N+1) = 0$. Since $(T_n \leq k) = (T_n \leq k - 1) \cup (T_n = k)$, it follows that

$$\mathbb{P}(T_n = k) = \frac{k^n - (k-1)^n}{N^n}, \quad \text{for } 1 \leq k \leq N.$$

Using Proposition 2.40, we can write

$$\mathbb{E}\, T_n = \sum_{k=0}^{N-1} \mathbb{P}(T_n > k) = \sum_{k=0}^{N-1}[1 - \mathbb{P}(T_n \leq k)] = N - \sum_{k=0}^{N-1}\left(\frac{k}{N}\right)^n.$$

2. If N is large enough, we have

$$\sum_{k=0}^{N-1} k^n \approx \int_0^N x^n dx = \frac{N^{n+1}}{n+1}.$$

Thus, $\mathbb{E}\, T_n \approx N - N/(n+1)$, or $\mathbb{E}\, T_n \approx nN/(n+1)$ that converges to N.
3. If the size of a population is unknown, the largest number in any observed sample is an asymptotically unbiased estimator of its size.   △

∇ **Exercise 5.4 (Sufficient Statistics)**

1. Let $(\Omega, \mathcal{F}, (\mathcal{U}(0, \theta))_{\theta \in \mathbb{R}_+^*})$ be a uniform statistical model. Give a sufficient statistic of the model for θ, and then an unbiased estimator of θ.
2. Let $(\Omega, \mathcal{F}, (\mathcal{U}(\theta - 1/2, \theta + 1/2))_{\theta \in \mathbb{R}_+^*})$ be another uniform model. Give a sufficient statistic and a free statistic of the model.
3. Consider $(\Omega, \mathcal{F}, (P_\theta)_{\theta \in \mathbb{R}_+^*})$, where P_θ is a translated exponential distribution with parameter 1, whose density is $f_\theta(x) = e^{-(x-\theta)}\mathbb{1}_{[\theta, +\infty[}(x)$. Give a sufficient statistic of the model, and then an unbiased estimator of θ.

Solution

1. We can write

$$L(\theta, x_1, \ldots, x_n) = \frac{1}{\theta^n} \mathbb{1}_{[0,\theta]}(x_{(n)}) \mathbb{1}_{\mathbb{R}_+}(x_{(1)}),$$

where $x_{(1)}$ and $x_{(n)}$ are the minimum and maximum values in $\{x_1, \ldots, x_n\}$. According to (3.5) p. 116, the density of $X_{(n)}$ is

$$f_{(n)}(t) = n \frac{t^{n-1}}{\theta^n} \mathbb{1}_{[0,\theta]}(t),$$

so we can write

$$L(\theta, x_1, \ldots, x_n) = f_{(n)}(t) \frac{1}{nt^{n-1}} \mathbb{1}_{[0,\theta]}(x_{(n)}),$$

and hence the maximum $X_{(n)}$ of the n-sample is a sufficient statistic of the model. Moreover,

$$\mathbb{E}_\theta[X_{(n)}] = \int_0^\theta n \frac{t^n}{\theta^n} dt = \frac{n}{n+1} \theta,$$

so $(n+1)X_{(n)}/n$ is an unbiased estimator of θ.
2. We can write $L(\theta, x_1, \ldots, x_n) = \mathbb{1}_{\{\theta+1/2 \le s \le t \le \theta+1/2\}}(s, t)$, where $s = x_{(1)}$ and $t = x_{(n)}$. The density of P_θ is $f_\theta(x) = \mathbb{1}_{[\theta-1/2,\theta+1/2]}(x)$ and its distribution function is given by

$$F_\theta(x) = (x - \theta - 1/2)\mathbb{1}_{[\theta-1/2,\theta+1/2]}(x) + \mathbb{1}_{[\theta+1/2,+\infty[}(x).$$

By applying Exercise 3.10, the density of $T_n = (X_{(1)}, X_{(n)})$ is

$$h_\theta(s, t) = n(n-1) f_\theta(s) f_\theta(t) (F_\theta(t) - F_\theta(s))^{n-2} \mathbb{1}_{\{s \le t\}}(s, t)$$

$$= n(n-1)(t-s)^{n-2} \mathbb{1}_{\{\theta+1/2 \le s \le t \le \theta+1/2\}}(s, t).$$

It follows that $L(\theta, x_1, \ldots, x_n) = h_\theta(s, t) g(x_1, \ldots, x_n)$, where $g(x_1, \ldots, x_n) = [n(n-1)(t-s)^{n-2}]^{-1}$, so T_n is sufficient.

We can write $X_i = U_i + \theta$, where $U_i \sim \mathcal{U}(-1/2, 1/2)$. Therefore, the distribution of the range $X_{(n)} - X_{(1)}$ is independent of the value of θ and is a free statistic of the model.
3. We have $L(\theta, x_1, \ldots, x_n) = e^{n\theta} e^{-\sum_{i=1}^n x_i} \mathbb{1}_{[\theta,+\infty[}(x_{(1)})$, and, according to (3.5) p. 116, the density of $X_{(1)}$ is $h_\theta(t) = ne^{-n(t-\theta)} \mathbb{1}_{[\theta,+\infty[}(t)$. So

$$L(\theta, x_1, \ldots, x_n) = h_\theta(t) \frac{e^{nt}}{n} e^{-\sum_{i=1}^n x_i},$$

and $X_{(1)}$ is sufficient.

Moreover,

$$\mathbb{E}_\theta X_{(1)} = \int_\theta^{+\infty} nt e^{-n(t-\theta)} dt = \theta + \frac{1}{n},$$

so $X_{(1)} - 1/n$ is an unbiased estimator of θ.   △

▽ **Exercise 5.5 (The Converse of Fisher's Theorem)** Let $(X_1, \ldots, X_n)$ be an n-sample of the standard Gaussian distribution P on $(\mathbb{R}, \mathcal{B}(\mathbb{R}))$.

1. We assume that S_n^{*2} and $\overline{X}_n$ are independent
   (a) Write S_n^{*2} as a function of $\sum_{i \neq j} X_i X_j$ and $\sum_{i=1}^n X_i^2$.
   (b) Compute $\mathbb{E}(X_i^k e^{it X_i/n})$ for $k = 0, 1, 2$, and then $\mathbb{E}(S_n^{*2} e^{it\overline{X}_n})$ as a function of the characteristic function ϕ of P and of its derivatives.
   (c) Show that $\mathbb{E}(S_n^{*2} e^{it\overline{X}_n}) = \phi(t/n)^2$, and that ϕ is solution of a differential equation with order 2.
   (d) Solve this equation.
2. Show that $\overline{X}_n$ and S_n^{*2} are independent if and only if P is a Gaussian distribution.

Solution
1.
   (a) We have

$$S_n^{*2} = \frac{1}{n-1} \sum_{i=1}^n X_i^2 - \frac{1}{n(n-1)} \left(\sum_{i=1}^n X_i \right)^2 = \frac{1}{n} \sum_{i=1}^n X_i^2 - \frac{1}{n(n-1)} \sum_{i \neq j} X_i X_j.$$

   (b) Since the random variables X_i are independent, Theorem 2.73 gives that

$$\mathbb{E}(e^{it\overline{X}_n}) = \phi(t/n)^n, \quad \mathbb{E}(X_i e^{it X_i/n}) = i\phi'(t/n), \quad \mathbb{E}(X_i^2 e^{it X_i/n}) = -\phi''(t/n).$$

   Therefore

$$\mathbb{E}(X_i^2 e^{it\overline{X}_n}) = \mathbb{E}(X_i^2 e^{it X_i/n}) \prod_{i \neq j} \mathbb{E}(e^{it X_j/n}) = -\phi''(t/n)\phi(t/n)^{n-1},$$

   and $\mathbb{E}(X_i X_j e^{it\overline{X}_n}) = -\phi'(t/n)^2 \phi(t/n)^{n-2}$. Finally

$$\mathbb{E}(S_n^{*2} e^{it\overline{X}_n}) = -\phi''(t/n)\phi(t/n)^{n-1} + \phi'(t/n)\phi(t/n)^{n-2}.$$

   (c) Since $\overline{X}_n$ and S_n^{*2} are independent,

$$\mathbb{E}(S_n^{*2} e^{it\overline{X}_n}) = \mathbb{E}(S_n^{*2})\mathbb{E}(e^{it\overline{X}_n}) = \phi(t/n)^2$$

holds true for any t, so

$$-\phi''(t)\phi(t) + (\phi'(t))^2 = (\phi(t))^2.$$

(d) Since $\phi(0) = 1$ and ϕ is continuous, we have, at least in a neighborhood of 0,

$$\frac{\phi''(t)}{\phi(t)} - \left[\frac{\phi'(t)}{\phi(t)}\right]^2 = -1.$$

The left-hand side of this equality is the second derivative of $\log\phi(t)$. Since $\phi'(0) = 0$, it follows that $\phi(t) = e^{-t^2/2}$ on this neighborhood, and hence on $\mathbb{R}$, since this function is analytical.
2. Using both 1 and Lévy's continuity theorem, we obtain that if S_n^{*2} and $\overline{X}_n$ are independent, then the distribution P is normal.

Fisher's theorem gives the converse. $\triangle$

∇ **Exercise 5.6 (Estimation for a Negative Binomial Distribution)** A coin is tossed. At each tossing, the probability of observing "head" is $p \in]0, 1[$. Let N be the number of tosses necessary for obtaining m "heads," where $m > 1$ is fixed.

1. Compute $\mathbb{E}\left[1/(N-1)\right]$ and give an unbiased estimator of p.
2. Show that, on the contrary, $\mathbb{E}\left(m/N\right) > p$.

Solution
1. Clearly, $N \sim \mathcal{B}_-(m, p)$, so $\mathbb{P}(N = m + k) = \binom{m+k-1}{m-1} p^m (1-p)^k$. We compute

$$\mathbb{E}\left[1/(N-1)\right] = \sum_{k \geq 0} \frac{1}{m+k-1} \mathbb{P}(N = m + k)$$

$$= \frac{p}{m-1} \sum_{k \geq 0} \binom{m+k-2}{m-2} p^{m-1} (1-p)^k = \frac{p}{m-1}.$$

Therefore, $\widehat{p} = (m-1)/(N-1)$ is an unbiased estimator of p.
2. We have

$$\mathbb{E}\left(\frac{m}{N}\right) = \mathbb{E}\left(\frac{m-1}{N-1}\right) + \mathbb{E}\left[\frac{N-m}{N(N-1)}\right].$$

The right-hand term of the sum is obviously positive, and the conclusion follows. $\triangle$

▽ **Exercise 5.7 (Comparison of Estimators)** Suppose that the number of purchases per day in a store of some kind of lamp is a random variable X with a Poisson distribution whose parameter λ is unknown.

Let N_j count the number of days during which j purchases have been made, for $j \in [\![0, J]\!]$, with $\sum_{j=0}^{J} N_j = n$, for n days of observation. Let X_i be the number of purchases for the i-th day of the observation.

(A)

1. Show that the empirical estimator $T_n = N_0/n$ is an unbiased estimator of $e^{-\lambda}$. Compute its variance. Is T_n a convergent estimator of $e^{-\lambda}$?
2. Let $\widehat{\lambda}_n$ be the maximum likelihood estimator of λ. Show that it is asymptotically unbiased, with variance converging to zero when n tends to infinity.
3. Set $T_n = e^{-\lambda_n^*}$, and let us compare λ_n^* and $\widehat{\lambda}_n$ as estimators of λ.

   Give an asymptotic equivalent of $\mathbb{V}\text{ar}\,\widehat{\lambda}_n / \mathbb{V}\text{ar}\,\lambda_n^*$. Show that $\widehat{\lambda}_n$ is a better estimator than λ_n^*. Hint: if (Z_n) is a random sequence converging to zero in probability, then $\mathbb{V}\text{ar}\,[\log(1 + Z_n)] \sim \mathbb{V}\text{ar}\,Z_n$ when n tends to infinity.

(B)

1. Deduce from $\widehat{\lambda}_n$ a plug-in estimator $\widetilde{p}_n(k)$ of $\mathbb{P}(X = k) = p(k)$ for any fixed k. Compute the mean and the variance of $\widetilde{p}_n(1)$ by using successive derivatives of the generating moments function of the Poisson distribution.
2. Show that the empirical estimator $\widehat{p}_n(k) = N_k/n$ is an unbiased and convergent estimator of $p(k)$. Compute its variance. Is $\widehat{p}_n(k)$ strongly convergent?

Solution

(A)

1. Clearly, $(X_1, \ldots, X_n)$ is an n-sample of the distribution $\mathcal{P}(\lambda)$. We have $T_n = \sum_{i=1}^{n} \mathbb{1}_{(X_i=0)}/n$ and $\mathbb{P}(X_i = 0) = e^{-\lambda}$, so $\mathbb{1}_{(X_i=0)} \sim \mathcal{B}(e^{-\lambda})$ and $N_0 \sim \mathcal{B}(n, e^{-\lambda})$. Therefore, $\mathbb{E}\,T_n = \mathbb{E}\,(N_0/n) = e^{-\lambda}$ and T_n is an unbiased estimator of $e^{-\lambda}$. Moreover,

$$\mathbb{V}\text{ar}\,T_n = \frac{1}{n^2}\mathbb{V}\text{ar}\,N_0 = \frac{1}{n^2}[ne^{-\lambda}(1 - e^{-\lambda})] = \frac{e^{-\lambda} - e^{-2\lambda}}{n}.$$

Chebyshev's inequality yields

$$\mathbb{P}(|T_n - e^{-\lambda}| \geq \varepsilon) \leq \frac{e^{-\lambda} - e^{-2\lambda}}{n\varepsilon^2}, \quad \varepsilon > 0,$$

so T_n converges in probability to $e^{-\lambda}$, and hence is a convergent estimator.

2. We compute $\widehat{\lambda}_n = \overline{X}_n$. By Theorem 5.16, the empirical mean is known to be an unbiased and convergent estimator of the mean of the distribution, here equal to λ. Its variance is λ/n, which converges to zero when n tends to infinity.

3. Set $T_n = e^{-\lambda} + B_n = e^{-\lambda}(1 + B_n e^{\lambda})$. We have $\lambda_n^* = \lambda - \log(1 + B_n e^{\lambda})$ and

$$\mathrm{Var}\,\lambda_n^* = \mathrm{Var}\,[\lambda - \log(1 + B_n e^{\lambda})] = \mathrm{Var}\,[\log(1 + B_n e^{\lambda})].$$

By 1, we know that B_n converges in probability to 0, so $\mathrm{Var}\,[\log(1 + B_n e^{\lambda})] \sim \mathrm{Var}\,(e^{\lambda} B_n)$, or $\mathrm{Var}\,\lambda_n^* \sim e^{2\lambda}\mathrm{Var}\,T_n$, that is $\mathrm{Var}\,\lambda_n^* \sim (e^{\lambda} - 1)/n$. Finally,

$$\frac{\mathrm{Var}\,\widehat{\lambda}_n}{\mathrm{Var}\,\lambda_n^*} \approx \frac{\lambda}{e^{\lambda} - 1}.$$

The function $f : \mathbb{R}^+ \to \mathbb{R}^+$ defined by $f(\lambda) = \lambda/(e^{\lambda} - 1)$ is decreasing and converges to zero rapidly when λ tends to infinity. Therefore, $\widehat{\lambda}_n$ is a better estimator: It is unbiased as is λ_n^* but converges more quickly to λ.

(B)

1. We get $\widetilde{p}_n(k) = (\widehat{\lambda}_n)^k \exp(-\widehat{\lambda}_n)/k!$. Set $S_n = \sum_{i=1}^{n} X_i$. We have $n\mathbb{E}\,[\widetilde{p}_n(1)] = \mathbb{E}\,(S_n e^{-S_n/n}) = M'_{S_n}(-1/n)$.

Since $S_n \sim \mathcal{P}(n\lambda)$, we compute $M_{S_n}(t) = \sum_{k \geq 0} e^{tk} e^{-\lambda} \lambda^k/k! = e^{\lambda(e^t - 1)}$ and $M'_{S_n}(t) = \lambda e^t e^{\lambda(e^t - 1)}$. Therefore $\mathbb{E}\,[\widetilde{p}_n(1)] = \lambda e^{1/n} e^{\lambda(e^{1/n} - 1)}/n$.

Similarly, $M''_{S_n}(t) = \lambda e^{\lambda(e^t - 1)} + \lambda^2 e^{2t} e^{\lambda(e^t - 1)}$, and

$$\mathbb{E}\,[\widehat{p}_n(1)^2] = \frac{1}{n^2}\mathbb{E}\,(S_n^2 e^{-2S_n/n}) = \frac{1}{n^2}M''_{S_n}(-2/n) = \frac{\lambda}{n^2}[1 + \lambda e^{-4/n}]e^{\lambda(e^{-2/n} - 1)},$$

from which the variance of $\widehat{p}_n(1)$ follows by König formula.

2.
(a) We have $\mathbb{E}\,[\widehat{p}_n(k)] = \sum_{i=1}^{n} \mathbb{P}(X_i = k)/n = p(k)$ and

$$\mathrm{Var}\,[\widehat{p}_n(k)] = \frac{1}{n^2}\sum_{i=1}^{n}\mathbb{E}\,[\mathbb{1}_k(X_i) - p(k)^2] = \frac{p(k) - p(k)^2}{n}.$$

This variance converges to 0 when n tends to infinity, so the estimator is strongly convergent.

$\triangle$

▽ Exercise 5.8 (Maximum Likelihood Estimators)

1. Let $(X_1, \ldots, X_n)$ be an n-sample of a log-normal distribution with parameters m and σ^2, with mean μ and variance τ^2.
   (a) Determine the maximum likelihood estimators $\widehat{\mu}_n$ and $\widehat{\tau}_n$ of μ and τ.
   (b) Show that $\widehat{\mu}_n$ is biased but asymptotically unbiased.

2. Let $(X_1, \ldots, X_n)$ be an n-sample of an inverse-Gaussian distribution with parameters μ and λ.
   (a) Determine the maximum likelihood estimators $\tilde{\mu}_n$ and $\tilde{\lambda}_n$ of μ and λ.
   (b) Show that $\tilde{\mu}_n$ is unbiased and give an unbiased estimator of $1/\lambda$. Hint: Fisher's theorem for inverse-Gaussian distributions states that $\overline{X}_n \sim \mathcal{N}^{-1}(\mu, n\lambda)$ and $\lambda \sum_{i=1}^n \left(1/X_i - 1/\overline{X}_n\right) \sim \chi^2(n-1)$ and are independent.

Solution

1.

(a) Clearly, $(U_1, \ldots, U_n) = (\log X_1, \ldots, \log X_n)$ is an n-sample of the distribution $\mathcal{N}(m, \sigma^2)$. According to Example 5.42, the maximum likelihood estimator of m is the empirical mean of the U_i, that is $\widehat{m}_n = \sum_{i=1}^n \log X_i$. Similarly, $\widehat{\sigma}_n^2 = \sum_{i=1}^n (\log X_i - \widehat{m}_n)^2/n$.

   We know that $\mu = \exp(m + \sigma^2/2)$, so $\widehat{\mu}_n = \exp(\widehat{m}_n + \widehat{\sigma}_n^2/2)$. Since $\tau = \mu(e^{\sigma^2/2} - 1)$, we get $\widehat{\tau}_n = \exp(\widehat{m}_n + \widehat{\sigma}_n^2/2)[\exp \widehat{\sigma}_n^2/2) - 1]$.

(b) We have $\mathbb{E}\,\widehat{\mu}_n = \mathbb{E}\,[\exp(\widehat{m}_n) \exp(\widehat{\sigma}_n^2/2)]$. Fisher's theorem says that $\widehat{m}_n \sim \mathcal{N}(m, \sigma^2/n)$ and $n\widehat{\sigma}_n^2/\sigma^2 \sim \chi^2(n-1)$ are independent. Therefore, on the one hand, $\exp \widehat{m}_n$ has the log-normal distribution with parameters m and σ^2/n, so $\mathbb{E}\,(\exp \widehat{m}_n) = \exp(m + \sigma^2/2n)$, and on the other hand, for $n > \sigma^2$,

$$\mathbb{E}\,[\exp(n\widehat{\sigma}_n^2/2)] = \frac{1}{2^{(n-1)/2}\Gamma((n-1)/2)} \int_{\mathbb{R}_+} e^{\sigma^2 s/2n} e^{-s/2} s^{(n-3)/2} ds$$

$$= \frac{1}{2^{(n-1)/2}\Gamma((n-1)/2)} \times \frac{\Gamma((n-1)/2)}{(1/2 - \sigma^2/2n)^{(n-1)/2}}$$

$$= (1 - \sigma^2/n)^{-(n-1)/2}.$$

Therefore, $\mathbb{E}\,\widehat{\mu}_n = (1 - \sigma^2/n)^{-(n-1)/2}\mu$. This estimator is biased, but

$$\log \mathbb{E}\,\widehat{\mu}_n = m + \frac{\sigma^2}{2n} - \frac{n-1}{2}\log\left(1 - \frac{\sigma^2}{n}\right) \longrightarrow m + \frac{-\sigma^2}{2}, \quad n \to +\infty,$$

so it is asymptotically unbiased.

2.

(a) We compute

$$\tilde{\mu}_n = \overline{X}_n \quad \text{and} \quad 1/\tilde{\lambda}_n = \frac{1}{n}\sum_{i=1}^n \left(\frac{1}{X_i} - \frac{1}{\overline{X}_n}\right).$$

(b) Therefore, $\tilde{\mu}_n$ is unbiased, and, since $n\lambda/\tilde{\lambda}_n \sim \chi^2(n-1)$, the estimator $n\lambda/(n-1)\tilde{\lambda}_n$ is an unbiased estimator of $1/\tilde{\lambda}_n$. $\triangle$

∇ **Exercise 5.9 (Fisher Information of Gaussian Models)**   Let us consider the
Gaussian statistical model $(\mathbb{R}, \mathcal{B}(\mathbb{R}), (\mathcal{N}(m, \sigma^2))_{(m,\sigma^2)\in\Theta})$.

1. Assume for this question that $\Theta = \mathbb{R}^* \times \{1\}$. Compute Fisher information of
   order 1 of the model.
2. Let now $\Theta = \mathbb{R}^* \times \mathbb{R}_+^*$. Compute Fisher information of order n of the model
   relative to m, and then to σ^2.
3. Give efficient estimators of m (if σ^2 is known) and of σ^2 (if m is known).

Solution
1. The density of the distribution $\mathcal{N}(m, 1)$ is $f_m(x) = e^{(x-m)^2/2}/\sqrt{2\pi}$. The model
   is regular, so Proposition 5.47 yields

$$\frac{\partial}{\partial m} \log L(m, x_1) = -x_1 + m$$

and

$$\mathcal{I}_1(m) = \int_{\mathbb{R}} \left(\frac{f_m'(x)}{f_m(x)}\right)^2 f_m(x)dx = \int_{\mathbb{R}} (m-x)^2 f_m(x)dx = \mathrm{Var}\, X = 1.$$

2. We have $\mathcal{I}_n(m) = n/\sigma^2$, because

$$\frac{\partial^2}{\partial m^2} \log L(m, \sigma^2, X_1, \ldots, X_n) = -\frac{n}{\sigma^2}.$$

   Fisher information increases when the variance decreases, which yields its
   interpretation in terms of information on the distribution.
3. It follows that the empirical mean is an efficient and unbiased estimator of the
   mean. Moreover,

$$\frac{\partial^2}{\partial(\sigma^2)^2} \log L(m, \sigma^2, x_1, \ldots, x_n) = \frac{n}{2\sigma^4} - \frac{1}{\sigma^6} \sum_{i=1}^{n} (x_i - m)^2,$$

   and $\mathbb{E}_{\sigma^2}[\sum_{i=1}^{n}(X_i - m)^2] = n\sigma^2$, so $\mathcal{I}_n(\sigma^2) = n/2\sigma^4$.

Since $nS_n^2 \sim \sigma^2\chi^2(n)$, we get $\mathrm{Var}\, S_n^2 = 2\sigma^4/n$ by Example 2.61. By Cramér–
Rao inequality, it follows that S_n^2 is an efficient estimator of σ^2.   △

∇ **Exercise 5.10 (Estimation for a Binomial Distribution)** Let X be a random
variable with distribution $\mathcal{B}(10, p)$, where $0 < p < 1$.

1.
(a) A 5-sample of the distribution of X is observed. The obtained values are: 3, 4, 4, 8, 10. What is the probability that these values (in whichever order) were observed?
(b) Compute the value p_0 for which this probability is maximum. What is the relation between p_0 and the mean of the observations?
2. Let $\overline{X}_n$ be the empirical mean of the 5-sample.
(a) What is its distribution when n is large enough?
(b) Give a bound for $p(1-p)$, and then a confidence interval I_α for p at level α.
(c) Give an estimator of $p(1-p)$, which is a function of X_n, and finally give another confidence interval for p at level α.
(d) Construct a test of the hypothesis H_0: "$p \leq p_0$" against H_1: "$p > p_0$" for a given $p_0 \in]0, 1[$.

Solution

1.
(a) Let A be the event "4, 4, 3, 8, 10 are observed, in any order." We have

$$\mathbb{P}(A) = 60\left[\binom{10}{4}p^4(1-p)^6\right]^2 \binom{10}{3}p^3(1-p)^7 \binom{10}{8}p^8(1-p)^2 p^{10}$$

$$= 71 \times 10^4 \times p^{29}(1-p)^{21}.$$

(b) We have $\log \mathbb{P}(A) = \log K + 29 \log p + 21 \log(1-p)$, so

$$\frac{d}{dp} \log \mathbb{P}(A) = \frac{29}{p} - \frac{21}{1-p} = 0$$

is satisfied for $p_0 = 29/50$, which indeed gives a maximum of $\mathbb{P}(A)$ because $\frac{d^2}{dp^2} \log \mathbb{P}(A) > 0$. Moreover, $\overline{x}_5 = 29/5 = 10p_0$, which confirms that the empirical mean is an unbiased estimator of the mean—by Theorem 5.16.

2.
(a) Set $S_n = \sum_{i=1}^n X_i$. We have $S_n \sim \mathcal{B}(10n, p)$ and $\mathbb{P}(\overline{X}_n = k/n) = \mathbb{P}(S_n = k)$, so

$$\mathbb{P}(\overline{X}_n = j) = \binom{10n}{10j} p^{10j}(1-p)^{10(n-j)}, \quad 0 \leq j \leq 10.$$

By the central limit theorem, $\sqrt{n}(\overline{X}_n - 10p)/\sqrt{10p(1-p)}$ converges in distribution to a standard Gaussian variable.
(b) For n large enough,

$$\mathbb{P}\left[-r_{1-\alpha/2} \leq \sqrt{n}\frac{\overline{X}_n - 10p}{\sqrt{10p(1-p)}} \leq r_{1-\alpha/2}\right] = 1 - \alpha,$$

where $r_{1-\alpha/2}$ is the quantile of order $1 - \alpha/2$ of the standard normal distribution.

For $0 < p < 1$, we have $0 < \sqrt{p(1 - p)} < 1/2$, so

$$\mathbb{P}\left(\frac{\overline{X}_n}{10} - \frac{r_{\alpha/2}}{20\sqrt{n}} \leq p \leq \frac{\overline{X}_n}{10} + \frac{r_{\alpha/2}}{20\sqrt{n}}\right) \geq 1 - \alpha,$$

and hence

$$I_\alpha = \left]\frac{\overline{X}_n}{10} - \frac{r_{\alpha/2}}{20\sqrt{n}}, \frac{\overline{X}_n}{10} + \frac{r_{\alpha/2}}{20\sqrt{n}}\right[$$

is a confidence interval for p at level α.

(c) By the method of moments, $(\overline{X}_n/10)(1 - \overline{X}_n/10)$ is known to be an unbiased estimator of $p(1 - p)$. Therefore,

$$\mathbb{P}\left(-r_\alpha \leq \sqrt{n}\,\frac{\overline{X}_n - 10p}{\sqrt{\overline{X}_n(1 - \overline{X}_n/10)}} \leq r_\alpha\right) = 1 - \alpha,$$

from which it follows that

$$I'_\alpha = \left]\frac{\overline{X}_n}{10} - \frac{r_{\alpha/2}\sqrt{\overline{X}_n(1 - \overline{X}_n/10)/10}}{\sqrt{10n}}, \frac{\overline{X}_n}{10} + \frac{r_{\alpha/2}\sqrt{\overline{X}_n(1 - \overline{X}_n/10)/10}}{\sqrt{10n}}\right[$$

is a confidence interval for p at level α.

(d) According to a., for n large enough,

$$\mathbb{P}(\overline{X}_n > r) = 1 - F\left(\frac{\sqrt{n}(r - 10p)}{\sqrt{10p(1 - p)}}\right),$$

where F is the distribution function of the standard normal distribution.

The function $p \to (r - 10p)/\sqrt{10p(1 - p)}$ decreases on $]0, 1[$ for all fixed $r \in [0, 10]$. The region $(\overline{X}_n > r)$ is a rejection region of H_0 at level α for r such that

$$F\left(\frac{\sqrt{n}(r - 10p_0)}{\sqrt{10p_0(1 - p_0)}}\right) = 1 - \alpha,$$

that is $r = (\sqrt{10p_0(1 - p_0)}q_{1-\alpha} + 10p_0)/\sqrt{n}$, where $q_{1-\alpha}$ is the quantile of order $1 - \alpha$ of the standard normal distribution. $\triangle$

Further Reading

Measure and Probability

Billingsley P. (2012) *Probability and measure*, 3rd edn., New York: Wiley.
Leadbetter R., Cambanis S., Pipiras V. (2014) *A Basic Course in Measure and Probability* Cambridge: Cambridge University Press.
Rudin W. (1987) *Complex and Real Analysis*, 3rd edn., New York: McGraw Hill.

Probability Theory and Statistics

Borkar V. (1995) *Probability theory, an advanced course* New York: Springer-Verlag.
Casella G., Berger R.L. (1990), *Statistical inference* Belmont, CA: Duxbury Press.
Chamond L., Yor M. (2003) *Exercises in probability*, Cambridge: Cambridge University Press.
Feller W. (1991) *An introduction to probability theory and its applications*, volumes 1 and 2, 2nd edn. New York: J. Wiley & Sons.
Fristedt B., Gray L. (1997) *A modern approach to probability theory* Boston: Birkhäuser.
Girardin V., Limnios N. (2014) *Probabilités, avec une introduction à la statistique*, 3rd edn., Paris: Vuibert.
Girardin V., Limnios N. (2014) *Probabilités, processus stochastiques et applications*, 3rd edn. Paris: Vuibert.
Girardin V., Limnios N. (2018) *Applied Probability: From Random Sequences to Random Processes*, Heidelberg: Springer International Publishing.
Gut A. (2013) *Probability: A graduate course*, 2nd Edition, New York: Springer.
Ibe O. C. (2005) *Fundamentals of applied probability and random processes* Amsterdam: Elsevier Academic Press.
Pinsky M., Karlin S. (2011) *An Introduction to Stochastic Modeling* 4th edn., London: Academic Press.
Resnick S.I. (1999) *A probability path*, Boston: Birkhäuser.
Roussas G. (2013) *Introduction to Probability* 2nd edn. London: Academic Press.
Shiryaev A.N. (1996) *Probability* 2nd edn., New York: Springer-Verlag.
Stirzaker D. (2003) *Elementary Probability* 2nd edn., Cambridge: Cambridge University Press.
Stoyanov J. (2013) *Counterexamples in probability* 3rd edn., Chichester: J. Wiley.
Tijms H. (2017) *Probability: A Lively Introduction*, Cambridge: Cambridge University Press.
Venkatesh S. (2012) *The Theory of Probability: Explorations and Applications* Cambridge: Cambridge University Press.
Young G.A., Smith R.L. (2010), *Essentials of statistical inference*, Cambridge: Cambridge University Press.

© The Author(s), under exclusive license to Springer Nature Switzerland AG 2022
V. Girardin, N. Limnios, *Applied Probability*,
https://doi.org/10.1007/978-3-030-97963-8

Applications

Barlow R., Proschan F. (1975), *Statistical theory of reliability and life testing* New York: Holt Rinehart Winston.

Cover L., Thomas J. (1991) *Elements of information theory* New-York: Wiley series in telecommunications.

DeCoursey W. (2003) *Statistics and probability for engineering applications* Amsterdam: Elsevier.

Evans M. J., Rosenthal J. S. (2004) *Probability and statistics* London: Freeman and Co.

Iosifescu M., Limnios N., Oprisan G. (2010) *Introduction to stochastic models* London: ISTE, J. Wiley.

Lange K. (2010) *Applied Probability* New York: Springer-Verlag.

Reza F. (1994) *Introduction to information theory* New York: Dover Publications.

Ross, S. (2010) *Introduction to probability models*, 10th edn, Burlington, MA: Elsevier and San Diego, CA: Academic Press.

Schinazi R. (2012) *Probability with statistical applications* Basel: Birkhäuser.

Index

Printed in the United States
by Baker & Taylor Publisher Services